Metals and Related Substances in Drinking Water

COST Action 637

Proceedings of the 4th International Conference
Metals and Related Substances in Drinking Water, METEAU

Kristianstad, Sweden, October 13-15, 2010

Editors

Prosun Bhattacharya
Ingegerd Rosborg
Arifin Sandhi
Colin Hayes
Maria Joäo Benoliel

List of Conference Sponsors

Royal Institute of Technology (KTH)
Stockholm, Sweden

Region Skåne
Kristianstad, Sweden

Malmberg Water AB
Åhus, Sweden

Kristianstads kommun
Kristianstad, Sweden

Vinnova
Stockholm, Sweden

Krinova Science Park
Kristianstad, Sweden

Published by **IWA Publishing**
Alliance House
12 Caxton Street
London SW1H 0QS, UK
Telephone: +44 (0)20 7654 5500
Fax: +44 (0)20 654 5555
Email: publications@iwap.co.uk
Web: www.iwapublishing.com

IWA Publishing
London • New York

First published 2012

Cover image: http:// www.vattenriket.kristianstad.se/eng/summary/index.htm

Disclaimer

British Library Cataloguing in Publication Data
A CIP catalogue record for this book is available from the British Library

Library of Congress Cataloging- in-Publication Data
A catalog record for this book is available from the Library of Congress

ISBN: 9781780400358

Contents

Section 6: Source waters

COST - "European CO-operation in the field of Scientific and Technical Research" - is the longest running and widest European intergovernmental mechanism for cooperation in research.

Founded in 1971, COST is an intergovernmental framework for European Cooperation in the field of Scientific and Technical Research, allowing the co-ordination of nationally funded research on a European level. COST Actions cover basic and pre-competitive research as well as activities of public utility. The goal of COST is to ensure that Europe holds a strong position in the field of scientific and technical research for peaceful purposes, by increasing European cooperation and interaction in this field.

Website: http://www.cost.esf.org/

COST Action 637 - METEAU, Metals and Related Substances in Drinking Water

The main objective of the Action is to stimulate better control of metals and related substances in drinking water and to minimize environmental impacts.

Major objectives

- To provide an on-going forum for knowledge exchange in connection with metals and related substances in drinking water.
- To promote good practice in the control of metals and related substances in drinking water.
- To more critically determine the environmental and socio-economic impacts of control measures through the sharing of practitioner experience.
- To stimulate relevant collaborative research and demonstration studies at the European scale.

Organizers / Core Committee

Dr. Colin Hayes
Swansea University, UK -Action Chair
cr.hayes@swansea.ac.uk

Dr. Maria Joäo Benoliel
EPAL- Empresa Portuguesa das Àquas Livres, SA, Lisbon, Portugal
mjbenol@epal.pt

Prof. George Pilidis
University of Ioannina, Greece
gpilidis@uoi.gr

Prof. Prosun Bhattacharya
Royal Institute of Technology (KTH), Stockholm, Sweden
prosun@kth.se

Dr. Ingegerd Rosborg
Royal Institute of Technology (KTH), Stockholm, Sweden
rosborg@spray.se

Dr. Peter Holm
Technical University Copenhagen, Denmark
peho@life.ku.dk

Dr Josef Klinger
TZW, Germany
klinger@tzw.de

Dr. Vladimira Nemcova
Ostrava Regional Health Authority, Czech Republic
vladimira.nemcova@zuova.cz

Dr. Matyas Borsanyi
National Institute of Environmental Health, Hungary
borsanyi.matyas@oki.antsz.hu

Dr. Larry Russell
Reed International
reedinternltd@aol.com

Local Conference Organization

Prof. Prosun Bhattacharya
Royal Institute of Technology, KTH, Stockholm, Sweden
prosun@kth.se

Dr. Ingegerd Rosborg
Royal Institute of Technology, KTH, Stockholm, Sweden
rosborg@spray.se

Local Conference Secretariet

Ms. Lollo Kruger
Krinova Science Park, 291 39 Kristianstad, Sweden,
Tel: +46(0)44204542, Fax: +46(0)44 2045 43
lollo.kruger@krinova.se

Technical Coordination

Dr. Maria João Benoliel
EPAL Empresa Portuguesa das Àquas Livres, SA, Lisbon, Portugal
mjbenol@epal.pt

Editors

Prof. Prosun Bhattacharya
Royal Institute of Technology, KTH, Stockholm, Sweden
prosun@kth.se

Dr. Ingegerd Rosborg
Royal Institute of Technology (KTH), Stockholm, Sweden
rosborg@spray.se

Arifin Sandhi
Department of Botany, Stockholm University, Stockholm, Sweden
sandhi@botan.su.se

Dr. Colin Hayes
Swansea University, UK -Action Chair
cr.hayes@swansea.ac.uk

Dr. Maria Joäo Benoliel
EPAL- Empresa Portuguesa das Àquas Livres, SA, Lisbon, Portugal
mjbenol@epal.pt

Summary of the 4th International COST Action 637 Conference, Kristianstad, Sweden, October 13-15, 2010

The conference was successfully accomplished with all different items on the program carried through in the best way. The main objective of the conference was to collate the understanding of the various factors which control metals and related substances in drinking water with an aim to minimize environmental impacts.The conference goals adequately fulfilled the goals of the COST Action 637 through:

- Sharing and exchange of knowledge on metals and related substances in drinking water.
- Promoting good practice in controlling metals and related substances in drinking water.
- Determining the environmental and socio-economic impacts of control measures through public participation
- Introducing the importance of mineral balance in drinking water especially when choosing treatment methods the sharing of practitioner experience.
- Strengthening relevant collaborative research and demonstration studies at the European as well as on a global scale.

The conference started in the evening, Wednesday 13 October, 2010 with a ceremonial public opening of the Drinking Water Well at Lilla Torg, Kristianstad. At 09.00 a.m. Thursday 14 October, 2010 the conference sessions started.

Session 1: Risk management and risk assessment. Bob Breach from UK was the keynote speaker. It was highlighted that the WSP relies on a partnership approach with the roles of the many different stakeholders being clearly defined, e.g. health and local authorities, national and/or local regulatory authorities, consumers and property owners, and plumbers.

Session 2: Health and aesthetic issues. Joby Boxall from Great Britain was the keynote speaker. Prof Boxall illustrated the management strategy regarding discoloration incidents from cast iron and other materials in pipes and installations should be holistic from source to consumer tap.

Session 3: Mineral balance in drinking water. Frantisek Kozisek from Czech Republic was the keynote speaker in this session. He made a concluding remark that scientific studies clearly indicate the importance of minerals from drinking water, and health effects from RO water are poorly studied.

Session 4: Treatment processes. The keynote speaker; Alberto Figoli from Italy, discussed about arsenic removal by membrane technologies. A consensus about the need of re-mineralisation after RO treatment was reached.

Session 5: Metal materials, testing and metal leaching. The keynote speaker Thomas Rapp from Germany, stated that a "Committee of Experts" is required to decide about the acceptance of materials according to EN 15664-1 and other significant data.

Session 6: Source water. David Polya from University of Manchester, UK was the keynote speaker. He demonstrated that exposure to arsenic through drinking water and rice may result in genetic and other damage in individuals.

Session 7: Bottled water. Peter Holm from Denmark gave the keynote paper and he concluded that the relation between taste and chemical content is poorly understood. Finally, Vice-Chair of Action - Maria João Benoliel summarized the conference.

There were nearly 27 poster presentations made during the second day of the conference and following this the Gala Conference dinner was held at the medieval Bäckaskog Castle.

MC meeting with chair of Action 637, Colin Hayes, Swansea University, ended the conference, which was a successful last event of COST Action 637, "Metals and related substances in drinking water". On Saturday 16 October, 2010, the last day of the conference, a field excursion at Vattenriket was organized.

The organizers of the COST 637 Conference 2010, are grateful for the support and sponsoring we received from VINNOVA, Kristianstad Municipality, Region Skåne, and Malmberg Water.

Ingegerd Rosborg and Prosun Bhattacharya
Department of Land and Water Resources Engineering, Royal Institute of Technology
SE-100 44 Stockholm, Sweden

Foreword from COST Action 637 Chair

With over 50 scientific presentations or posters on metals and related substances in drinking water, the fourth International Conference of COST Action 637, in Kristianstad, Sweden on 13-15 October 2010, was a resounding success.

For a range of metals and metalloids, presentations focused on,

- Risk assessment and management
- Water and health
- Mineral balance
- Treatment processes
- Natural and bottled waters

This conference will be of interest to water managers and scientists throughout Europe and elsewhere.

Since this conference was held, the research network (funded by COST until November 2010) has been incorporated as a Specialist Group within the International Water Association, and continues to be active. If you wish to participate in our on-going activities, please visit the IWA Water Wiki at www.iwahq.org - you will find us via “Group Spaces”.

Dr. Colin Hayes
Chair, COST Action 637

Section 1

Risk Management and Risk Assessment

How water safety plans can help to address risks from metals in drinking water

Bob Breach
Water Quality and Environmental Consultancy, UK
Corresponding author e-mail: bob.breach@btinternet.com

Abstract
The launch in 2004 of the 3rd edition of the WHO guidelines for drinking water quality [1] jointly with the IWA Bonn Charter for safe Drinking Water [2] marked the culmination of a huge shift in worldwide advice on water quality management that had been evolving for some years. This redressed the balance of effort away from reliance solely on end product testing towards a more proactive, risk prevention approach known as a Water Safety Plan (WSP). In essence this is a regularly repeated cycle of assessing and managing water quality risks from catchment to consumer. This paper will briefly review the WSP approach to managing drinking water quality with a particular focus on metals in drinking water and describe how the principles in the Bonn Charter can be used to develop effective partnerships to minimise such risks.

1. Introduction

Since the launch of the WHO guidelines and Bonn Charter, there has been a welcome and massive increase in information for utilities and other stakeholders on how to adopt such a risk based approach for water quality management. Central to this is the valuable guidance which continues to be produced by WHO and IWA [3, 4], as well as a wealth of conferences, papers and other material. Despite this, progress globally with WSP adoption is patchy and many utilities still do not appreciate their benefits or have major difficulty in their adoption.

The primary goal of the Bonn Charter is the provision of:

"Good safe drinking water which has the trust of consumers"

To secure this goal the Charter provides a set of high level principles which are universally applicable to both developed and developing countries across the world. At the heart of the process is the development of a risk based water safety plan from catchment to consumer. But the document also covers a number of other important areas including the setting of water quality standards, roles and responsibilities, institutional arrangements, communication and water pricing and financing. Central to the idea of the Bonn Charter is that whilst the principles of good water quality management are universal, the way they are applied locally will depend on a range of factors including cultural, legal, institutional, and socioeconomic. The Bonn Charter recognises that although water utilities are pivotal in delivering good safe drinking water through the adoption of WSPs, this can only be truly effective if a wide range of other stakeholders are also fully involved in the process. This applies particularly to the sometimes overlooked issue of metals in drinking water.

2. The WSP approach

It is now widely accepted that managing drinking water quality solely through end product testing is invariably "too little and too late". By contrast a risk based approach yields many benefits including:

1) Significantly reduced risk of incidents which impact public health
2) Improved compliance with regulatory standards and other statutory requirements
3) Improved consumer trust through more reliable water quality and improved water acceptability
4) Improved confidence of key stakeholders
5) More cost effective operation and more targeted capital investment planning
6) Improved staff commitment

Although the concept of a WSP is quite simple, in practice implementation is a long term process which requires not only good management procedures but also full support and commitment from senior management and a change in culture and behaviour across the whole organisation through training and awareness raising.

A WSP is essentially a structured and documented plan that:
1) Identifies all risks from catchment to consumer
2) Puts in place controls
3) And verifies their effectiveness

But it is also important to recognise that a WSP is a continuous cycle of review and improvement. It is impossible to assess and address all risks in the first cycle and thus the improvement programme that results should be prioritised to first address the key risks and "quick wins" with other risks being addressed in subsequent cycles. And since water suppliers normally only have responsibility for treatment and distribution this must inevitably also involve a wide range of other stakeholders.

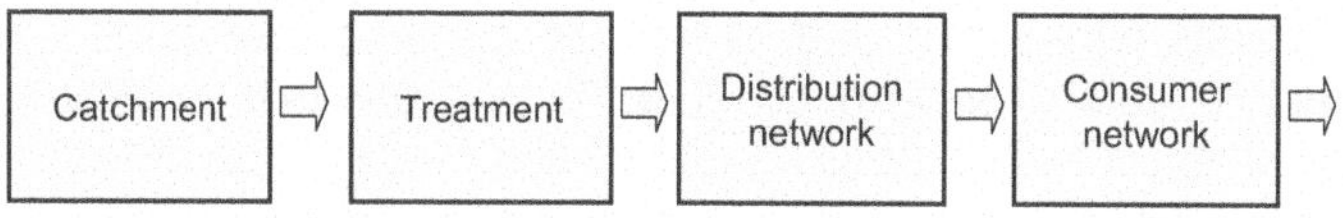

Figure 1. The catchment to consumer approach

The approach can be summarised in the following diagram, which is based on that developed by WHO.

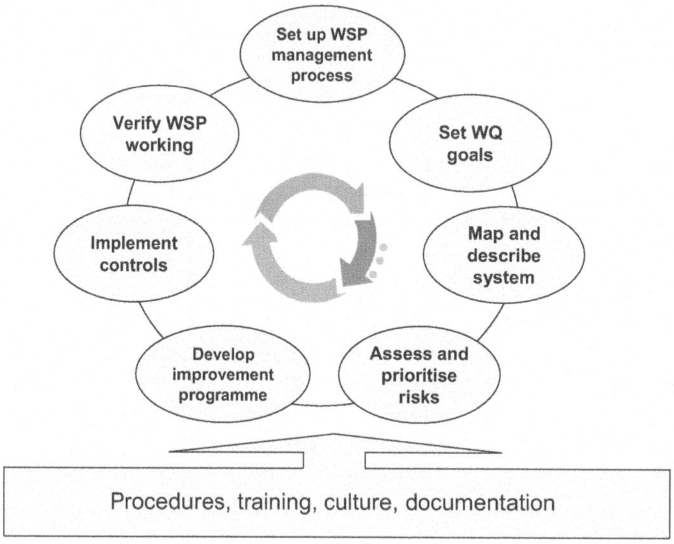

Figure 2: The basic Water safety plan cycle

Much more information on WSPs can be found on the websites of both WHO and IWA and also in the WHO water safety plan manual [3, 4, 5].

By definition the primary goal of a water safety plan is to ensure that the water delivered to consumers protects public health and is safe to drink at all times. However in moving towards such goals, the specific standards or health targets applied may legitimately vary from country to country and over time as recognised in the continuous review process for the WHO drinking water guidelines.

But the Bonn Charter also recognises that to meet the goal of "Good safe drinking water which has the trust of consumers" the water supplied has to meet objectives which go beyond simply compliance with vital health and statutory standards. It reminds us that there are a three related fundamental objectives to which all those involved in the supply of drinking water should strive:

1) "Access to good, safe, and reliable drinking water. This is one of the most basic needs of human society. In many areas water quality may already be high and continuing to improve. In others, where waterborne disease or other quality deficiencies are still prevalent, the basic provision of safe and good supplies is vital;

2) Water that is not just safe to drink but considered of good aesthetic quality by consumers; and

3) Water supplies in which consumers have confidence"

Water safety plans can address all these aspects, although in implementing such plans the priority must first and always be given to public health protection. It is for water suppliers in conjunction with the relevant authorities and their consumers to clearly determine the goals of their water safety plan. However, these would normally be considered under three main headings as identified in the Bonn Charter. But in the context of metals in drinking water all three are potentially relevant.

3. Hazards from metals in drinking water

Hazards from metals in drinking water can potentially arise at all stages of the supply process from catchment to consumer, and if not properly controlled can impact on health, consumer acceptability and/or consumer trust. The most likely ones are summarised in the diagram below.

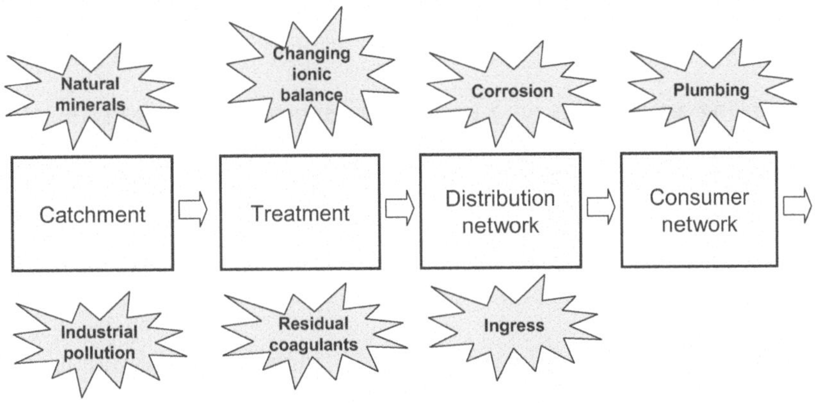

Figure 3: Typical water quality hazards from metals

Many of these issues are covered in much more detail in other papers from this conference, but in summary:

1) Industrial pollution

a. In some countries poor control of both solid and liquid industrial waste within the catchment can remain a hazard

2) Natural minerals

a. There are a range of naturally occurring minerals which can impact water quality whether; hazardous to health e.g. arsenic/uranium/lead; impact on consumer acceptability e.g. calcium; maybe both e.g. Magnesium

3) Changing ionic balance

a. Treatment which changes the mineral balance of the water e.g. ion exchange, desalination, can have both positive and negative consequences.

4) Residual coagulants

a. Poor control of residual coagulants coupled with inadequate removal of naturally occurring metals and other material can lead to deposits in the network and subsequent water discolouration

5) Corrosion

a. Corrosion of iron mains coupled with metals from residual metals after treatment can lead to deposits in the network and subsequent water discolouration

b. Risks of Calcium leaching from cement structures if alkalinity is low

6) Ingress

a. Ingress of pollution when mains pressure is low or zero can be a major risk, particularly for microbiological hazards but also metals

7) Plumbing

a. Leaching of metals from plumbing materials due to a range of factors including:
b. Inadequate corrosion control
c. Poor plumbing
d. Poor manufacturing

The actual risk from each of these will vary considerably between utilities and assessment of the specific risks can only be undertaken locally based on full knowledge of the design and operation of whole supply system.

4. A partnership approach

Control of metals in drinking water relies on a partnership approach with the roles of the many different stakeholders being clearly defined. The precise arrangements will vary from country to country but can include:

1) Water utilities - who have a variety of responsibilities including treatment of water to a high standard, including minimising the risk of plumbing metal dissolution and advising consumers on what they need to do to reduce risks

2) Health and local authorities - who may have a range of responsibilities including defining acceptable levels of metals in water and also advising consumers

3) National and/or local regulatory authorities - who may set standards for consumer pipework and plumbing equipment in contact with drinking water and also establish certification/training arrangements for plumbers and plumbing suppliers

4) Catchment authorities who have responsibility for managing risks in drinking water catchments

5) Builders, plumbers and plumbing suppliers - who must make sure they follow best plumbing practice and only use approved materials and fittings

6) Consumers and property owners themselves - who ultimately are responsible for the condition of plumbing in their properties.

Water utilities need to identify all key external WSP stakeholders and prioritise development of effective partnerships with them both at local level and nationally in collaboration with other water suppliers. However in every case it will be important to consider:

1) The messages to be conveyed
2) The best way to communicate these
3) What resources will be needed and how long will it take
4) The most realistic outcome that can be secured

5. Summary and conclusions

There is widespread agreement that WSPs are at the heart of good water quality management. WSPs can cover all types of water quality risk including water acceptability.

Metals hazards can be a significant in drinking water and can arise at all points in the supply system with both health and consumer acceptability impact. However the actual risks and the most cost effective way to mitigate these risks will vary considerably between utilities.

Control of the risks associated with drinking water can be difficult and has to take account of a range of complex interactions and involve a wide range of external partnerships to be successful.

References

[1] World Health Organisation Guidelines for Drinking Water Quality 3rd edition, 2004
[2] International Water Association: Bonn Charter for Safe Drinking Water, 2004
[3] International Water Association website on water quality
http://www.iwahq.org/Home/Themes/Water_and_health/Drinking_water_quality
[4] World Health Organisation website on drinking water quality
http://www.who.int/water_sanitation_health/dwq/WSP/en/index.html
[5] Bartram, J., Corrales, L., Davison, A., Deere, D., Drury, D., Gordon, B., Howard, G., Reinehold, A., Stevens, M. 2009 Water Safety Plan Manual. Step-by-Step Risk Management for

QC/QA scheme applied for monitoring of metals concentrations in water intended for human consumption sampled from the area of Warsaw performed by ICP-MS and ICP-OES techniques

Sławomir Garboś, Dorota Święcicka

National Institute of Public Health - National Institute of Hygiene, Department of Environmental Hygiene, 24 Chocimska Str., 00-791 Warsaw, Poland
Corresponding author e-mail: dswiecicka@pzh.gov.pl

Abstract

Monitoring of metals concentrations in water intended for human consumption sampled from the area of Warsaw was performed within DWM/N176/COST/2008 project financed by Polish Ministry of Science and Higher Education. Several metals which are listed in EU Directive 98/83/EC (Al, As, Cd, Cr, Cu, Fe, Mn, Ni, Pb) and additionally Zn were determined in 100 tap water samples collected from the area of Warsaw. The part of Warsaw supplied in drinking water by Central Water Supply System was chosen as control area. *Random Day Time* (RDT) monitoring based on taking 1 l of water directly from the tap used for consumption water drawing at a time randomly chosen within the day during normal office hours was applied for collection of tap water samples. For the determination of Al, As, Cd, Ni and Pb ICP-MS was applied while for the determination of rest of metals ICP-OES was used. During determinations of elements by ICP-MS and ICP-OES QA/QC applied scheme included: maintenance of optimal performance of spectral measurements, calibration control, procedural and on-field blanks levels control, precision control for analysis of double samples, check sample and certified reference material control. Appropriate control charts were prepared in order to assure of adequate quality control for achieved analytical results. Additionally Laboratory of Physicochemical Analysis of the Environment has participated in interlaboratory comparison organized by Technical University of Cracow (April 2010).

1. Introduction

Quality assurance is defined as all those planned and systematic actions necessary to provide adequate confidence that a product or service will satisfy given requirements for quality. National Institute of Public Health - National Institute of Hygiene (which consists of group of several laboratories) has been accredited according to PN-EN ISO/IEC 17025:2005 international standard by Polish Centre of Accreditation since 2004 (certificate Nr AB 509 - Figure 1). As first laboratory in Poland this laboratory achieved flexible range of research which provides the possibility of addition of analytes to previously developed research procedures and modification of determination range of analytes. Twenty six general procedures including PO-11 "Validation of chemical methods" and PO-13 "Chemical research quality control" are used in laboratories of NIPH-NIH. Among this big structure of laboratories the main tasks of Laboratory of Physicochemical Analysis of the Environment (part of the Laboratory of the Environmental Hygiene) are spectral and chromatographic analyses of water and the air. Analytical work in Laboratory of Physicochemical Analysis of the Environment is based on: several research procedures (developed according to PN-EN ISO international standards) for spectroscopic methods and chromatographic methods and 72 instructions (29 related own instructions and 43 instruction manuals). During measurements several actions are applied including quality control which consists of internal quality control IQC and external quality control EQC (mainly based on participation in interlaboratory comparisons and ring tests).

POLSKIE CENTRUM AKREDYTACJI
POLISH CENTRE FOR ACCREDITATION
PCA
CERTYFIKAT AKREDYTACJI
LABORATORIUM BADAWCZEGO
ACCREDITATION CERTIFICATE OF TESTING LABORATORY
Nr AB 509
Potwierdza się, że / This is to confirm that
NARODOWY INSTYTUT ZDROWIA PUBLICZNEGO
– PAŃSTWOWY ZAKŁAD HIGIENY
ul. Chocimska 24, 00-791 Warszawa
Certyfikat akredytacji ważny do dnia 14.07.2012 r.
Akredytacji udzielono dnia 15.07.2004 r.
DYREKTOR
KAROL HAUPTMANN

Fig. 1. The certificate Nr AB 509 issued for National Institute of

2. Materials and Methods

2.1 Spectroscopic techniques applied for elemental analysis of water. Validation of analytical methods

Simultaneous inductively coupled plasma optical emission spectrometer - IRIS Advantage DUO ER/S (Thermo Jarrell Ash, USA) is used for determinations of several metals in water intended for human consumption. Sample introduction system of ICP-OES spectrometer consists of: four channel peristaltic pump, glass concentric nebulizer, cyclone spray chamber, horizontal DUO plasma torch, axial and radial observation systems.

Additionally inductively coupled plasma mass spectrometer with collision cell technology - XSeries II (Thermo Electron Corporation, UK) is used for trace elemental analysis of drinking water. Sample introduction system of ICP-MS spectrometer consists of: three channel peristaltic pump, glass concentric nebulizer, Peltier cooled conical spray chamber ("impact-bead" type), quartz plasma torch equipped with silver screen (in order to achieve the best sensitivity), nickel sampling cone and nickel skimmer cone applied for construction of MS sector interface.

The operating conditions for ICP-OES and ICP-MS measurements are given in Table 1 and Table 2, respectively. Following validation parameters were established for analytical methods applied for elemental analysis of water by ICP-OES and ICP-MS techniques:

- selectivity,
- calibration functions used for construction of calibration graphs,
- linear ranges, linearity (range of correlation coefficients),
- sensitivities,
- detection limits and quantification limits (LOD and LOQ),
- repeatability,
- reproducibility,
- trueness,
- recoveries (in the presence of drinking water matrix),
- expanded uncertainties (assessment of uncertainty budgets).

All established during validation detection limits, precisions (as repeatability) and trueness met the requirements listed in Directive 98/83/EC.

Table 1. Operating conditions for ICP-OES measurements

Sample introduction system / parameter - ICP-OES	**Type / value**
Plasma torch	Quartz, horizontal duo
Spray chamber	Cyclone
Nebulizer	Glass concentric
RF frequency	27.12 mhz
Forward power	1150 w
Argon flow rates: - plasma - intermediate - optics interface - purging optics - purging cid detector - nebulizer pressure	 15 l/min 1 l/min 4 l/min 4 l/min 80 units 26 psi
Sample pumping flow rate Waste pumping flow rate	110 rpm (approx. 2 ml/min) 110 rpm
Rinsing time	60 s
No. Replicates/sample	4
Integration time in the range of Wavelengths: 175 - 275 nm Including: Cr - 206.149 nm; ni - 231.604 nm; Cu - 224.700 nm; pb - 220.353 nm; Fe - 238.204 nm; zn - 206.200 nm; Mn - 257.610 nm	50 s (axial observation system)

Table 2. Operating conditions for ICP-MS measurements.

Sample introduction system / parameter - ICP-MS	**Type / value**
Plasma torch	Quartz, equipped with silver screen
Spray chamber	"impact-bead" type (cooled to 2°c with peltier system)
Nebulizer	Glass concentric
R.F. frequency	27.12 mhz
Forward power	1400 w
Argon flow rates: Cool / auxiliary / nebulizer	13 l/min / 0.72 l/min / 0.95 l/min
Target analyte isotopes monitored	^{27}al, ^{75}as, ^{114}cd, ^{60}ni, ^{208}pb
Internal standard	^{89}y
Number of points per peak (channels per mass)	1
Dwell time per isotope	10 ms
Sweeps per run	230
Acquisition time - main run	30 s - peak jumping
No. of runs per sample	3
Sample pumping flow rate	Approx. 0.8 ml/min
Uptake and wash times	60 s

3. Results and discussion

3.1 Internal quality control scheme (IQC)

Internal quality control (IQC) scheme included:

- maintenance of optimal performance of ICP-OES and ICP-MS spectrometers, - calibrations of ICP-OES and ICP-MS spectrometers,
- composition of sequence of analytical batch (run) for obligatory measurements,
- determination of elements in certified reference materials.

3.1.1 Maintenance of optimal performance of ICP-OES and ICP-MS spectrometers

For assurance of optimal performance of ICP-MS spectrometer several actions are performed e.g. checking:

- sensitivities (have to be higher than established previously acceptable minimum levels),
- precisions for metals measured in tune solution at concentration levels of 1 μg/l (≤ 2 %),background at 220 amu (equal or lower than 1 cps),
- oxide ions level (CeO^+/Ce^+ equal or lower than 2 %),
- double charged ions level (Ba^{2+}/Ba^+ equal or lower than 2.5 %),
- pulse counting voltage,
- cross calibration between analog detection mode and pulse counting mode (Figure 2),
- mass calibration (Figure 3),
- fluctuation of laboratory temperature (equal or lower than 2°C/h during first 0.5 h, then equal or lower than 1°C/h).

For assurance of optimal performance of ICP-OES spectrometer several actions are performed e.g. checking:

- correctness of emission lines imaging,
- symmetrical coverage of emission peaks integration (integration of 2×3 pixels or 3×3 pixels),
- sensitivities (higher than previously established acceptable minimum levels),
- precisions for metal measured in tune solution at concentration levels of 0.5 mg/l (≤ 2 %),analog indication of argon flow rate used for purging optics interface (equal 4 l/min),
- analog indication of nebulizer pressure (equal 26 psi),

- digital indication of optics temperature (90.0°F ± 0.3°F),
- fluctuation of laboratory temperature (equal or lower than 2°C/h during first 0.5 h, then equal or lower than 1°C/h).

For optimization of the ICP-MS spectrometer performance tune solution is applied - Analityk-CAL-40 (Inorganic Ventures, USA) consists of Ba, Be, Bi, Ce, Co, In, Li, Ni, Pb and U.

For optimization of the ICP-OES spectrometer performance special set of tune solutions is applied - "ICP Multi Element Standard Solution IV CertiPUR" (Merck, Germany) consists of Ag, Al, B, Ba, Bi, Ca, Cd, Co, Cr, Cu, Fe, Ga, In, K, Li, Mg, Mn, Na, Ni, Pb, Sr, Tl and Zn, "Arsenic ICP Standard CertiPUR" (Merck, Germany) and "Antimony ICP Standard CertiPUR" (Merck, Germany)

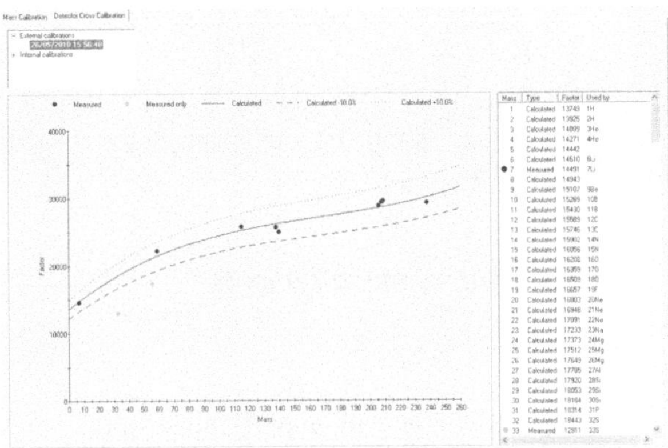

Fig. 2. Cross calibration between analog detection mode and pulse counting mode.

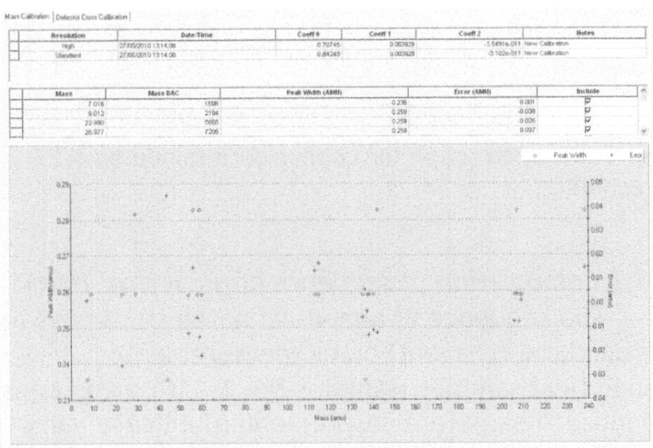

Fig. 3. Mass calibration of ICP-MS spectrometer.

3.1.2 Calibrations of ICP-OES and ICP-MS spectrometers

At the beginning of each measurement day calibrations of ICP-OES and ICP-MS spectrometers are performed. For calibrations of ICP-OES and ICP-MS spectrometers calibration solutions based on reference material (RM) solutions are applied - "ICP Multi Element Standard Solution IV CertiPUR" (Merck, Germany) consists of Ag, Al, B, Ba, Bi, Ca, Cd, Co, Cr, Cu, Fe, Ga, In, K, Li, Mg, Mn, Na, Ni, Pb, Sr, Tl and Zn, "Arsenic ICP Standard CertiPUR" (Merck, Germany) and "Antimony ICP Standard CertiPUR" (Merck, Germany).

All concentrations of elements listed in RM description were traceable to NIST standard reference materials. For preparations of adequate calibration solutions UltraPUR concentrated nitric acid (60 %, Merck, Germany) and deionized water achieved in Simplicity 185 system (Millipore, USA) were used. In the cases of ICP-OES and ICP-MS techniques borosilicate calibration flasks class A and PMP calibration flasks class A were applied, respectively. Only calibratedmicropipettes (calibration every three months or when it is necessary) with adequate disposable micropipette tips were used for dosage of stock RM solutions and nitric acid.

Typical calibration graphs for Cu determined by ICP-OES (calibration solutions in the range of 0.2 - 1.4 mg/l) and for Al determined by ICP-MS (calibration solutions in the range of 0.5 - 30 µg/l) were presented in Figure 4 and Figure 5, respectively. Weighted linear regression was applied for ICP-OES calibration while ordinary linear regression with intercept drawn "through blank" was applied in the case of ICP-MS calibration. Correlation coefficients better than 0.9999 were usually achieved in both cases - ICP-OES and ICP-MS calibrations.

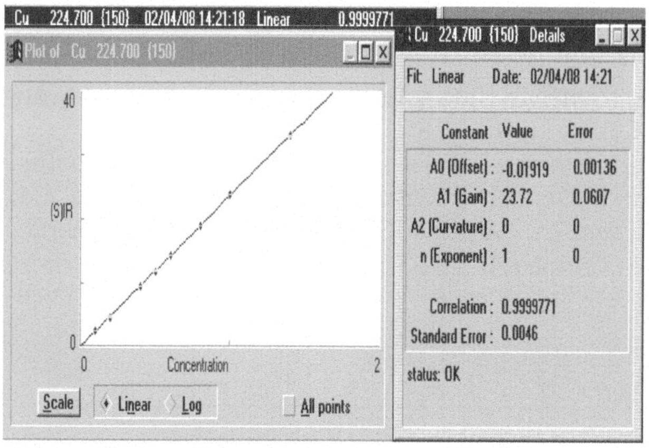

Fig. 4. Calibration graph for copper determined by ICP-OES technique.

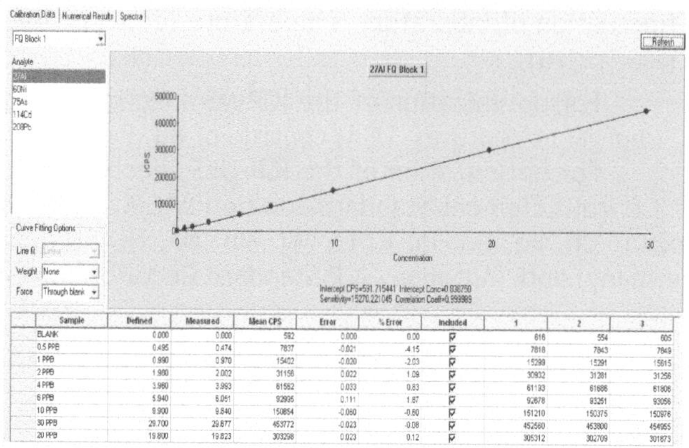

Fig. 5. Calibration graph for aluminium determined by ICP-MS technique.

3.1.3 Composition of sequence of analytical batch (run) for obligatory measurements

The sequence of analytical batch (run) is as follow:

a) procedural (reagent) blank measurement,
b) analytical samples (samples No. 1 - 10) measurements,
c) duplicate sample measurement (duplicate sample is correlated to one of analytical samples No. 1 - 10),
d) "on-field" blank measurement,
e) check standard measurement with appropriate concentrations of analytes and additions of matrix elements - Ca (100 mg/l), Na (60 mg/l), Mg (20 mg/l) and K (20 mg/l).

It means that each sequence of analytical batch (run) consists of 14 sample measurements (or lower measurements for shorter series with lower number of analytical samples) with obligatory measurements presented in points a, c, d, e:

- one procedural (reagent) blank per one batch of 10 analytical samples,
- duplicate sample per one batch of 10 analytical samples,
- one on-field blank per one batch of 10 analytical samples,
- one check standard (reference material - control material) per one batch of 10 analytical samples.

3.1.3.1 Procedural (reagent) blank and "on-field" blank measurements

For preparation of procedural (reagent) blank 0.5 ml of UltraPUR concentrated nitric acid (Merck, Germany) was added into calibration flask with the volume of 100 ml and then deionized water achieved from Simplicity 185 system (Millipore, USA) was added up to 100 ml. This solution is sometimes called as "reagent blank" but it tests more than the purity of reagents (e.g. concentrated nitric acid and deionized water). For example it is capable of detecting contamination of the analytical system originating from any source, for example:

- glassware (borosilicate and PMP calibration flasks),
- micropipette tips,
- atmosphere of laboratory,
- carry-over between samples (especially between samples characterized with different concentrations of analytes).

Nevertheless the main task connected with measurements of procedural blank is monitoring detection limit (LOD) stability. Therefore for each element adequate control chart is prepared with the use of data concerning procedural blank measurements (x_1, x_2, x_3, ... x_n; $n=10$). All results of procedural blank measurements have to lie in the range ±LOD. In the case when a result for procedural blank measurement is outside above mentioned range analytical batch is stopped and then the reasons have to be identified.

After filling full control chart with 10 results of procedural blank measurements (which are in the range ±LOD) appropriate statistical calculations are performed. Q-Dixon test is applied in order to eliminate outlier data from the set of 10 results of procedural blank measurements. For this purpose coefficients Q_a and Q_b are established after arranging data (x_1, x_2, x_3, ... x_n; $n=10$) in ascending order (x_{min} - x_{max}). Q_a and Q_b are calculated according to Equation 1 and Equation 2, respectively:

$$Q_a = \frac{x_2 - x_1}{x_{n-1} - x_1} \quad (1)$$

$$Q_b = \frac{x_n - x_{n-1}}{x_n - x_2} \quad (2)$$

and then compared with suitable critical value $Q_{critical}$. When Q_a and Q_b are equal or lower than $Q_{critical}$ then all 10 results are used for estimation of detection limit from the current control chart. When Q_a and/or Q_b are higher than $Q_{critical}$ appropriate outlier result is erased from the set of 10 results. Then after taking into account only set of 9 results of procedural blank measurements Q-Dixon test is applied once again. If the result of test is "PASS" then the set of 9 results is applied for the calculation of detection limit from the current control chart.

After estimation of LOD from the current control chart in general two cases could be observed:

- estimated detection limit from the current control chart is equal or lower than the detection limit estimated during validation process. In such case any action is performed and the detection limit achieved during validation process is used during reporting.
- estimated detection limit from the current control chart is higher than the detection limit estimated during validation process. In such case F-test has to be applied in order to state if the difference between calculated detection limit from the current control chart and the detection limit achieved during validation process is statistically essential or not. F-test is based on the comparison of standard deviation SDblank estimated for the set of 10 results of procedural blank measurements (or for lower number of results due to application of Q-Dixon test) and standard deviation SDblank_val estimated during validation process. Coefficient Fest is calculated after taking into account adequate variances (Equation 3):

$$F_{est} = \frac{SD^2_{blank}}{SD^2_{blank\_val}} \quad (3)$$

and then compared with suitable critical value $F_{critical}$. When F_{est} is lower than $F_{critical}$ any action is performed and the detection limit achieved during validation process is used during reporting. When F_{est} is higher than $F_{critical}$ then new higher detection limit achieved from the current control chart has to be applied during reporting.

The same above described process has to be applied for data concerning "on-field" blanks (n=10) which additionally provides possibility of monitoring possible contamination originating from HDPE containers, micropipette tips, concentrated nitric acid, atmosphere of on-field sampling.

The examples of control charts achieved with the use of data concerning procedural blank (n=10) and of data concerning "on-field" blank measurements (n=10) for cadmium determined by ICP-MS and for manganese determined by ICP-OES are presented in Figure 6 and Figure 7, respectively.

All results of procedural and "on-field" blanks measurements for cadmium and manganese are lied in the range ±LOD. ICP-OES detection limits estimated during validation process and detection limits calculated after taking into account: data based on procedural blanks (n=10) and data based on "on-field" blanks (n=10) were listed in Table 3. Higher detections limits based on data concerning procedural blanks in comparison to detection limits achieved during validation process were calculated in the cases of chromium, nickel and zinc determined by ICP-OES. Additionally higher detections limits based on data concerning "on-field" blanks in comparison to detection limits achieved during validation process were calculated in the cases of copper, nickel and lead determined by ICP-OES. But differences between LODs estimated during validation process and LODs calculated from currently finished control charts are not statistically essential.

ICP-MS detection limits estimated during validation process and detection limits calculated after taking into account: data based on procedural blanks (n=10) and data based on "on-field" blanks (n=10) were listed in Table 4.

As can be seen for all elements determined by ICP-MS (with exception of aluminium) LODs estimated during validation process are higher than those calculated from currently finished control charts for procedural and "on-field" blanks measurements. In the case of aluminium higher detection limit based on data concerning "on-field" blank measurements in comparison to detection limit achieved during validation process was calculated, however, any action was performed because of positive result achieved for F-test.

3.1.3.2 Duplicate sample measurements

Tap water sample with the volume of 1 l was mixed directly after collection and two sub-samples (one analytical sample and one duplicate sample) with the volumes of 100 ml were transferred into 125 ml HDPE containers. Then sub-samples were acidified with 0.5 ml of concentrated nitric acid (UltraPUR, Merck, Germany), marked and transported to laboratory in refrigerator at 4ºC ± 2.5ºC. One of above described sample is analyzed as analytical sample while second one is typical IQC sample called duplicate sample. Thus duplicate sample measurement is performed once per one batch (run) of 10 analytical samples. It is capable of detecting contamination originating from several sources, for example: HDPE containers, micropipette tips, concentrated nitric acid, atmosphere of the environment of on-field sampling. Additionally possible adsorption process of analytes on internal walls of HDPE containers could be taken into account.

Table 3. ICP-OES LODs estimated during validation process and LODs calculated after taking into account: data based on procedural blanks (n=10) and data based on "on-field" blanks (n=10).

Element	**LOD estimated during validation process [µg/l]**	**LOD estimated using data concerning procedural blank [µg/l]**	**LOD estimated using data concerning "on-field" blank [µg/l]**
Cr	1.1	1.1 [F-test: $F_{est}=1.04 < F_{critical}=3.13$]	1.0
Cu	1.5	0.92	1.9 [F-test: $F_{est}=1.59 < F_{critical}=3.13$]
Fe	0.71	0.68	0.68
Mn	0.19	0.14	0.07
Ni	1.0	1.3 [F-test: $F_{est}=1.71 < F_{critical}=3.13$]	1.1 [F-test: $F_{est}=1.12 < F_{critical}=3.13$]
Pb	8.0	6.8	8.3 [F-test: $F_{est}=1.07 < F_{critical}=3.13$]
Zn	0.49	0.56 [F-test: $F_{est}=1.3 < F_{critical}=3.13$]	0.41

For duplicate sample analyses additional table is filled with following data: determined concentrations of element in sub-samples (in analytical sample and in correlated duplicate sample), average concentration, standard deviation of mean concentration and relative standard deviation. Adequate control charts for duplicate sample analyses are constructed for controlling precisions of determinations expressed as relative standard deviation values (RSD; %).

The analytical batch was accepted when RSDs for the determinations of all elements in sub-samples were below 5 %. However, this requirement is only applied for determined average concentrations (calculated for duplicate sample and adequate analytical sample analyses) equal or higher than: 1/10 of maximum admissible concentration levels listed in Directive 98/83/EC for Al, As, Cd, Cr, Fe, Mn, Ni and Pb; 25 μg/l for Cu and Zn.

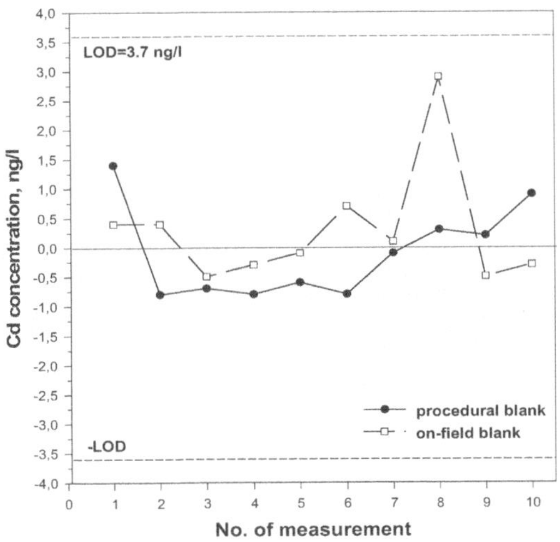

Fig. 6. Control charts for procedural (reagent) and -field blanks measurements of cadmium determined by ICP-MS.

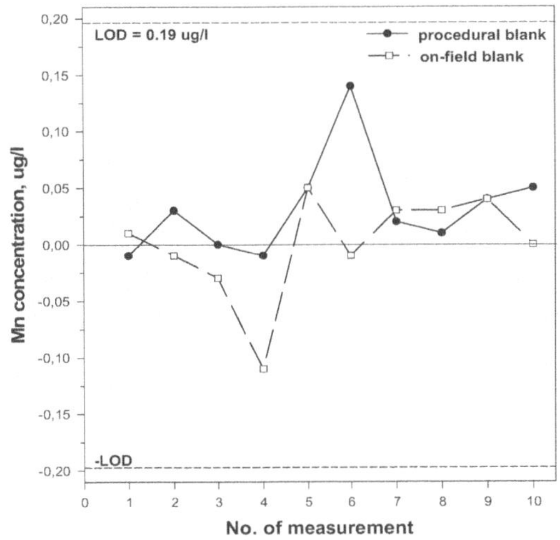

Fig. 7. Control charts for procedural (reagent) and on-field blanks measurements of manganese determined by ICP-OES.

Table 4. ICP-MS LODs estimated during validation process and LODs calculated after taking into account: data based on procedural blanks (n=10) and data based on "on-field" blanks (n=10).

Element	LOD estimated during validation process [ng/l]	LOD estimated using data concerning procedural blank [ng/l]	LOD estimated using data concerning on-field blank [ng/l]
Al	43	23	48 [F-test: $F_{est}=1.24 < F_{critical}=3.13$]
Ni	24	11	8
As	43	22	28
Cd	3.7	2.4	1.3
Pb	34	11	4

The ranges of determined average concentrations of elements for duplicate samples and threshold values above which requirement "RSD ≤ 5 %" is applied are presented in Table 5.

Table 5. The ranges of determined average concentrations of elements for duplicate samples and threshold values above which requirement "RSD ≤ 5 %" is applied.

Element	Range of determined average concentrations for duplicate samples [µg/l]	Threshold value (when average concentration is ≥ this value then the requirement "RSD ≤ 5 %" is applied) [µg/l]
Al	2.11 - 8.10	20
As	0.34 - 0.56	1.0
Cd	0.02 - 0.61	0.50
Cr	<1.1	5.0
Cu	1.70 - 1460	25
Fe	11.0 - 655	20
Mn	0.65 - 22.8	5.0
Ni	1.77 - 32.5	2.0
Pb	0.34 - 4.62	1.0
Zn	37.2 - 1085	25

Control charts of RSDs calculated for adequate average concentrations determined in duplicate samples for As, Al, Cd, Cu, Fe, Mn, Ni, Pb and Zn are presented in Figure 8 - 16. All calculated RSDs were equal or lower than 5 % for all determined average concentrations of Al, Cd, Fe, Ni, Pb and Zn existed in duplicate samples at concentration levels even below threshold values listed in Table 5. In the cases of As, Cu and Mn all calculated RSDs were equal or lower than 5 % for all determined average concentrations higher than adequate threshold values - 1.0 µg/l, 25 µg/l and 5.0 µg/l, respectively.

3.1.3.3 Check standard measurements

At the end of sequence of analytical batch (run) check standard measurements are performed. Check standard solution consists of appropriate concentrations of analytes and additions of matrix elements - Ca (100 mg/l), Na (60 mg/l), Mg (20 mg/l) and K (20 mg/l). Nominal concentrations of Cr, Cu, Fe, Mn, Ni, Pb and Zn in check standard solution analyzed by ICP-OES were at the levels of 0.2 mg/l while nominal concentrations of Al, As, Cd, Ni and Pb in check standard solution analyzed by ICP-MS were at the levels of 4 µg/l. For preparation of check standards solutions the same type of reference materials like in the case of calibrations were used but with different number of series. Check standard measurements are performed in order to control the stability of currently applied calibrations in the presence of simulated water matrix (similar to that which is present in natural tap water samples) and additionally errors connected with contamination and/or preparation of calibration standards could be indicated. The results of determinations of elements in analytical samples achieved within analytical batch were accepted when determined concentrations of elements in check standards were in the range 95-105 % of nominal concentrations. All results of determinations for above mentioned elements achieved during 10 analytical runs were within the range 95 - 105 % of normal concentrations:

Cr, Cu, Fe, Mn, Ni, Pb and Zn determined by ICP-OES - within the concentration range 0.19 to 0.21 mg/l and Al, As, Cd, Ni and Pb determined by ICP-MS - within the concentration range 3.8 - 4.2 µg/l. Thus all results derived from 10 analytical runs were accepted because trueness for check standards measurements for all elements was in the range ±5 %. Adequate control charts (achieved within 10 analytical runs) for Cr, Cu, Fe, Mn, Ni, Pb and Zn present in check standard solution at the levels of 0.2 mg/l determined by ICP-OES and for Al, As, Cd, Ni and Pb present in check standard solution at the levels of 4 µg/l determined by ICP-MS are presented in Figure 17 and Figure 18.

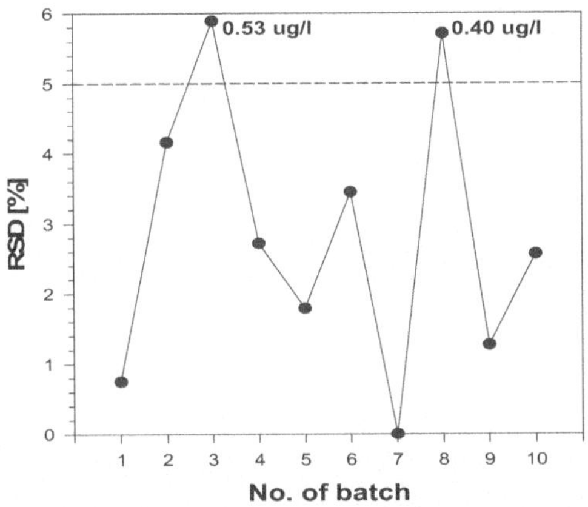

Fig. 8. Control charts of RSD calculated for adequate average concentrations of As determined by ICP-MS in duplicate samples.

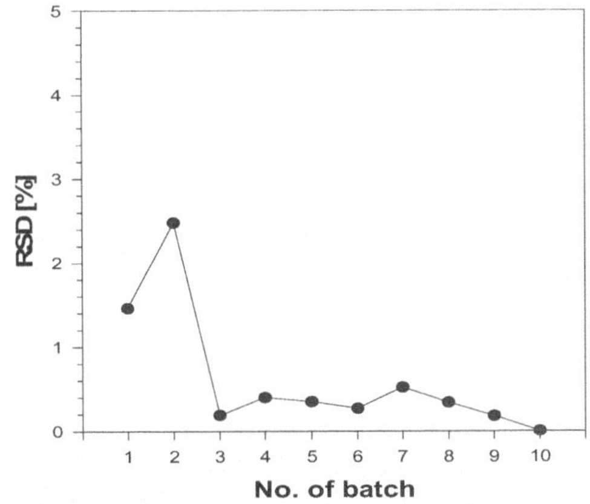

Fig. 9. Control charts of RSD calculated for adequate average concentrations of Al determined by ICP-MS in duplicate samples.

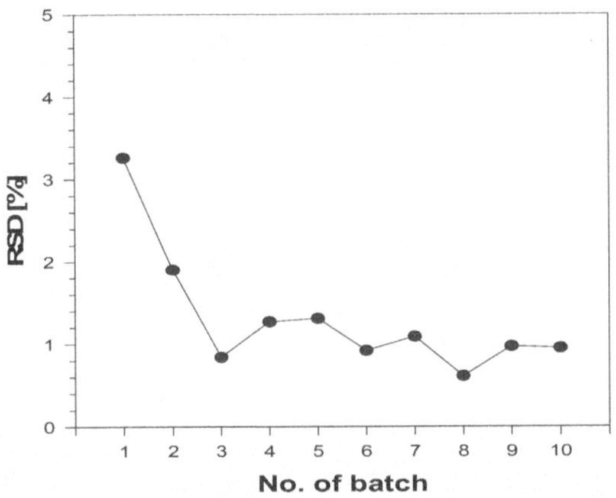

Fig. 10. Control charts of RSD calculated for adequate average concentrations of Cd determined by ICP-MS in duplicate samples.

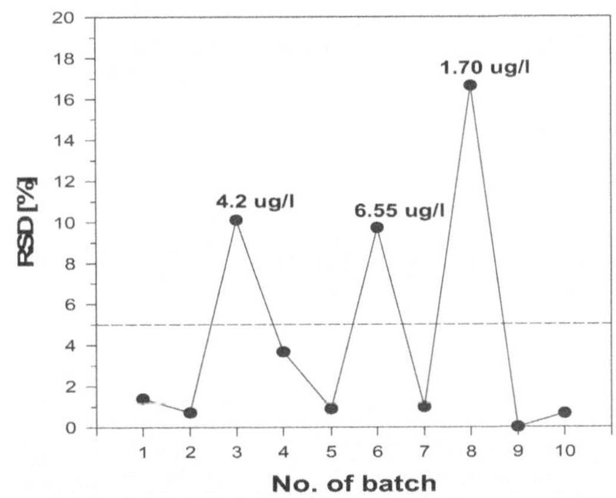

Fig. 11. Control charts of RSD calculated for adequate average concentrations of Cu determined by ICP-OES in duplicate samples.

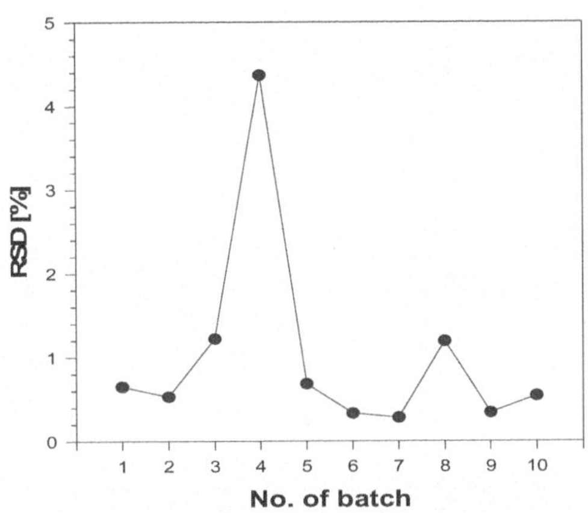

Fig. 12. Control charts of RSD calculated for adequate average concentrations of Fe determined by ICP-OES in duplicate samples.

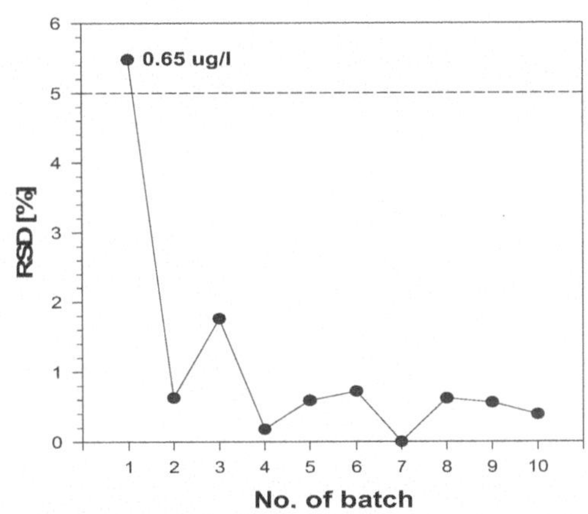

Fig. 13. Control charts of RSD calculated for adequate average concentrations of Mn determined by ICP-OES in duplicate samples.

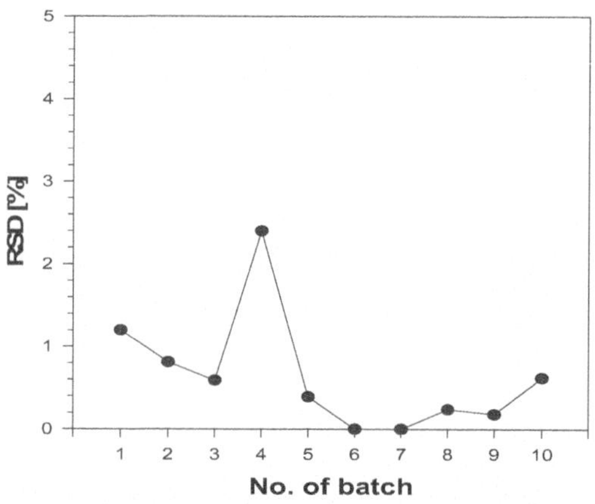

Fig. 14. Control charts of RSD calculated for adequate average concentrations of Ni determined by ICP-OES in duplicate samples.

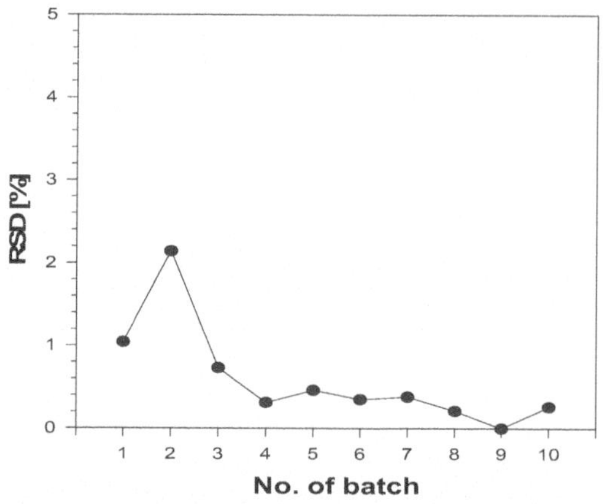

Fig. 15. Control charts of RSD calculated for adequate average concentrations of Pb determined by ICP-MS in duplicate samples.

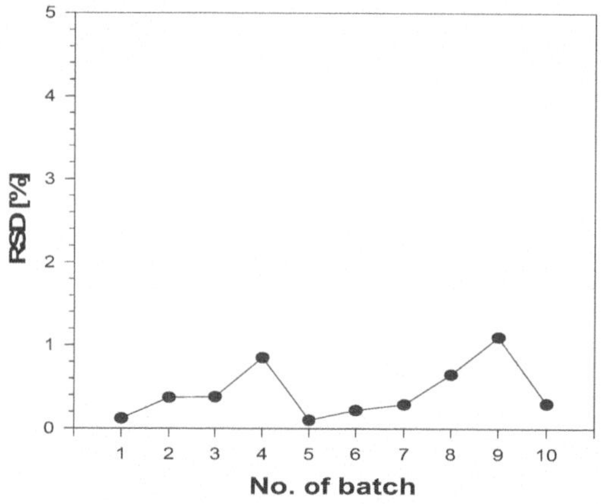

Fig. 16. Control charts of RSD calculated for adequate average concentrations of Zn determined by ICP-OES in duplicate samples.

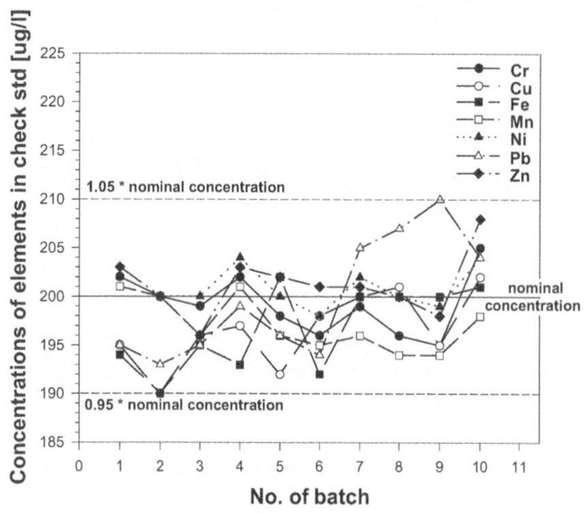

Fig. 17. Concentrations of Cr, Cu, Fe, Mn, Ni, Pb and Zn determined by ICP-OES in check standard solution (nominal concentrations of elements = 0.2 mg/l).

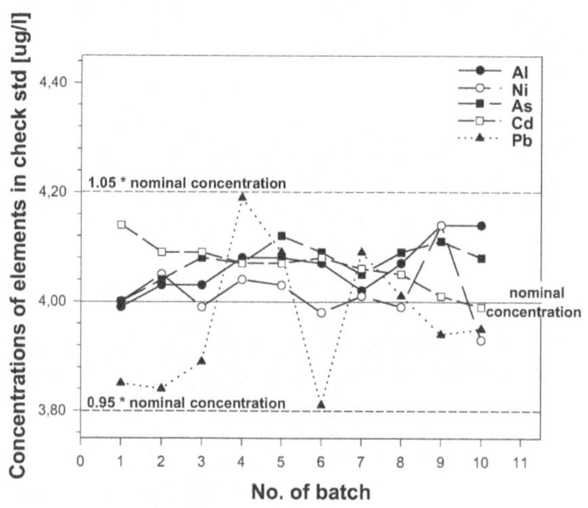

Fig. 18. Concentrations of Al, Ni, As, Cd and Pb determined by ICP-MS in check standard solution (nominal concentrations of elements = 4 µg/l).

3.1.3.4 Determination of elements in certified reference materials - trueness control

During each measurement day certified reference materials were analyzed in order to check the trueness of metals determinations in the presence of typical water matrixes. For this purpose several certified reference materials including: CRM TMDA-51.3 *"A high level fortified standard for trace elements"* (Environment Canada) and SRM 1643e *"Trace Elements in Water"* (National Institute of Standards & Technology, USA) were applied.

The example of results concerning determinations of metals in SRM 1643e *"Trace Elements in Water"* (National Institute of Standards & Technology, USA) by ICP-OES technique (including achieved trueness) are presented in Table 6. All results were achieved with satisfactory accuracy - trueness of determinations was better than 10 % for all determined elements.

Table 6. The results concerning determinations of metals in SRM 1643e by ICP-OES technique.

Element	True value [mg/l]	Uncertainty [mg/l]	Determined concentration [mg/l]	Trueness [%]
Cr	0,02040	0,00024	0,0212	3,9
Zn	0,0785	0,0022	0,0766	-2,4
Cd	0,006568	0,000073	0,0069	5,1
Mg	8,037	0,098	7,76	-3,4
Mn	0,03897	0,00045	0,0370	-5,1
Cu	0,02276	0,00031	0,0216	-5,1
Ni	0,06241	0,00069	0,0644	3,2
Na	20,740	0,260	20,4	-1,6
Ca	32,300	1,100	30,7	-5,0
Fe	0,0981	0,0014	0,0978	-0,31

3.2 Participation in interlaboratory comparisons - trueness control

Laboratory of Physicochemical Analysis of the Environment participated in interlaboratory comparison organized by Institute of Chemistry and Inorganic Technology of Technical University of Cracow in April 2010. The example of results concerning determinations of metals in water sample by ICP-OES technique including achieved Z-scores and trueness are presented in Table 7. All results were achieved with satisfactory accuracy. All achieved Z-scores were equal or lower than 2 and additionally trueness was better than 10 % for all determined elements.

Table 7. The results concerning determinations of metals in water sample by ICP-OES technique within interlaboratory comparison organized by Technical University of Cracow (April 2010).

Element	True value [mg/l]	Determined concentration [mg/l]	Trueness [%]	Z-score
Cr	0,046	0,044	-4,3	-0,54
Zn	0,068	0,0687	1,0	0,13
Cd	0,0043	0,0045	4,7	0,56
Cu	0,014	0,014	0,0	0
Mg	7,1	6,91	-2,7	-0,34
Mn	0,061	0,059	-3,3	-0,36
Ni	0,013	0,014	7,7	0,96
Pb	0,022	0,023	4,5	0,44
Na	51,8	51,4	-0,8	-0,23
Ca	76,5	76,0	-0,7	-0,31
Fe	0,48	0,444	-7,5	-2,0

4. Conclusions

The developed research methods for the determinations of elements in drinking water by ICP-OES and ICP-MS techniques and achieved results within applied internal quality control and external quality control met requirements described in Directive 98/83/EC and in Polish Decree of Minister of Health from 29 March 2007 (with further changes) on the quality of water intended for human consumption. All achieved detection limits are lower than 1/10 of maximum admissible concentration levels for elements listed in Directive 98/83/EC. During performed IQC actions no statistically essential changes for the detection limits established during validation process were observed. All results of determination of metal concentrations in certified reference materials and in water analyzed within interlaboratory comparison were achieved with trueness better than 10 %. Stability of calibrations within analytical runs was sufficiently correct. All achieved precisions of determinations expressed as RSDs were lower than 10 % for concentrations equal or higher than established threshold concentration values.

5. Acknowledgments

The work was done within DWM/N176/COST/2008 project financed by Polish Ministry of Science and Higher Education.

Drinking water quality in the city of Belgrade and health risks from domestic use of filters with reverse osmosis

Ivana Ristanovic-Ponjavic, Marina Mandic-Miladinovic, Sezana Vukcevic
Public Health Institute, Belgrade, Bulevar despota Stefana no. 54a
11000 Belgrade, Serbia

Corresponding author e-mail: ivana.ristanovic@zdravlje.org.rs

1. Background

Public Health Institute of the City of Belgrade performs an audit control of drinking water from the Belgrade waterworks in terms of public health protection. Drinking water monitoring program includes laboratory analysis of about 6500 water samples per year. Despite good water quality from the City waterworks, we have registered an increase of demands for analysis of water filtrated through filters with reverse osmosis for domestic use. Prolonged consumption of low mineralized or demineralized water could have adverse health effects, such as homeostatic disorders, disorders connected with low calcium and magnesium intake, low intake of essential elements and micronutrients, etc.

2. Method

In 2009. we have analyzed 6650 water samples from waterworks. 14 basic (indicator) parameters were analyzed in 6.158 samples, 49 parameters in 360 samples and 124 parameters in 132 water samples. Indicator parameters, cations and anions were analyzed in 30 samples of filtrated water.

3. Results

The results of physical and chemical analysis of water samples from waterworks showed high level (over 98%) of compliance to the Serbian Regulation on drinking water quality, as well as to the Council Directive on water intended for human consumption 98/83/EC. Turbidity and iron concentration were the most frequent causes of the incompliance. The values of these parameters were not of concern in terms of influence on human health. Values of heavy and toxic metals, polynuclear aromatic hydrocarbons (PAH), polychlorinated biphenyls (PCBs), pesticides, phenols, cyanides, mineral oils and disinfection by-products were either bellow upper limits, or bellow detection limits. Cations and anions ranges were: calcium (Ca: 54.1-77.8 mg/L), magnesium (Mg: 10.2-25.8 mg/L), potassium (K: 1.3-2.1 mg/L), sodium (Na: 6.9-18.0 mg/L), bicarbonates (HCO3: 159.4-335.8 mg/L), total dissolved solids (TDS: 243.3-386.1). Total hardness ranged from 10.6 to 16.9 °dH, depending on water source. All samples of filtrated water had very low total dissolved solids (TDS: < 50.0 mg/L), calcium (Ca: < 2.0 mg/L), bicarbonates (HCO3: < 30.0 mg/L), and were bellow WHO recommendations for demineralized water (Nutrients in Drinking Water, WHO, 2005).

4. Conclusion

Upon results of water quality monitoring program there is no reason for domestic use of filters with reverse osmosis. Very low mineralization of water from these filters could have adverse health effects. Because of this, the setting of minimum values for the content of the essential elements or TDS in drinking water regulations should be considered.

Consumer concerns about drinking water in an area with high levels of naturally occurring arsenic in groundwater, and the implications for managing health risks

J. Leventon [1,2] , S. Hug [3]

[1] *Central European University, Budapest, Hungary,*
[2] *Technical University of Crete, Chania, Greece,*
[3] *EAWAG, Zurich, Switzerland*

Corresponding author e-mail: leventon_julia@phd.ceu.hu

Abstract

The aim of this exploratory study was to examine and explain the role local populations play in the successful management of drinking water. In Békés County, Eastern Hungary, 98% of the population receive piped water in their households from supplies of moderately treated (usually chlorinated) geothermal groundwater. Meeting the EU Drinking Water Directive parameters for a number of geogenic elements including arsenic and boron is a significant challenge for water managers. This mixed method research includes interviews, questionnaires and water samples. The results demonstrate distinct differences in the way policy and people construct quality.These differences are important because policy shapes management, whereas people as consumers dictate use; if they do not correspond, management cannot be successful.

1. Introduction

This paper presents part of a larger PhD research project addressing the governance in the EU of drinking water high in naturally occurring arsenic. Natural arsenic in groundwater poses a management challenge in areas where groundwater is relied on as a source of drinking water. Groundwater is a source of bacterially-clean drinking water. Arsenic is a known carcinogen [1] which is linked to a range of illnesses including diabetes and ischemic heart disease [2,3]. Arsenic in drinking water has a wide impact in parts of Southeast Asia, including Cambodia, Vietnam and Bangladesh. In Bangladesh alone, between 35-77 million people are exposed to chronic arsenic poisoning [4]. Parts of the European Union (EU) have aquifers with high levels of naturally occurring arsenic [5]. In some countries, including Slovakia, Romania and Hungary, these aquifers have historically been used as drinking water sources [6]. In these cases, preventing exposure to arsenic through drinking water, and protecting human health in these areas is shaped by the legal system of the EU. Under the EU's Drinking Water Directive (DWD), drinking water must not exceed 10 ppb As.

This paper examines the role of consumers in water governance in a Hungarian case study. In Hungary, municipalities have the primary responsibility to deliver safe drinking water under the Water Act, Act LXIII of 1994. Municipalities are controlled by residents via the electoral system, and therefore residents are important stakeholders in water governance. In Békés county, Southeastern Hungary, drinking water supplies rely on geothermal aquifers. Besides being high in arsenic, water is also high in organic matter, boron and manganese. The area is known for its geothermal water and has a number of thermal spas fed by the same waters. Approximately 65% of the county's population receive drinking water that breaches this requirement. In response, the central Hungarian government has launched the DARFU Drinking Water Improvement Program. Under the program municipalities and companies must form associations and instigate changes to technology and infrastructure[1]. In addition, in some of the towns there are publicly accessible wells, giving people access to untreated geothermal water. Previous research shows that consumer perceptions of drinking water are based on organoleptic properties, knowledge and information, and trust in water suppliers [7]. However, these factors are context specific. The aim of this paper is to

[1] More information on the DARFU Drinking Water Improvement project can be found at: http://www.ivovizunk.hu/index.php (Hungarian only)

identify the factors which contribute to consumer perceptions of water quality, how they vary, and what influence this has on drinking water management.

2. Methods

In order to understand how consumer quality perceptions changed with water chemistry, we collected data in four towns. In the Békés county study area, each distribution system delivers water with different chemical properties. Each is also managed in a different way: either through a localised distribution system which is municipally run; a localised distribution system which is contracted to a central water company; or a central company, central distribution system. There are currently 11 companies operating 40 water distribution systems, delivering water to approximately 203,000 people in 75 towns and villages. By comparing between four towns with different management approaches and water properties, we could explore how consumer perceptions changed, and the extent to which they were motivated by changes in the management approach and water properties. The four towns chosen, along with management information is shown in Table 1, columns A - C.

Table 1. The study towns

A: Town	B: Population	C: Distribution Type	D: Water Samples	E: Street Interviews	F: Questionnaires
A	20,647	Central company, system serves many towns	10	32	23
B	3,960	Municipal country, single town system	6	29	26
C	32,016	Municipal country, single town system	4	81	31
D	9,465	Central company, system serves two towns	5	29	28

In each town, data on resident's perceptions were collected in two ways. Initially, people were stopped on the street and asked to talk about their opinions on concerns on tap water and their own drinking water habits. The number of people interviewed in this way is shown in Table 1, column E. These interviews were non-structured, beyond the initial question "what do you think about tap water?". Follow up questions were formulated in response to answers given. In order to explore the opinions and motivations of well users, similar interviews were conducted by visiting the wells at various times of the day. The wells were located in Gyula (1) and Békés (2). At least 50 people at each well were interviewed. The starting question for well users was "Why do you use this well?". In both the street and well interviews, responses were noted by hand. They were analysed and coded using an iterative system of refinement and revisiting. There were no pre-defined codes, but patterns were spotted and coding was refined in response. Using the results from this coding exercise, a standardised questionnaire was formulated with multiple choice answers. These were placed in libraries in each of the towns, and were collected 2 months later. The number of completed questionnaires for each town is shown in Table 1, column F. Using questionnaires allowed a quantitative exploration of concerns. Questionnaire data and interview data could not be combined for analysis because the questionnaires were prompted and the interview data not; often the exact same questions were not asked in both, nor were answers given in the same form. However, questionnaire data identifies a trend or pattern, and the interview data explains it.

Tap water samples were taken in order to examine how well physical properties of water correlated with resident's concerns. Samples were collected from the point of consumption; this includes taps in people's kitchens and on-street tap-water fed pumps. The tap or pump was run for one minute and the water was checked to be running cold before the sample was collected. We collected samples from all over the distributions system, not limited to the studied town, were collected in order to ensure that water in the system was fairly uniform. The number of samples reflected the size of the distribution network and can be seen in Table 1, column D. At each sample site, Free and Total chlorine were measured using SenSafe™ test strips. A basic smell and taste test were conducted, using the smell-taste wheel to identify likely compounds [8]. This was conducted by the lead researcher, and repeated separately by an assistant, and then notes compared and discussed. A sample of the water was collected for ICP-MS analysis. We analysed for B, Na, Mg, Si, P, K, Ca, Mn, Fe, As and Pb. All samples were collected in July/August of 2009, during the same period as the interview and questionnaire data.

3. Results

a) Street Interviews

Street interview respondents can be split according to their drinking water behaviour. Respondents either stated that they drank only bottled, mineral water, only tap water, only well water, or a combination of these options. This is shown in Table 2. The use of well water was higher in the towns which had wells than in the towns without. However, for the other water behaviours, proportions are similar across all towns. Sample sizes are not large enough to state whether or not there is any real and significant difference between drinking water behaviour in each town, and it is suggested that the behaviours are similar across all towns. Between 39 and 48% of respondents in each town rely entirely on non-tap sources (well and mineral water).

Whether or not a street interviewee spoke positively or negatively about tap water was influenced by their drinking water behaviour. Their responses to interview questions were coded in order to determine if they had a generally positive or negative perception of particular water sources. This was not whether or not they said it was bad or good quality, but the overall impression they had of it. Indifferent opinions and no expressed opinions were both deemed to be indifferent, on the basis that if strong opinions were held they would be voiced. Figure 1 (a, b, and c) shows the perception of each water source categorised by the primary drinking water source of respondents. They are collective graph for all towns together. While the proportion negative, indifferent and positive varies between towns, the graphs for each town individually are very similar. As there are not large enough samples to know whether or not these differences are significant, the overall graph for them collectively is given. They show that the mineral water drinkers are mainly indifferent (or have no opinion) about well water and mineral water, but are negative about tap water. The well water drinkers are unique in their responses in that they are not only negative about tap water (similar to the mineral water drinkers), but they are also mainly positive about well water. This positive opinion of their chosen water source separates them from the mineral water drinkers who are indifferent to their chosen water source. Of those who drink tap water, either wholly or as one of their mixed sources, there are some positive responses to tap water. However the majority remain indifferent to all water source options.

Table 2. Drinking water behaviours by count (percentages in brackets)

Town	Tap water (%)	Well water (%)	Mineral water (%)	Mixed (%)
A	10 (32)	4 (13)	8 (26)	9 (29)
B	11 (38)	0 (0)	10 (34)	8 (18)
C	9 (31)	0 (0)	14 (48)	6 (21)
D	16 (20)	4 (5)	35 (43)	25 (32)
TOTAL	46 (27)	8 (5)	67 (40)	48 (28)

The issues that shaped people's perceptions of tap water were largely to do with the content of the water and the perceived nature of the distribution system. Issues to do with water content included particles or suspected bacteria. A number of people stated that they had been made sick by the tap water. Chlorine added to the drinking water was a popular concern; almost 20% of the interviewees, unprompted, raised the issue of chlorine in a negative way. This links to concerns related to the distribution and treatment process. Besides the addition of chlorine, people were concerned about the maintenance of the pipes, and the state of the water towers and storage facilities. A number of people stated that they did not like the smell, taste or colour of the tap water. However this was often coupled with other reasons. For example, the smell of chlorine was often mentioned as a concern because of the linked assertion that chlorine is unnatural. These concerns were repeated in all towns.

The reasons given for well use by the well users include both the benefits of well water, and the disadvantages of tap water. Almost all well users prefer the taste of well water. In addition, the well water is seen as being natural and clean, and without 'chemicals'. Reasons of habit and tradition are given, including reference to previous generations of the same family also having used the same well. Some of the older respondents find that their regular well visits serve a social function. Many of the respondents were happy to talk for a long time about the well and their memories that were connected with the well. The water provided by it is seen as a resource to be proud of and to protect.
Arsenic was not a large concern amongst well users or tap water users. Even without being directly asked, respondents would mention arsenic in their water source. However, this was often as a closing or passing comment, and could rarely be classed as a concern. It was seen as being a fact that everyone knew about. Well users tended to believe that tap water and well water were equally affected by arsenic. The level of risk people associated with the arsenic was linked to experience. For example, one person for whom arsenic was a worry, said that in the 1980's the town was provided with water from a tanker because of the arsenic. This had made her wary of it. However, many people (both well respondents and street respondents) said that elderly people who had drunk well water all their lives were suffering no ill health effects and lived to be very old. This was given as the reason that arsenic was not concerning to them. There was some level of ignorance around what arsenic was; statements such as "I am not an expert" and "what is arsenic anyway?" were common.

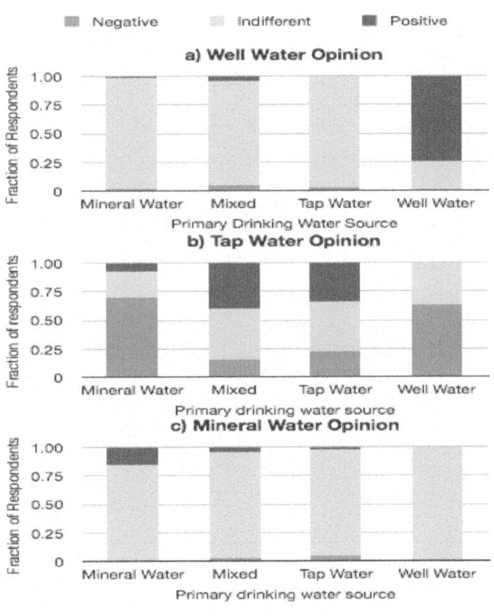

Figure 1. Expressed overall opinion of water sources grouped by respondent's primary drinking water source.

b) Questionnaires

The opinions collected through the questionnaires are shown as percentages in Table 3. It can be seen that fewer people claim to have a negative opinion of tap water quality when directly asked in the questionnaire. However, higher numbers have specific concerns about smell or taste, or specific chemicals than have a negative opinion.

Table 3. Percentage of questionnaire respondents with concerns.

Town	Cl concern	Smell concern	Taste concern	As concern	Negative opinion of quality
A	22	13	13	50	0
B	27	38	38	39	11.5
C	16	29	6	67	10.6
D	46	43	39	46	6.5

c) Water Samples

The water chemistry varies between distribution systems, but is constant within each single system system. For a selection of the individual elements tested, the concentrations with error bars are shown in Figure 2, a and b. The figure has been split into two parts only for ease of display. The errors do not include the instrumental error of the ICP-MS. For each individual element, the error bars are small, showing that the variation within the distribution system is low. There is little overlap between the means of single elements in each town, showing that there is a real difference in the concentrations between towns. It is very difficult to convert each chemical analysis into a statement on the quality of the water. In terms of policy, if water fails a parameter, or a collection of parameters, then it is bad quality. It is therefore an absolute measure, and not scalar. Water in town A fails to meet the policy standard for Mn. Town B meets the standards for all those parameters we measured. Town C fails on As, B and Na. Town D fails on As.

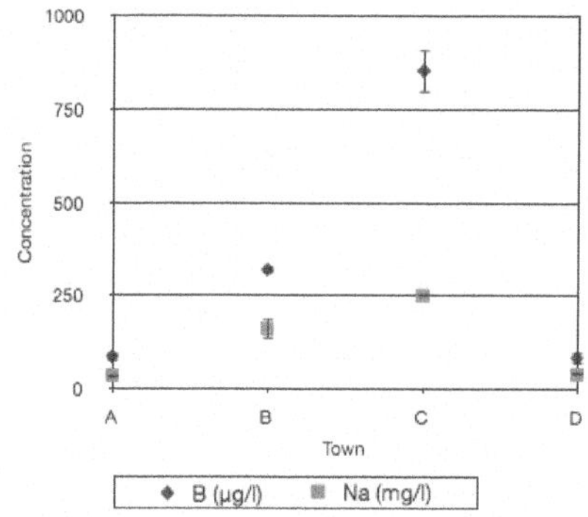

Figure 2 (a). Mean concentrations of B and Na in samples from distribution systems serving test towns.

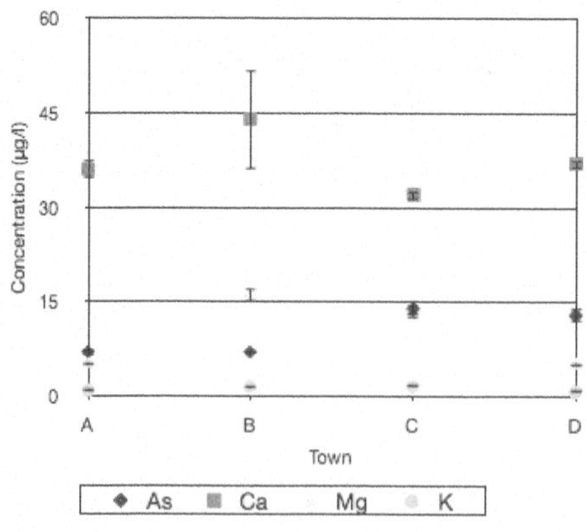

Figure 2 (b). Mean concentrations of As, Ca, Mg and K in samples from distributions systems serving test towns.

The smells and taste tests indicate the presence of phenols in the water supplies of all the towns. There was no colour, sediment or particles in the samples taken. All householders confirmed that the sample could be considered an example of how their water usually is. This smell is described by the householders as a chlorine smell. The smells detected were 'medicinal' with no particular mouth feel or taste. According to the Suffet et al., (2004) smell and taste wheel, this smell represents chlorophenols, iodophenols and/or bromophenols. This would be expected where chlorine is added to water with high organic matter, as it is in this area. No attempt was made to indicate the strength of the smell as this could not be done objectively.

4. Discussion

People's concerns are based mainly on their experiences and perceptions. In tap water, arsenic is of low concern to residents because it is not observable, and because residents are unclear about what its negative impact is. Chlorine is a common concern because it is observable (smell) and associated with an 'industrial' process. The industrial, non-natural nature of tap water is seen by residents as a bad thing. It may be delivering good quality water, but it is a negative characteristic. Conversely, the natural state of the well water is a positive characteristic of the well water for users. The lack of concern about its known arsenic content is because it is natural, and because users have little knowledge of what arsenic is. Additionally, they have no direct experience of it harming anyone. Indeed, much of their experience indicates that people live to old-age by drinking this water. Furthermore, well use has a traditional and social aspect to it; It is a rejection of the tap water, but also an embracing of the well water for its naturalness and the tradition it represents.

These experience-based perceptions of water do not influence the overall opinion of water quality. The questionnaires reveal that people can have a non-negative opinion of tap water whilst still remaining concerned by issues such as smell, taste, chlorine and arsenic. In addition, the questionnaires reveal a much lower percentage of people rating their water as bad quality than would be expected by studying the street interviews. This suggests that when asked directly to rate quality, they consider it good or alright. However, they will still talk negatively about the water when considering it for their own purposes or tastes. This indicates that quality is based on something other than the tangible concerns that residents have. Instead, to the residents, water quality is an abstract and undefined concept; they trust that the tap water quality is good, but still have concerns or dislikes.

In contrast policy bases water quality on tangible and measurable criteria. Even though the water chemistry in each town is different, it is not possible to determine the difference in the policy defined quality. It is only possible to say how many parameters they fail, but how these failing relate to each other is subjective. It is therefore impossible to examine how public opinion of water quality correlates with policy-defined quality. This is important because it demonstrates how differently policy and the public define water quality.

This in part explains why there is little variation in opinions between towns with different water chemistry. Similar experiences and opinions are expressed throughout all the towns, and at both wells. Yet water chemistry and the management approach are different. However, such characteristics are not necessarily detectable by residents. They have little influence over perceptions such as the non-natural status of water. The characteristics that are detectable to residents, and that might influence their perception, such as the smell, are constant between all towns. Therefore management approach and water chemistry do little to influence public perception of water.

5. Conclusions

Policy and people do not define water quality in the same way. Residents are concerned mainly by experiential aspects of their water. They do not equate their concerns with water quality. Quality is an abstract concept that is not easily broken into measurable indicators and often bears little relationship to whether or not a person has a negative opinion of the water. On the other hand, policy characterises quality only in measurable indicators. These measurable indicators do not represent the same concerns that residents have.

This has clear consequences for the way water is managed and controlled in the study area. People aren't going to change their behaviour and stop well use, or support a project, without a clear understanding of why. Telling residents that their water quality will be improved will garner little support in an area where concerns are not linked to quality perceptions. Instead, information needs to be clear and relevant, and it should work with the knowledge and experiences that people already have. Information must directly tackle their experiences and knowledge. For example, explaining in real terms what this arsenic concentration means, and how it affects people, thus enabling them to put their observations into context.

Acknowledgments

We acknowledge funding from the European Commission (AquaTRAIN MRTN-CT-2006-035420).

References

[1] IARC.1990. Arsenic and arsenic compounds, summaries and evaluations. Monographs on the Evaluation of Carcinogenic Risks to Humans

[2] Smith, Allan H and Craig M Steinmaus. 2009. Health effects of arsenic and chromium in drinking water: Recent human findings. Annual Review of Public Health 30 (1): 107-122.

[3] US Agency for Toxic Substances and Disease Registry. 2007. Toxicological profile for arsenic.

[4] Argos, Maria, Tara Kalra, Paul J Rathouz, Yu Chen, Brandon Pierce, Faruque Parvez, Tariqul Islam, et al. 2010. Arsenic exposure from drinking water, and all-cause and chronic-disease mortalities in bangladesh (HEALS): A prospective cohort study. Lancet 376 (9737): 252-8.

[5] Smedley, P L and D G Kinniburgh. 2002. A review of the source, behaviour and distribution of arsenic in natural waters. Applied Geochemistry 17 (5): 517-568.

[6] Lindberg, Anna-Lena, Walter Goessler, Eugen Gurzau, Kvetoslava Koppova, Peter Rudnai, Rajiv Kumar, Tony Fletcher, et al. 2006. Arsenic exposure in hungary, romania and slovakia. J Environ Monit 8 (1): 203-8.

[7] Doria, Miguel de Franca. 2010. Factors influencing public perception of drinking water quality. Water Policy 12 (1): 1-19.

[8] Suffet, I H, L Schweitze, and D Khiari. 2004. Olfactory and chemical analysis of taste and odor episodes in drinking water supplies. Reviews in Environmental Science and Biotechnology 3 (1): 3-13.

Section 2

Health and Aesthetic Issues

Discolouration in water supply, the role of metals

J.B. Boxall

Pennine Water Group, University of Sheffield, United Kingdom
Corresponding author e-mail: j.b.boxall@sheffield.ac.uk

Abstract

Water distribution systems were originally conceived to safeguard public health. However, the long term development and operation of most systems has been dominated by water quantity issues, leading to systems that are large and complex to provide hydraulic capacity and resilience. Such water quantity biased systems often have a variety of associated water quality issues, from low velocities and hence high residence times upwards. Discolouration is the most obvious and directly attributable water quality issue experienced by customers, often accounting for over one third of all contacts received by water suppliers. Traditionally discolouration has been addressed in a reactive manner, triggered when contacts pass a supplier specific threshold, number of contacts per thousand population. The most common remedial intervention has being flushing, this is a relatively simple option requiring the opening of fire hydrant(s) to discharge dirty water and attempt to cleanse the system. However, this has been shown to provide only short term, local amelioration. With more emphasis on customer service, tighter regulation and conservation of water, understanding to inform system management and operation to control discolouration has been improving. This paper explores the state of the art understanding of discolouration in potable water distribution systems, based on an active research theme at the University of Sheffield since 2001. Summary data from a notable number of extensive field studies is presented, focusing on the role of metals in discolouration. Overall it is shown that while discolouration incidents and customer contacts are often associated with local effects the management strategy should be holistic from source to tap and that better understanding of the role of, and mechanisms affecting, metals behaviour throughout the systems is of great importance.

1. Introduction

Discolouration is a major and ongoing issue facing water supply companies. In England and Wales in 2007 a total of 154,985 customer contacts where reported with 124,671 (80%) relating to discoloured water (DWI, 2008). While many of these are attributable to local effects, including customer plumbing, discolouration events also occur at a larger scale. Major water quality incidents occurring in England and Wales are investigated and formally reported to the Drinking Water Inspectorate, the breakdown of such reported investigations in 2006 is shown in Figure 1. These levels of discolouration incidents occur despite significant investment in asset renewal and rehabilitation, specifically targeting water quality, since the water supply sector in England and Wales was privatised in 1989. Discolouration is an important issue for water supply companies and better knowledge, understanding and tools are needed.

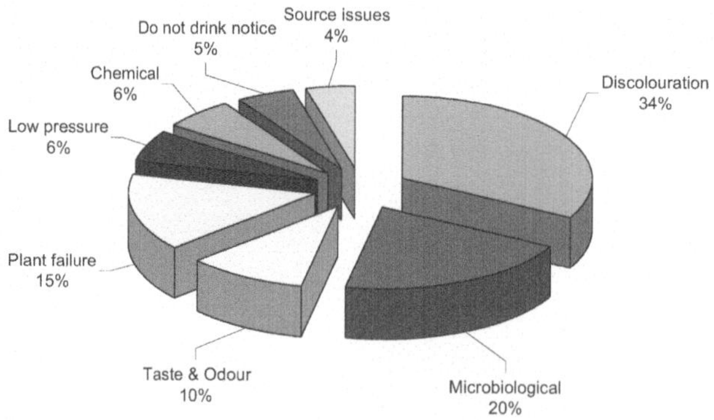

Figure 1. Breakdown of DWI investigated incidents in England and Wales in 2006 (source data from DWI 2007, image from Husband et al 2010)

2. Discolouration material

Traditional engineering judgement has attributed corrosion of cast iron pipes as a major factor in the cause and occurrence of discolouration, captured by the often used alternative expression 'red water' reflecting the dominance of oxidised ferric compounds in discoloured water samples. The material responsible for causing discolouration is particulate in nature, and hence turbidity is a good measure of the aesthetic issues experienced by customers. This particulate nature has lead to traditional management approaches seeking to use classical mathematical description of individual loose particle behaviour, governed by the interaction of self weight, drag and hydraulic forces, to derive management tools and guidelines. However, the size of particles found in water distribution systems is generally small, as shown in figure 2. For such size fractions many addition forces and mechanisms will have an effect on particle behaviour and should be considered in the understanding of discolouration.

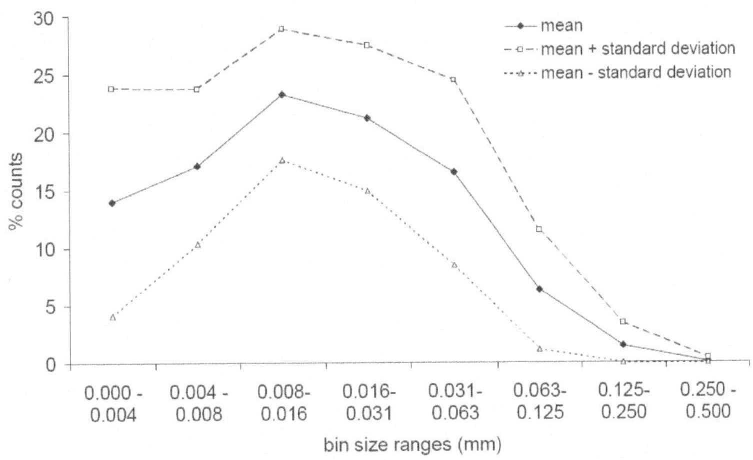

Figure 2. Particle size distribution from flushing samples, 30 samples with consistent analysis (Boxall et al 2003)

At the University of Sheffield a programme of research has been underway since 2001 to develop modelling tools to help Predict and understand Discolouration in Distribution Systems (PODDS). The PODDS approach (Boxall et al 2001) is based on 'cohesive' behaviour of fine sediments (Parchure and Mehta 1985 and Mehta and Lee 1994) that effectively retains material at the pipe wall. In the PODDS model discolouration behaviour is described through consideration of the interaction of hydraulic (shear stress) forces and the cohesive strength of the material layers. The PODDS model has been widely validated through fieldwork across England (Boxall and Saul 2005, Husband and Boxall 2010) and internationally (Boxall and Prince 2006) and in the laboratory (Husband et al 2008). While the empirical PODDS model has proven practicable predictive capabilities for the mobilisation of discolouration material it does not provide a direct understanding of the processes and mechanisms leading to the accumulation of material or retention of material at the pipe wall. This paper will explore the role of metals in discolouration from the field sampling and investigations undertaken over the course of and in association with the PODDS projects.

3. Metal composition of discoloured water

Much of the field work undertaken to validate the PODDS model has involved the careful planning and monitoring of flushing operations. These flushing operations used a uni-directional approach, design to affect a single length of pipe material and diameter with turbidity monitoring at inlet, to confirm no upstream effects, and at the flushing point where discrete samples were also collected. Discrete samples analysis data as reported here was generally undertaken by the related Water Company's usual certified laboratory. Figures 3a and b show typical examples of results from flushing of pipes in a distribution system and part of a trunk main system respectively. From these figures it can be seen that there is a strong correlation over the duration of the flushing, mobilisation event, between turbidity and key metal parameter including iron, manganese and aluminium. It is interesting to note that the temporal form and association between metals concentrations and turbidity is consistent between distribution system pipes and trunk main pipes with iron as the dominant metal.

From the data shown in Figure 3a Boxall et al (2003) attempted a conversion from turbidity to suspended solids and hence, by assuming uniform material layers along the pipe length, to a depth of material per unit surface area of pipe. The result of this is that the discolouration shown in figure 3a resulted from mobilisation of only 0.145 mm of material per unit surface area of pipe. The relationship for conversion from turbidity to suspended solids has appreciable scatter, hence this number should be considered indicative rather than absolute. The low material depth per unit surface area of pipe is contrary to traditional engineering judgement which associated discolouration with the breaking and mobilisation of complete corrosion tubercles. This analysis suggest that it is the surface of material layers that are most important in discolouration as experienced by customers.

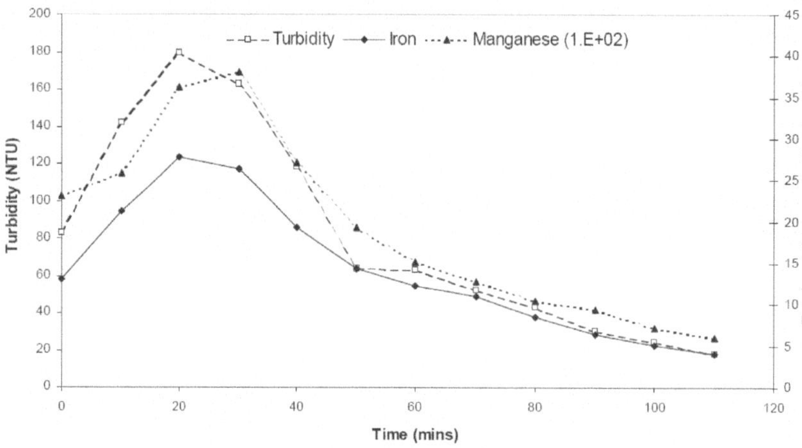

Figure 3a. Turbidity, iron and manganese response observed from flushing a ~1.6km ~75mm diameter cast iron distribution pipe (Boxall and Saul 2005)

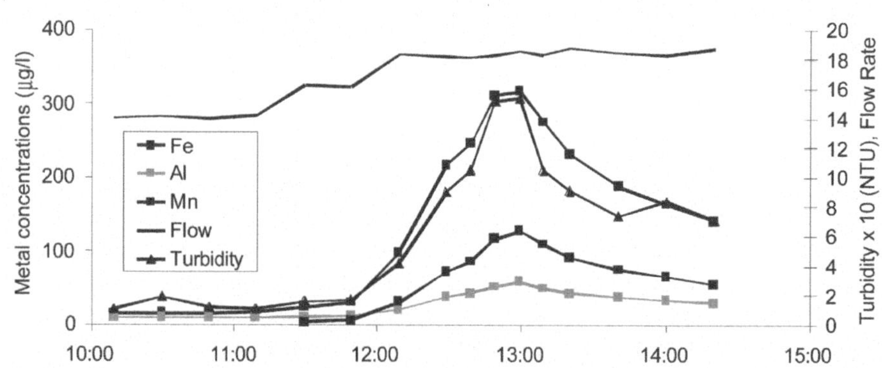

Figure 3b. Turbidity, iron and manganese response observed from flow increase in a ~3.7km 440mm diameter lined ductile iron trunk main (Seth et al 2009)

Iron is the dominant material found in discoloured water samples in England hence it seems logical to expect an association with corrosion of iron pipes and fittings within distribution systems. Figure 4 shows the results of metals analysis of discrete samples from a series of pipes in a looped network area. Flush numbers 3, 4, 6 and 7 affected cast iron pipes, the remainder being either PVC or MDPE pipe material. From this it can be seen that within a system with different pipe materials, including cast iron, it is not necessarily the cast iron pipes that produce the greatest concentrations of iron, and other metals, as discolouration products. Boxall and Saul (2005) showed that the degree of discolouration resulting from these pipes is related to the normal, or conditioning, shear stress experienced by the pipes, an important attribute for managing discolouration and one of principles of the PODDS approach.

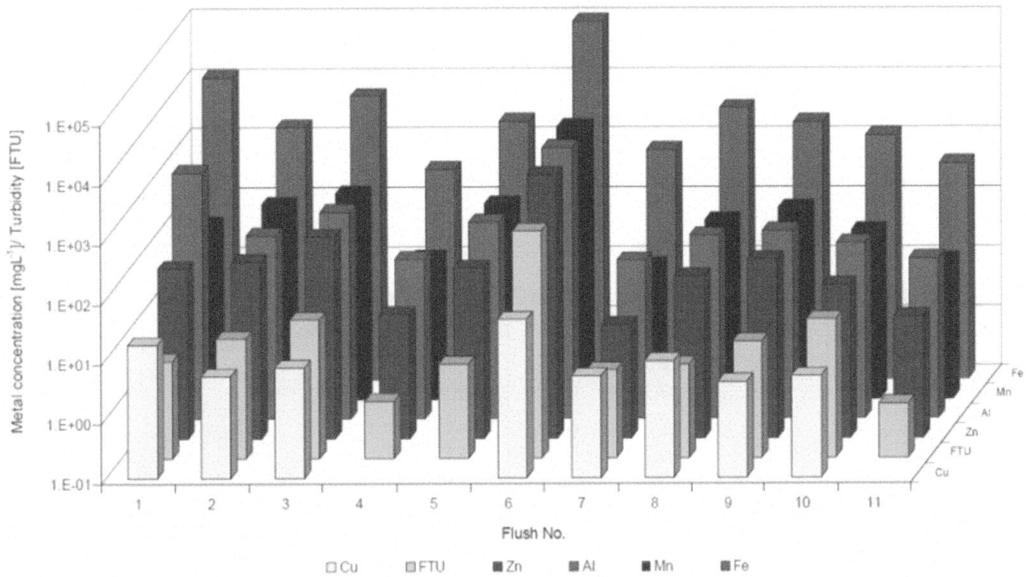

Figure 4. Metals observed in flush samples from looped network area, note: y-axis is log scale (Seth 2007)

Figure 4 shows that a 'cocktail' of different metals had accumulated within the system tested and that this cocktail was subsequently mobilised by the hydraulic disturbance of the flushing operations. These sampled showed similar particle size distributions across the different pipe materials (Seth 2007). Similar analysis across three Distribution Management Areas (DMA) supplied from the same treatment works showed strong (R^2 generally >0.9) linear correlations between turbidity and metals concentrations and between metals concentrations, including turbidity, iron, manganese, aluminium, zinc and copper (Seth 2007). However wider investigations have shown that the relative concentrations and even presence of certain metals does vary between different networks, for example as a function of source water, treatment train and trunk main system characteristics.

In parallel with the extensive flushing studies, such as reported above, Seth (2007) undertook analysis of material accumulations within pipe samples retrieved from the field as part of ongoing rehabilitation investment. This was to enable direct evaluation of the materials, corrosion products in particular, present within water distribution systems. Table 1 presents some summary results from this seeking to evaluation the difference in metal concentrations in material accumulation in different pipe types and at different locations around cast iron pipes. Table 1 again shows iron to be the dominant metal present in both cast iron and non-cast iron pipes, with a cocktail of other metals present in significantly lower concentrations. The data does not show significant change in metal composition with location around the cast iron pipes, as might be expected if gravity processes where leading to enhanced deposition of certain metals to the pipe invert. Average values for invert and crown are shown in table 1, any trend that it may be tempting to seek in the table is an artefact of the averaging and was not present for individual pipe samples. Other locations around the pipes were analysed, again not revealing any repeatable trends.

Table 1. Average concentrations of elements in samples from different orientations on cast iron pipes compared to non-cast iron pipes and example images showing variation in appearance and extent of corrosion of cast iron pipes examined (Seth 2007).

	Relative abundance (mg/g)			
	Surface	Invert	Crown	Non CI
Fe	232	631	694	159
P	5.05	10.93	10.06	3.80
Al	3.35	12.39	5.56	4.99
Mn	9.44	21.39	3.66	4.53
S	2.45	13.42	7.98	3.43
Ca	8.26	9.35	3.71	25.03
Zn	2.19	7.00	1.00	1.61
Mg	1.33	5.73	1.20	10.91
K	1.16	0.75	0.10	0.60
Na	0.68	1.43	1.07	1.12
V	0.70	0.20	< 0.10	0.41
Cr	0.83	0.10	< 0.10	/
Cd	/	0.01	0.01	/
Pb	/	0.05	0.05	/
Cu	/	0.05	0.02	1.04

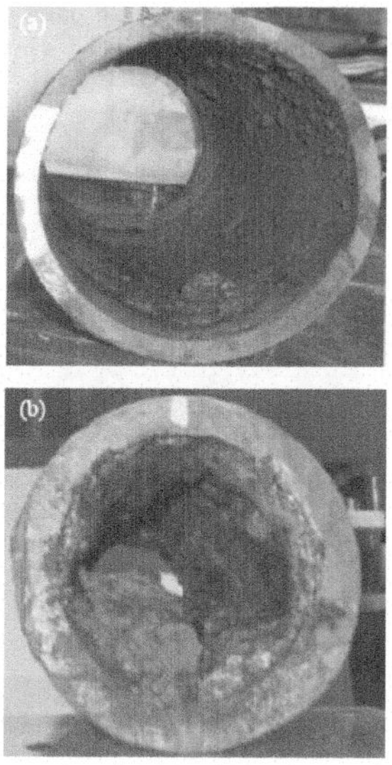

The difference in metal concentrations between the 'surface' and 'crown / invert' analysis of the cast iron pipes in table 1 is of interest when considered with respect to the low material depth calculations presented earlier. The metal concentrations, iron in particular, for the surface material are lower than in the crown or invert, in fact the values are closer to those of the non cast iron pipe samples. This supports the observation from figure 4 that there is not a marked difference in the metallic composition of discolouration material with pipe material, and that the material mobilised from cast iron pipes is not solely due to the corrosion of the pipes themselves. These observations lead to the conclusion that it is primarily the surface of corrosion products that are of interest and concern with respect to discolouration rather than the complete, complex structure of corrosion tubercles.

Husband and Boxall (2010) presented a comprehensive analysis of metals concentrations from flushing operations seeking to establish if there was a difference in the materials from cast iron and other pipes. They present a table which shows that the iron pipes tested had accumulated more iron than manganese or aluminium than was present in the bulk water, however in the plastic pipes the accumulation of metals in the discoloration layers was proportional to the bulk water. While this trend was apparent overall, there was significant variation for the individual pipes, some of which could be due to the discrete nature of samples collected during flushing as well as variability and low sample numbers available to derive some regulatory values. However they went on to suggest two simplified material accumulation processes: Mechanism 1 as a result of the corrosion of iron pipes and fittings occurring in cast iron pipes and fittings only; and Mechanism 2 occurring due to an accumulation of material from the background concentrations in the bulk water occurring in all pipes.

4. Accumulation of discolouration material / asset deterioration

While tools such as the PODDS model provide predictive capabilities for system response to hydraulic disturbances and other changes and the analysis of material composition may provide insight into the source of materials, there remains a relative lack of understanding of the rates of material 'regeneration' or asset deterioration and the associated influential factors. Such deterioration information is a key requirement for operators of water distribution systems so that they can plan the interval between maintenance operations and evaluate this against other investment options such as upgrading treatment work or pipe replacement / relining.

Regeneration, or asset deterioration, has been investigated through the field work of the PODDS projects by undertaking repeat visits to sites. Such repeat testing has sought to induce physically identical flushing and monitoring operations, generally at a 12 month interval. Figure 5 shows discrete sample analysis data from repeat visits to sites, initially tested in 2008 and repeat testing in 2009. All sites had no known operational or maintenance interventions for prolonged periods prior to or between testing. While there is scatter in each of the correlation plots, it is apparent that some degree of regeneration / deterioration occurred at all sites and there was consistency in the relative amounts of turbidity, iron, manganese and aluminium, suggesting repeatable, consistent long term processes. The results from such testing show ubiquitous but varying degrees of regeneration / deterioration and has started to provide insight into the rates of regeneration / deterioration and the impact of investment options.

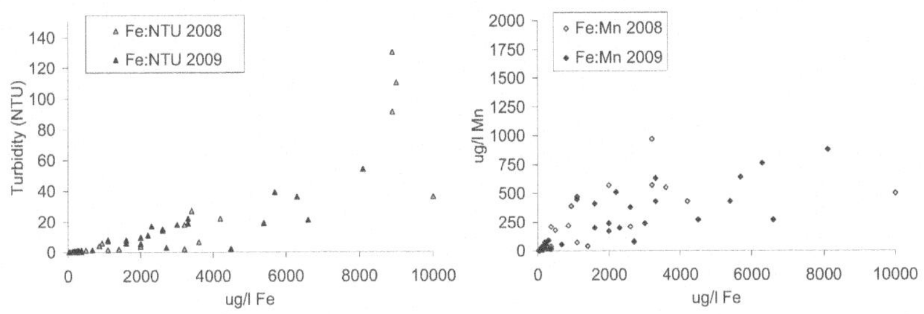

Figure 5. Correlation of turbidity and metal concentrations during repeat flushing fieldwork operations.

Table 2 shows further analysis of repeat flushing tests of field sites across England, after Husband and Boxall (2010b). The table shows the average, mean and coefficient of variation for the percentage of material regenerated in 12 months, assuming initial testing represented the total amount of material feasible to be retained by a given pipe, primarily dictated by the usual hydraulic conditions within that pipe. A detailed exploration of the implications of the full data set partially presented in table 2 is available in Husband and Boxall (2010b). Focusing on the mean regeneration percentages (first column of data) comparing cast iron and plastic pipe materials (first two data rows) it is apparent that on average cast iron pipes had regenerated ~50% of their initial material while for plastic pipes this is only ~25%. While there is significant scatter around these values, and the table shows that many other factors influence regeneration processes, this is interesting particularly in the light of observations about material composition: while the same materials may accumulate in both plastic and cast iron pipes the rate of accumulation in plastic pipes is about half that in cast iron pipes. Further evaluation of the table shows the need for consideration of cumulative effects from source water to the pipe of interest, for example the relative impact of different source waters can be seen, together with the impact of different coagulation (or lack of) processes in treatment works as well as the presence of unlined cast iron upstream of the pipe of interest. All these factors directly influence mechanism 2, the bulk water quality at a given pipe, while the pipe material itself dictates mechanism 1, direct corrosion of cast iron pipes and fittings.

As shown previously iron is usually the dominant metal present in discolouration samples. Iron can also be identified as a common factor in table 2: from source water, from coagulation processes, from upstream unlined cast pipes and from each pipe itself. It is therefore potentially interesting and informative to examine bulk water iron concentration as a single measure capturing the dominant influences on the bulk water quality. Hence Husband and Boxall (2010b) plotted the bulk water iron concentrations for each pipe length against percentage regeneration, reproduced here as figure 6. It should be noted that bulk water iron (not the concentrations exiting treatment works) was determined using data from historic regulatory sampling in the DMAs studied. Although not conclusive with a proportional variance R^2 of 0.57, figure 6 indicates that a relationship exists between bulk water iron concentration and material regeneration. The general trend is for regeneration to increase with iron levels, whilst of operational benefit it appears a greater discolouration risk is likely when concentrations rise above 25 to 40 µgFe/l. This reinforces the findings that water quality is a key factor governing cohesive layer development

Table 2. Discolouration material regeneration figures for different potential influential factors (after Husband and Boxall 2010b)

Factor	Regeneration per annum, %		
	μ	σ	C_v
CI pipe (all sites)	52	30.7	0.6
Plastic pipe	28	15.7	0.6
CI / surface water	66	32.0	0.5
CI / ground water	31	11.4	0.4
Surface water	55	32.5	0.6
Blended water	24	14.2	0.6
Ground water	32	10.5	0.3
Iron coagulation	49	33.2	0.7
Aluminium coagulation	40	29.2	0.7
No coagulation	32	10.5	0.3
Upstream unlined CI pipes	54	30.1	0.6
No unlined CI pipes	27	14.2	0.5
Average C_v			**0.5**

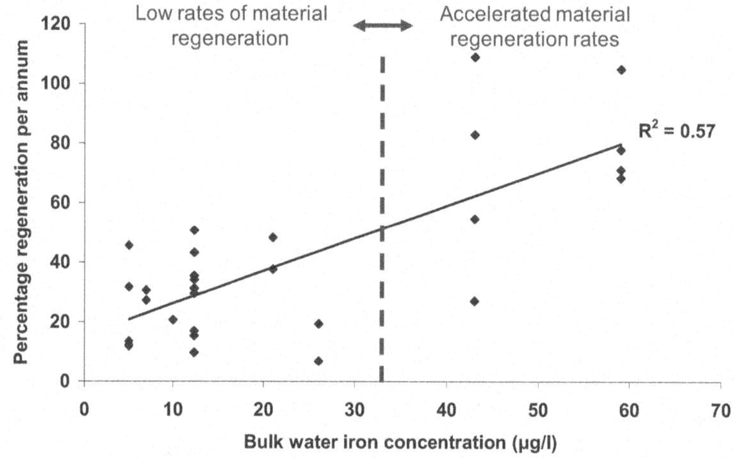

Figure 6. Influence of bulk water iron concentration on annual discolouration material regeneration (after Husband and Boxall 2010b)

Husband et al 2010a specifically investigated the role of trunk mains in discolouration. They found that cleaning of the trunk main resulted in a reduction in the inherent discolouration risk inferred from turbidity measurements and tentatively from customer contacts. They also found that cleaning of the trunk main resulted in reduced rates of material regeneration in downstream DMAs, particularly in non-corroding parts of the network. Thus they concluded that cleaning of trunk mains can reduce inherent discolouration risk as well as providing downstream benefits in terms of reduced rates of asset deterioration.

5. Summary

The material responsible for discolouration is particulate in nature, predominately of size less than 100μm. Once present at the pipe wall this material generally displays resistive forces and behaviour inconsistent with gravity driven sedimentation processes alone. The mobilisation behaviour of such discolouration material may be described by the PODDS model, based on a cohesive transport approach.

The material mobilised by flushing shows strong correlation, over the duration of the flushing, between turbidity and key metal parameter including iron, manganese and aluminium. This is consistent between distribution system and trunk main pipes with iron as the dominant metal. While iron is the dominant material a cocktail of substances is generally mobilised including increased levels of sulphur, chlorine, phosphorous, silicon and calcium, plus a cocktail of other metals. There is no appreciable difference in the metallic composition of the material mobilised from different pipe types within given networks, indeed there is correlation in material between areas supplied from the same source water and treatment works.

The metallic composition of corrosion products within cast iron pipes does not show consistent variation with location around the pipe perimeter, suggesting gravity effects are not influential. These corrosion products are dominated by iron with a cocktail of other metals. The dominance of iron is significantly lower in the near surface layer of corrosion products in cast iron pipes, approaching those found in non cast iron pipes. It is primarily the surface of corrosion products that are of interest and concern with respect to discolouration rather than the complete, complex structure of corrosion tubercles.

Studies into material regeneration (asset deterioration) have shown regeneration to be endemic and ubiquitous. Iron pipes tested were found to have accumulated more iron than manganese or aluminium than was present in the bulk water, however in the plastic pipes the accumulation of metals in the discoloration layers was proportional to the bulk water. Two conceptual material sources are suggested: Mechanism 1 as a result of the corrosion of iron pipes and fittings, occurring in cast iron pipes and fittings only; and Mechanism 2 occurring due to an accumulation of material from the background concentrations in the bulk water, occurring in all pipes. This highlights that corrosion of iron pipe themselves is not the only material source, and is often not the dominant source. While the same materials may accumulate in both plastic and cast iron pipes the rate of accumulation in plastic pipes is on average about half that in cast iron pipes, although there are many other influential factors. The need for consideration of cumulative effects from source water to the pipe of interest is shown and it is suggested that bulk water iron concentrations may be a useful first catch all indicator for regeneration rates.

Acknowledgments

Particular acknowledgement and thank to Dr Stewart Husband and Dr Allyson Seth, who were responsible for carrying out many of the studies and facilitating much of the data reported here. Acknowledgments are also to the support of the collaborating water companies and the Engineering and Physical Science Research Council - Anglian Water, Northumbrian Water, Severn Trent Water, Southern Water, Thames Water, United Utilities, Veolia Water, Wessex Water, Yorkshire Water.

References

Boxall, J. B., Saul, A. J. and Skipworth, P. J. (2001). A Novel Approach to modelling sediment movement in distribution mains based on particle characteristics. Water Software Systems: v. 1: Theory and Applications (Water Engineering & Management). B. Ulanicki, B. Coulbeck and J. P. Rance, Research Studies Press, Hertfordshire, UK. 1: 263-273

Boxall, J.B., Saul, A.J., Gunstead, J.D. and Dewis, N. (2003) 'Regeneration of discolouration in distribution systems' Proc. ASCE, EWRI, World water and environmental resources conference, 23-26 June, Philadelphia, USA

Boxall, J. B. and Saul, A. J. (2005). "Modelling Discolouration in Potable Water Distribution Systems." Journal of Environmental Engineering ASCE 131(5): 716-725

Boxall, J.B. and Prince, R.A. (2006) 'Modelling discolouration in a Melbourne (Australia) potable water distribution System' Journal of Water Supply: Research and Technology - AQUA. Vol. 55, No. 3, pp. 207-219

DWI (2007). Drinking Water 2006; Drinking Water in England and Wales 2006. A report by the Chief Inspector

Husband, P. S., Boxall, J. B. and Saul, A. J. (2008). "Laboratory Studies Investigating the Processes Leading to Discolouration in Water Distribution Networks." Water Research 42(16): 4309-4318

Husband, P.S., Whitehead, J. and Boxall, J.B. (2010) 'The Role of Trunk Mains in Discolouration' Proc. of the Institution of Civil Engineers, Water Management Vol 163 Issue WM8 pp 397-406

Husband, P.S. and Boxall J.B. (2010a) 'Field Studies of Discolouration in Water Distribution Systems: Model Verification and Practical Implications' J. Water Resources Planning and Management ASCE Jan, Vol. 136. Vol. 136, No.1, pp 86-94

Husband P.S. and Boxall, J.B. (2010b) 'Field Studies To Inform The Management Of Discolouration Risk In Water Distribution' Water Research (in press August 2010)

Mehta, A. J. and Lee, S.-C. (1994). "Problems in Linking the Threshold Condition for the Transport of Cohesionless and Cohesive Sediment Grain." Journal of Coastal Research 10(1): 170-177

Parchure, T. M. and Mehta, A. J. (1985). "Erosion of soft cohesive sediment deposits." Journal of Hydraulic Engineering 111(10): 1308-1326

Seth, A., Bachmann, R.T., Boxall, J.B., Saul, A.J. and Edyvean, R. (2004) 'Characterisation of materials causing discolouration in potable water systems' Water Science & Technology Vol. 49, No 2 pp. 27-32

Seth, A. (2007) 'An investigation into materials causing water discolouration and characterisation of corrosion products in water distribution system' PhD Thesis, University of Sheffield

Seth, A.D., Husband, P.S. and Boxall, J.B. (2009) 'Rivelin trunk main flow test' Integrating Water Systems. Proceedings of the 10th Computing and Control for the Water Industry, Boxall and Maksimovic (eds), Taylor and Francis, pp 431-434, ISBN 978-0-415-54851-9

Metals and related substances in drinking water - from source to the tap. Krakow tap survey 2010

A. Postawa, E. Kmiecik, K. Wator

Faculty of Geology, Geophysics and Environmental Protection, AGH University of Science and Technology. 30-059 Krakow, Poland

Corresponding author e-mail: Postawa@agh.edu.pl

Abstract

The tap survey in Krakow was conducted as a part of screening tap survey, performed in Poland in 2010 for the purpose of joined research project: "Metals and related substances in drinking water in Poland". Sampling points were randomly selected on the base of regular geographical grid. All samples were collected using RDT (random daytime) sampling protocol. Obtained results show significant variations of all analyzed metals and metalloids concentrations along a way "from source to the tap". Concentrations of metals and metalloids in tap water samples also vary significantly. The highest variations show Zn - from 21 to 2845 µg/L, Cu - from 2 to 640 µg/L, and Fe -from 56 to 560 µg/L. 3% of samples fail to comply with 10 µg/L lead standard. 17% of samples fail to comply with Fe standard.

1. Introduction

Krakow, former capital of Poland, is one of the largest and oldest cities in Poland, with population of over 750,000 permanent inhabitants plus 500 000 of students, tourists and people not living within city limits but employed here. Krakow is situated on the Vistula River banks in southern Poland. It is now the capital of the Malopolska Province.

First waterworks in Krakow become operational in 1901. Since then water supply system has been continuously developed and in 2009 total length of distribution network exceeded 2015 km. Annual water consumption reaches over 57 millions cubic meters. The area of Krakow is divided into water-supply zones, which are supplied by 4 treatment works: "Bielany", "Rudawa", "Raba" - (from surface water catchments) and "Dlubnia" (surface water and groundwater). "Raba" treatment works is the largest one with, daily production of nearly 200 thousand cubic meters that covers approximately 54% of Krakow water demand. Some parts of the city are supplied with mixed water from "Raba and "Bielany" or "Raba" and "Dlubnia" treatment works. Raw waters, treated water and waters from distribution network are regularly sampled and analyzed by Central Laboratory of Krakow Municipal Waterworks and Sewer Enterprise - MPWiK SA. The results of water quality monitoring, performed by MPWiK SA, revealed some problems with high concentrations of metals in drinking water in Krakow, therefore for better recognition of the extent of this problem a tap survey was undertaken.

2. Materials and Methods

2.1 Sampling sites

Krakow tap survey 2010 comprised southern part of the Krakow agglomeration - Debniki and Podgorze districts. This water supply zone is supplied with water from Raba River (Dobczyce reservoir) via "Raba" treatment works. Sampling points within water supply zone were randomly selected on the base of regular orthogonal grid with the cell size of 250 m by 250 m. Target sampling point locations were set at the centres of grid cells. Samples were collected as close to the target point as possible, but in some cases, when householder was absent or uncooperative, real sample collection points were relatively far from required ones.

2.2 Sampling and analytical methods

During 2010 sampling campaign total number of 101 samples were collected using RDT (random daytime) sampling protocol [3, 4]. Additionally 26 control samples were collected for QA/QC program purposes. Samples were preserved with concentrated nitric acid to reduce sample pH to the value below 2 and delivered to a laboratory within 6-8 hours from collection. The samples were stored in a lightproof container and cooled down to a temperature 4-5 degrees Celsius.

Where applicable, electric conductivity, pH, and water temperature were measured on site. Householders were also interviewed about age of buildings, age and materials used for connection pipes, internal installations and appliances within their premises.

Samples were analyzed using ICP-MS method for metals and related substances: Al, As, Ca, Cd, Cr, Cu, Fe, K, Mg, Mn, Na, Ni, Pb, Zn. Concentrations of chlorides, sulphates and alkalinity were additionally determined.

3. Results and Discussion

3.1. Raw water

Raw waters and waters pumped into distribution network are regularly sampled, in accordance to a drinking water directive [1], and analyzed by Central Laboratory of Krakow Municipal Waterworks and Sewer Enterprise - MPWiK SA for selected parameters, including metals and metalloids. 3.5 % of raw water samples from Raba River show concentrations of iron higher than parametric value set for drinking water [1] while nearly 37 % of samples fail to comply with standard for manganese.

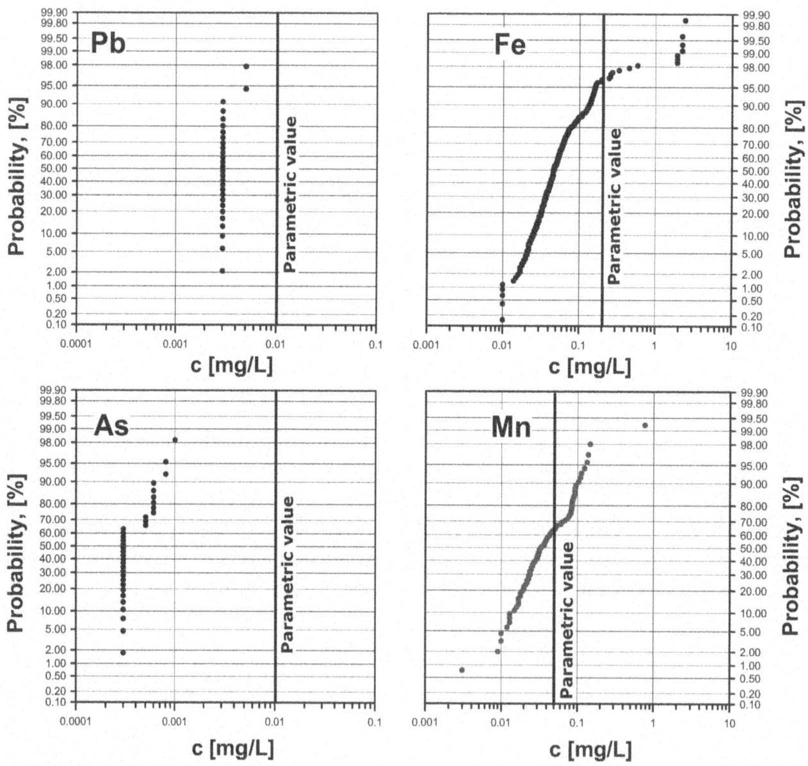

Figure 1. Probability plots for selected metals and metalloids concentrations in raw water from Raba River

Naturally high concentration of iron and manganese in raw water is quite easy to remove during treatment and it is not considered as an important problem. No failures in respect to other metals and metalloids were encountered (Figure1). Water pumped into the distribution network after treatment meets all quality standards.

3.2. Water in distribution network

Water samples from distribution network are collected from fixed sampling points and analyzed by MPWiK SA Laboratory. Within investigated water supply area 29 sampling points are located. All samples are being collected as "full flush" samples. Unfortunately only limited number of parameters is regularly analyzed (Fe, Cd, Cu, Ni, Pb, Hg).

Table 1 shows summarized results of metals and metalloids monitoring in distribution network. All concentrations except for iron are significantly lower than respective parametric values. The occurrence of iron concentrations exceeding parametric value seems to be the most important drinking water quality problem in Krakow, especially from aesthetic point of view because it causes many users complains.

Table 1. Metals in distributed water within "Raba" supply area

	Fe	Cd	Cu	Ni	Pb	Hg
	μg/l					
min	<25	<1	<5	<6	<5	<0.2
AVG	72	-	9	-	5.3	-
max	287	-	46	-	6	-

It is very difficult to solve iron problem in distributed water since 36% of mains and 32% of distribution pipes in Krakow are made of cast iron moreover nearly 60% of pipes are older than 20 years. There is no evidence of using lead pipes as mains or connections. Lead connecting pipes presence was reported by owners or administrators in a few buildings, usually constructed before or during World War II. Unfortunately, in general, knowledge of pipes age and materials amongst householders is very limited. 65% of them declare no knowledge of connection pipes material.

3.2. Tap waters

All samples were collected according to RDT protocol usually from kitchen taps. Obtained results show that drinking water quality within investigated water supply zone generally meets quality standards however some problems in respect to metals were revealed.

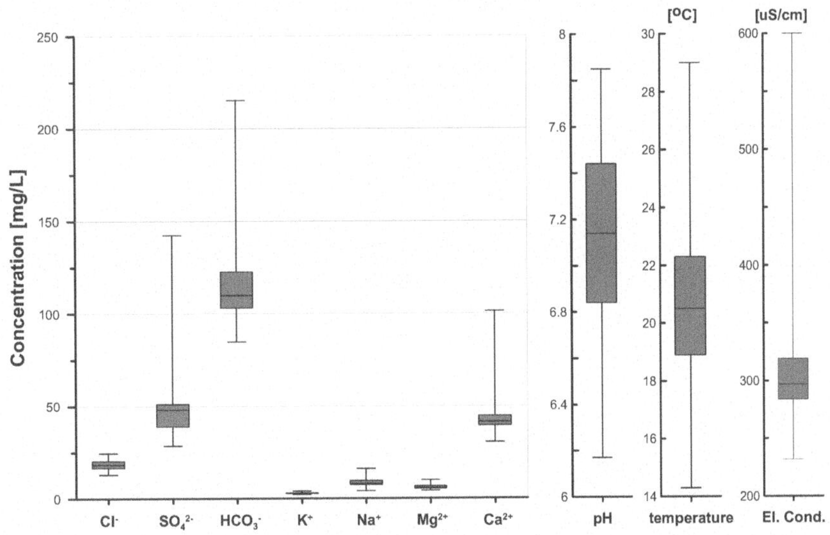

Figure 2. Summary on drinking water composition. Krakow tap survey 2010.

As it is shown on figure 2, tap waters within investigated water supply zone are of medium mineralization with average pH value practically neutral. Average tap water temperature was close to 20 degrees Celsius. Dominating anions are hydrocarbons and sulphates. Dominating cation is Ca^{2+}. Such a chemical composition is typical to surface waters in Poland. The chemical composition unfortunately makes water favourable to metal solvency. Concentrations of all analyzed metals and metalloids except for iron and lead are much lower than parametric values, usually by one order of magnitude (Table.2).

Table 2. Metals and metalloids in tap waters within "Raba" supply area

	Al	As	Cd	Cr	Cu	Fe	Mn	Ni	Pb	Zn
	μg/l									
Minimum	13.9	0.2	0.03	2.0	2.1	32.5	0.9	0.4	0.1	21.5
AVG	29.4	0.7	0.3	4.9	40.1	134.8	4.1	2.3	1.8	513.8
Maximum	67.4	1.5	3.9	15.5	640.2	559.1	36.6	19.1	27.3	2845.5
param. value	200	10	5	50	2000	200	50	20	10	-
% of failures	0	0	0	0	0	**17**	0	0	**3**	0

3% of samples failed to lead standard of 10 micrograms per litre but only in one case the concentration exceeded 25 micrograms per litre (present standard). The sample was collected from a kitchen tap localised in an over 30 years old building. Householders declared that internal installations are made of galvanized iron pipes and that the tap is less than 10 years old. Considering the age of the building, that was rather more than 50 years, despite householder's information, there is a possibility that lead pipes were used as connection. Lead pipes presence was reported in some buildings, usually constructed before or during World War II. Sometimes lead pipes are replaced during renovation but it happens that only end part of pipe is replaced in order to adapt it to new fittings. In many cases even owners of buildings have no knowledge of pipe material used in their properties especially these who bought it recently. 56% of interviewed householders declared no knowledge on internal pipe work in their properties and 92% of them know practically nothing about what their taps are made of.

Concentrations of iron, close to parametric value or exceeding it, are the most common problems with tap water quality in Krakow. 17% of samples showed concentrations up to 559 micrograms per litre. This phenomenon is quite understandable since cast iron is common material used for main and distribution pipes. A black iron and zinc alloyed iron pipes were the most popular material used for internal installations for many years. Many modern homes use PVC pipes, which are cheaper and easier to work with, or copper pipes.

Zinc concentrations are relatively high. Maximum concentration is over 2.8 mg/litre. Zinc is not listed among parameters potentially harmful to human health but is considered as potentially unacceptable to consumers since water containing zinc at concentrations in excess of 3-5 mg/litre may appear opalescent and develop a greasy film on boiling [2]. Increased zinc concentrations may serve as an indicator of poor condition of internal installation or low quality of brass used by taps manufacturer.

4. Conclusions

Krakow tap survey 2010 on metals was generally successful and allowed to improve knowledge on drinking water quality in one of the largest and most populated cities in Poland. Quality of drinking water in Krakow in respect to metals and metalloids is generally satisfying. The main problems are high concentrations of iron (17% of samples) and some potential problems with keeping 10 micrograms per litre lead standard (3% of failures). Relatively high concentrations of zinc suggest generally poor condition of internal installations in investigated buildings. Random daytime protocol (RDT) proved to be effective for screening and operational purposes.

Acknowledgments

This investigation was carried out by support from Ministry of Science and Higher Education (projects No. 28.28.140.7013 and No. 11.11.140.139)

References

[1] DWD, **1998**, *Council directive on the quality of water intended for human consumption*. Official Journal L 330, 05/12/**1998** p. 0032 - 0054.

[2] Guidelines for drinking water quality, 3rd edition, World Health Organisation WHO, Geneva, **2004**.

[3] Hoekstra E.J., Hayes C.R., Aertgeerts R., Becker A., Jung M., Postawa A., Russell L., S. Witczak, ***2009***, *Guidance on sampling and monitoring for lead in drinking water, JRC Scientific and Technical Reports*

[4] Van den Hoven Th.J.J., Buijs P.J., Jackson P.J., Miller S., Gardner M., Leroy P., Baron J.,Boireau A., Cordonnier J., Wagner I., Marecos do Monte H., Benoliel M.J., Papadopoulos I.,Quevauviller Ph., **1999**, *Developing a new protocol for the monitoring of lead in drinking water*; European Commission, BCR Information, Chemical Analysis, EUR 19087 EN

Relation between arsenic in drinking water and carcinoma of urinary bladder: Data from Municipality of Zrenjanin

Dragana Jovanovic[1], Zorica Rasic-Milutinovic[2], Gordana Perunicic-Pekovic[2], Snezana Zivkovic- Perisic[1], Tanja Kneževic[1], Dragan Miljus[1], Miroslav Radosavljevic[1], Katarina Paunovic[3]

[1] *Public Health Institute "Dr Milan Jovanovic Batut", Belgrade, Serbia*
[2] *University Hospital Zemun, Belgrade, Serbia*
[3] *Institute of Hygiene and Medical Ecology, School of Medicine, Belgrade, Serbia*

Corresponding author e-mail: dragana_jovanovic@batut.org.rs

Abstract

The tap survey in Krakow was conducted as a part of screening tap survey, performed in Poland in 2010 for the purpose of joined research project: "Metals and related substances in drinking water in Poland". Sampling points were randomly selected on the base of regular geographical grid. All samples were collected using RDT (random daytime) sampling protocol. Obtained results show significant variations of all analyzed metals and metalloids concentrations along a way "from source to the tap". Concentrations of metals and metalloids in tap water samples also vary significantly. The highest variations show Zn - from 21 to 2845 µg/L, Cu - from 2 to 640 µg/L, and Fe -from 56 to 560 µg/L. 3% of samples fail to comply with 10 µg/L lead standard. 17% of samples fail to comply with Fe standard.

1. Introduction

Arsenic in drinking water is known to cause cancers of the urinary bladder, lung and skin in humans [1], but there is limited evidence for development of cancers of the kidney, liver and prostate [2]. A wide variety of adverse health effects including skin, bladder, and internal cancers have been associated with chronic arsenic exposure [3,4], with the toxic effects most evident in regions where the groundwater contains high arsenic concentrations.Previous ecologic studies in Argentina, northern Chile and southwestern Taiwan reported increased bladder cancer mortality rate associated with arsenic concentrations above 200 µg/l. Contrary to this, the effects of lower levels of arsenic remain uncertain.The municipality of Zrenjanin is located in the north-eastern region of Serbia, and lies on quaternary sedimentary aquifers within the Pannonian Basin, which are known to contain high concentrations of naturally occurring arsenic [5, 6]. Groundwater sources in Zrenjanin supply water from the depth of around 100 meters. Water is formed in interaction with sedimentary rocks (clays, marls) and rich in arsenic. The composition of this water is variable, with wide range of arsenic concentration and high concentration of natural organic matter. Systematic water quality monitoring in Serbia shows that 82% of Zrenjanin population is exposed to arsenic concentrations above 2 µg/l. The majority of Zrenjanin population is supplied by 24 public water supply systems. The largest water supply system in this region was established in Zrenjanin town during 1950s. The first measurements of arsenic in drinking water were carried out in 1990, but regular monitoring has been implemented in 1999.The aims of this study were to compare incidence and mortality rates of urinary bladder cancer between exposed and unexposed populations and to determine whether low to moderate level of arsenic exposure from drinking water supply systems in Zrenjanin municipality is associated with increased risk of bladder cancer.

2. Materials and Methods

The research was designed as retrospective ecological study. Two study populations were included in the study: the exposed population included people living in Zrenjaninin municipality, known to contain drinking water arsenic concentrations ranging from 2 to 349 μg/l; and the unexposed population that included people living in Pirot municipality, known to contain drinking water arsenic below 0.5 μg/l.
Arsenic concentrations in drinking water were obtained from National water quality monitoring programs from 2004 to 2008. Incidence and mortality data for bladder cancer were obtained from National Cancer Register, supported by the CanReg3 programme package (Department of Descriptive Epidemiology, IARC, Lyon, France, 2002-2005) [7]. These data were available for the same five-year period (2004 to 2008).
Standardized incidence (SIR) was calculated by direct standardization, with the World (ASR-W) standard population. In addition, rate ratio and relative risk for the occurrence of urinary bladder cancer were calculated. STATISTICA software was used for all data analyses (Version 6, StatSoft Inc., Tulsa, OK, USA).

3. Results and Discussion

Standardized incidence rates for urinary bladder cancer in Zrenjanin and Pirot municipalities from 2004 to 2008 in men and women were presented in Figures 1 and 2. SIR for bladder cancer was higher in Zrenjanin municipality in comparison to Pirot municipality in the whole investigation period, both in men and in women. The highest SIR was reported in 2006, being 21.8 per 100.000 in men and 6.6 per 100.000 in women.

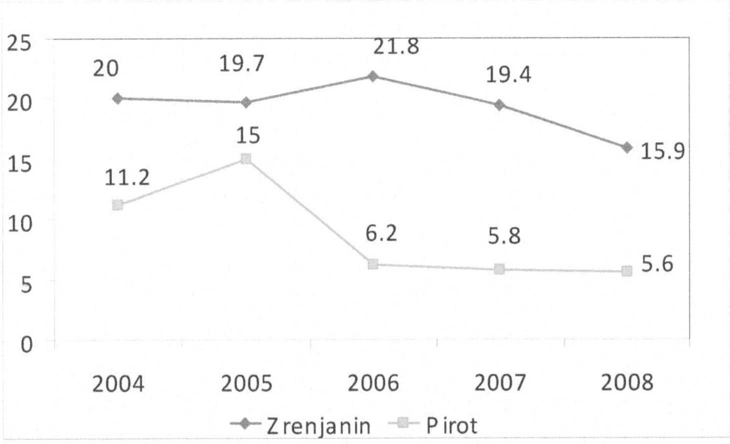

Figure 1 Bladder cancer standardized incidence rates for men in Zrenjanin region compared with Pirot region from 2004-2008

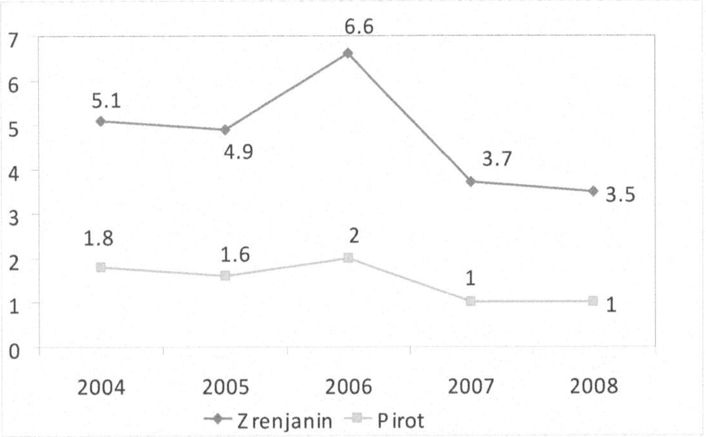

Figure 2. Bladder cancer standardized incidence rates for women in Zrenjanin region compared with Pirot region from 2004-2008

Rate ratios in men and women among exposed and unexposed population are presented in Table 1. Among men, the highest rate ratio was observed in 2007 and 2008, and in 2004 among women. The rate ratio ranged from 1.73 to 6 among men, and from 2 to 4.5 among women.

Table 1. Bladder cancer rate ratios among exposed and unexposed male and female population (5-years comparison)

Year	Rate ratio in men	95% Confidence interval	Rate ratio in women	95% Confidence interval
2004	2.63	1.62-4.01	4.50	2.05-8.54
2005	1.73	1.07-2.65	3.50	1.53-6.92
2006	4.60	2.98-6.79	3.67	1.92-6.37
2007	5.50	3.55-8.19	3.00	1.21-6.23
2008	6.00	3.67-9.30	2.00	0.63-4.82

Relative risk for the occurrence of bladder cancer in men and women was presented in Table 1. In general, men from the exposed municipality had up to three times more risk to develop urinary bladder cancer, compared to those form the unexposed population. Exposed women were up to 2 times more likely to develop urinary bladder cancer, compared to unexposed women.

Table 2. Bladder cancer relative risk among exposed and unexposed male and female population (5-years comparison)

Year	Relative risk in men	95% Confidence interval	Relative risk in women	95% Confidence interval
2004	1.31	0.58-2.95	2.22	0.48-10.30
2005	0.86	0.41-1.81	1.62	0.34-7.82
2006	2.30	0.87-6.05	1.70	0.47-6.11
2007	2.75	0.95-7.98	1.39	0.28-6.90
2008	3.00	0.88-10.18	2.00	0.17-5.07

We found significantly higher standardized incidence rates for bladder cancer for both gender in the exposed population compared to unexposed population in the observed five-year period. The highest standardized incidence rates were observed in year 2006 for both men and women. Standardized incidence rates were on average 4 times higher for men and 3 times higher for women in exposed population in comparison to unexposed population. However, relative risk suggested a rising trend of risk for bladder cancer for men during a short period of time, starting from year 2005.

Conflicting results have been obtained in other studies on arsenic and bladder cancer conducted in areas with low arsenic concentrations in drinking-water. A Finnish case-cohort study (based on 61 cases) reported an increased risk for bladder cancer in association with exposure to arsenic [8], which was significant for short latency exposure only. In contrast, a study in Denmark, based on 214 cases, showed no increase in bladder cancer risk [9]. Studies carried out in the United States did not find an elevated risk for bladder cancer with increasing arsenic exposure in areas with arsenic concentrations in drinking-water ranging between 0.5 and 160 µg/L [10,11,12].

Our study shares limitations common to similar ecological studies, including the lack of data on individual arsenic exposure and the lack of data on the presence of other risk factors for bladder cancer in both exposed and unexposed populations, such as dietary habits, water consumption, smoking and occupational exposures.

4. Conclusions

This ecological study showed that standardized incidence rates for bladder cancer for both gender were significantly higher in the exposed population compared to unexposed population in the observed five-year period with peak in 2006. Standardized incidence rates were on average 4 times higher for men and 3 times higher for women in exposed population in comparison to unexposed population. Relative risk suggested a growing trend of bladder cancer for men during a short period of time. Further studies are needed in order to assess individual exposure to arsenic from drinking water.

References

[1] IARC. Monographs. Evaluation on Carcinogenic to Humans Some Drinking-Water Disinfectants and Contaminants, Including Arsenic. Vol. 84. 512 pp. International Agency for Research on Cancer, Lyon, 2004.

[2] IARC. A review of human carcinogens. C. Metals, arsenic, dusts and fibres. IARC Monographs 100. Lyon, International Agency for Research on Cancer, 2010.

[3] H.R. Guo, H.S. Chiang, H. Hu, S.R. Lipsitz, R.R. Monson, Epidemiology 8 (1997) 545-550.

[4] C. Hopenhayn-Rich, M.L. Biggs, D.A. Kalman, L.E. Moore, A.H. Smith. Environ. Health Perspect 104 (1996) 1200-1207.

[5] I. Varsanyi, L.O. Kovacs. Appl Geochem 21 (2006) 949-963.

[6] K. Koppová, E. Fabiánová, K. Slotová, P. Bartová, M. Drímal, Arsenic health risk assessment and molecular epidemiology project in Slovakia. In: K.C. Donnelly, L.H. Cizmas (Eds.), Environmental Health in Central and Eastern Europe. Springer, Dordrecht, the Netherlands, 2006, pp. 53-160.

[7] CanReg3 programme package. Department of Descriptive Epidemiology, IARC, Lyon, France, 2002-2005.

[8] P. Kurttio, E. Pukkala, H. Kahelin, A. Auvinen, J. Pekkanen. Environ Health Perspect 107 (1999) 705-710.

[9] R. Baastrup, M. Sorensen, T. Balstrom, K. Frederiksen, C.L. Larsen, A. Tjonnenland et al. Environ Health Perspect 116 (2008) 231-237.

[10] M.N. Bates, A.H. Smith, K.P. Cantor. Am J Epidemiol 141 (1995) 523-553.

[11] S.H. Lamm, A. Engel, M.B. Kruse, M. Feinleib, D.M. Byrd, S. Lai, et al. J Occup Environ Med 46 (2004) 298-306.

[12] C. Steinmaus, Y. Yuan, M.N. Bates, A.H. Smith. Am J Epidemiol 158 (2003) 1193-1201.

Blood pressure and drinking water's magnesium level in some Serbian Municipalities

Zorica Rasic-Milutinovic[1], Gordana Perunicic-Pekovic[2], Dragana Jovanovic[3], Ljiljana Bokan[4], Milce Cankovic-Kadijevic[5]

[1]Department of Endocrinology, Zemun Clinical Hospital, Belgrade, Serbia
[2]Department of Clinical Nephrology and HD Unit, Zemun Clinical Hospital, Belgrade, Serbia
[3]Institute for Public Health, Belgrade, Serbia
[4]Biochemical Laboratory, Clinical Hospital, Belgrade, Serbia
[5]Institute of Blood Transfusion, Belgrade, Serbia

Corresponding author e-mail: zoricar@eunet.rs

Abstract

Chronic exposure to lower level of magnesium (Mg) in drinking water increases risk of magnesium deficiency and its potential association with Hypertension. The aim of the study was to assess the effect of mineral contents in drinking water on blood pressure in healthy population. The study was cross-sectional, recruited 90 healthy blood donors, 20 to 50 years age, from tree municipalities. Area of one, Pozarevac, had four times higher mean Mg level in drinking water (42 mg/l), than other, Grocka, (11mg/l). Diastolic blood pressure was the lowest in subjects from Pozarevac. Serum Mg was significantly highest and Ca/Mg lowest in subjects from Pozarevac, and after adjustment for confounders (age, gender, BMI), only total cholesterol and serum Mg level were independent predictors of diastolic blood pressure in subjects from these tree municipalities. Therefore, Mg supplementation in area of lower magnesium level in drinking water may be an important tool in prevention of Hypertension.

1. Introduction

Hypertension is the leading cause of cardiovascular morbidity and mortality of individuals worldwide. Although the exact etiology is unknown, the fundamental hemodynamic abnormality in hypertension is increased peripheral resistance, due primarily to changes in vascular structure and function. Obesity and dietary macronutrients clearly play a role in the risk for hypertension, but the role of micronutrients in this process is not clear.

Several epidemiologic studies suggest a close relation between water hardness, and risk for cardiovascular disease (CVD) [1, 2, 3, 4, 5]. Regarding individual minerals, several studies have been reported where hypertensive subjects were treated orally with nutritional doses of Mg [6, 7, 8]. The results suggested a dose-dependent reduction in blood pressure from the Mg intervention, as well as supplementation of Mg together with other minerals, among persons with a low body burden of Mg and Ca [9, 10].

Magnesium is an essential element that has numerous biological functions in the cardiovascular system. At the subcellular level, Mg regulates contractile proteins, modulates transmembrane transport of Ca, Na and K, and acts as an essential cofactor in the activation of ATPase, controls metabolic regulation of energy-dependent cytoplasmic and mitochondrial pathways [11]. Small changes in extracellular Mg levels and/or intracellular free Mg concentration have major effects on cardiac excitability and on vascular tone, contractility and reactivity [12, 13]. Decreased Mg levels enhance reactivity of arteries to vasoconstrictor agents, attenuate responses to vasodilators, promote vasoconstriction and increase peripheral resistance, leading to increased blood pressure [13, 14]. Elevated Mg levels have opposite effects leading to vasodilatation, reduced vascular tone and decreased blood pressure. Thus magnesium may be physiologically important in blood pressure regulation.

The hardness of ground water, defined by concentrations of Ca and Mg, is different in variety parts of Serbia. Until now, there have not been conducted any research about the relation between drinking water quality and cardiovascular disease, or hypertension, particularly.

In this cross-sectional, epidemiological study, we intended to explore association between water hardness and the level of blood pressure in different parts of country.

2. Materials and Methods

2.1 Study design

The study was randomized, cross-sectional, epidemiological, with three groups of subjects, 90 healthy blood donors, aged 18-58 years (mean 34.57±10.56), 30 from each of 3 municipalities of Serbia (Banovci, Grocka and Pozarevac). One of the municipalities, Pozarevac, was characterized by hard water, and the other two by softer water, particularly Grocka. The present study was designed to determine whether a relationship exists between water hardness and blood pressure in healthy people, and to evaluate potential mechanisms leading to hypertension.

Descriptions of hardness correspond roughly with ranges of mineral concentrations, and total water hardness according to the scale of degree of General Hardness (dGH) (defined as 10 milligrams of calcium oxide per liter of water) could be from: 0-4 dGH (very soft), 4-8 dGH (soft), 8-12 dGH (slightly hard), 12-18 dGH (moderately hard), 18-30 dGH (hard), and > 30 dGH (very hard).

Total water hardness, magnesium and calcium concentrations, electroconductivity and total dissolved solids were measured in water samples from public water supply systems as part of the National Monitoring Programme of Drinking Water Quality from Public Water Supply Systems from 2003 to 2004. Water samples from individual wells were not taken into consideration.

Sampling and chemical analyses of drinking water from water supply systems were performed at the following laboratories: at the Institute of Public Health in Belgrade, the Institute of Public Health in Sremska Mitrovica and the Institute of Public Health in Pozarevac. All laboratories were accredited and authorized according to SRPS ISO/IEC 17025 and SRPS ISO 9001 standards. Laboratory procedures for sample management, analytical methods, and quality control measures (accuracy, precision, and detection limits) were standardized by Serbian laws (Book of Regulations for Water Sampling 87/33 1987; Book of Regulations on the Hygienic Correctness of drinking water 98/42 1998). Following these protocols, water Ca and Mg levels were analyzed by inductively coupled plasma optical emission spectrometry (ICP-OES) [15]. Total water hardness was measured by gravimetric methods [16]. Water conductivity was measured directly using a conductivity probe [17]. Total dissolved solids are determined gravimetrically [18]. Descriptions of hardness correspond roughly with ranges of mineral concentrations.

2.2 Subjects

Subjects were recruited by referrals from primary care physicians. They were evaluated and recruited by physicians of Institute of blood transfusion in Belgrade. A complete history and physical exam were performed. Height and weights of participants were measured in centimeters and kilograms and body mass index, (BMI) kg/m^2 was calculated.

2.3 Blood pressure

Blood pressure was recorded by standard mercury sphygmomanometer, before the blood samples were taken. Two separate recordings were made after 5 minutes of supine rest. The blood pressure, systolic, diastolic and mean arterial pressure, was reported as the average of these recordings.

2.4 Blood samples.

Blood samples were taken after overnight fast, for at least 8h, to measure serum concentration of Mg, Ca, Na, K, P, creatinine, glucose, lipids, insulin, red blood cells, white blood cells, platelets, and haematocrite values.

2.5 Laboratory tests

The analyses were performed at Biochemical laboratory of Clinical Hospital Zemun, Belgrade. Serum minerals Na, K, Ca, were measured by ion-meter AVL- 988-3, and P and Mg were measured with colorimetric assay by IL 650 analyzer. Glucose level, total cholesterol, and triglycerides were measured by commercial enzymatic tests. Insulin level was measured with immunofluorescence assay by IMMULITE 1000.

2.6 Statistical analysis

Data are expressed as means ± SD. Differences between groups were analyzed by general model ANOVA and *post hoc* multiple comparisons were performed using LSD test when ANOVA testing was significant ($p < 0,05$). For variables with skewed distribution, the values were log-transformed and a normal distribution was confirmed by the Komolgorov-Smirnov goodness of-fit test ($p>0.15$). Correlation analysis was performed by calculating Pearson's correlation coefficient. By multivariate linear regression analysis we evaluate the relative importance of factors possibly contributing to the variation in risk factor levels. All statistical analyses were done with the SPSS statistical software package (version 15.0; SPSS, Chicago, USA).

3. Results

The total hardness of water, defined as the sum of Ca and Mg, the levels of Ca and Mg separately, as well as the ratio of Ca/Mg, of three municipalities is presented in Table 1. The water from the water supply from Pozarevac municipality had the highest degree of hardness. The median content of Ca in this water supply system was 99.79 mg/L, almost twice than in water supply system from Grocka, or Banovci. The median content of Mg in water supply from Pozarevac was significantly higher, and the ratio of Ca/Mg was significantly lower than the level in water supply from Grocka. The median content of Mg in water supply system did not differ between Pozarevac and Banovci, as well as the ratio of Ca/Mg (Table 1).

Table 1. The hardness of drinking water from three municipalities

	Pozarevac	Grocka	Banovci
Total hardness(dGH)	23.71	11.15*	13.87*
Calcium (mg/L)	99.79	58.85*	49.7*
Magnesium (mg/L)	42.25	11.8*	54.92
Ca/Mg	2.36	4.98*	0.9
Water conductivity	761	752.7	667.2
Total dissolved solids ($t=105\ ^0C$)	546.0	503.1	434.6

There were no significant differences between groups of subjects according the age, gender, or nutritional status. The subjects were not obese. The mean systolic blood pressure did not differ between groups. There were significant difference between groups for diastolic blood pressure, it was the lowest in subjects from Pozarevac (Table 2).

There were no differences between groups for mean levels of serum Ca^{2+}. The mean serum Mg was the highest in group of Pozarevac, and the ratio Ca^{2+}/Mg was the lowest in serum of the same subjects (Table 2). The mean values of serum triglycerides, as well as creatinine were the lowest in the subjects from Pozarevac (Table 2). Table 3 shows the Pearson's correlation between all clinical and laboratory data of subjects from three geographic area. We observed inverse association between diastolic blood pressure and hardness of drinking water, as well as the mean level of serum Mg. Diastolic blood pressure directly correlated with the ratio of serum Ca^{2+}/Mg, total cholesterol, triglycerides, and creatinine (Table 3).

Serum level of sodium (Na) did not correlate with diastolic blood pressure, but correlated directly with systolic blood pressure. Blood pressure correlated directly with age and BMI, as we expected (Table 3).

In multivariate regression analysis 23% of the variation in diastolic blood pressure was explained by the variation in serum Mg, and cholesterol, after adjustment for age, gender, BMI, the ratio of serum Ca^{2+}/Mg, Na, triglycerides, and creatinine (Table 4).

Table 2. Characteristics of healthy blood donors from three municipalities

Variables	**Pozarevac** mean±SD	**Grocka** mean±SD	**Banovci** mean±SD	Significance p_1; p_2
Age (year)	40.88±9.47	39.33±4.84	37.21±3.85	0.42; 0.08
gender (male/female)	23/7	19/11	22/8	
BMI (kg/m^2)	26.65±4.72	25.44±5.37	25.91±3.67	0.54; 0.29
SBP (mmHg)	124.50±7.80	125.00±5.30	126.96±5.72	0.87; 0.49
DBP (mmHg)	78.30±4.87	81.66±5.86	82.50±4.33	**0.03**; 0.08
MAP	93.70±4.68	96.11±5.17	97.08±4.50	0.18; 0.05
s-Ca^{2+} (mmol/L)	1.06±0.04	1.10±0.02	1.07±0.05	0.13; 0.38
s-Mg (mmol/L)	0.87±0.09	0.71±0.05	0.73±0.05	**0.01; 0.04**
s-Ca/Mg	1.23±0.13	1.54±0.10	1.48±0.12	**0.02**; 0.06
s-Na (mmol/L)	137.52±2.29	141.11±0.42	141.84±2.4 5	**0.04**; 0.96
s-Cholesterol (mmol/L)	5.47±0.95	5.62±1.05	5.22±0.93	0.70; 0.97
s-Triglycerides (mmol/L)	1.20(1.05-1.95)	2.00(0.95-3.60)	1.40(1.00-1.80)	**0.04**; 0.05
s-Creatinine (μmol/L)	72.96±6.98	78.66±12.46	84.10±8.93	**0.01; 0.03**

P_1 = difference between the mean (median) value of variables from Pozarevac and Grocka;
p_2= difference between the mean (median) value of variables from Pozarevac and Banovci

Table 3. Correlations (Pearson correlation coefficient *r*) between hardness of drinking water, serum calcium, magnesium, sodium, cholesterol, triglycerides and blood pressure

	age	BMI	sCa	sMg	sCa/Mg	sNa	sCh	sTg	sCreatinine	SBP	DBP	MAP
age	1	.438**	.145	-.106	.161	-.272	.344**	.227*	.240*	.225*	.400**	.401**
BMI		1	.104	-.099	.135	.161	.244*	.353**	.318**	.412**	.293**	.404**
sCa			1	-.256**	.369**	.100	.175	.046	-.011	.137	.128	.158
sMg				1	-.698**	-.415**	.149	-.205*	-.326**	-.226*	-.262**	-.287
sCa^{2+}/Mg					1	.377**	.214*	.175	.256*	.287*	.254*	.319**
sNa						1	-.083	.020	.197	.224*	.098	.174
sCh							1	.324**	.137	.216*	.358**	.365**
sTg								1	.342**	.198	.291*	.307**
sCreatine									1	.249*	.404**	.414**
SBP										1	.352**	.712**
DBP											1	.908**
MAP												1

*$p<0.05$; ** $p<0.01$

Table 4. Independent predictors of diastolic blood pressure in healthy blood donors from three municipalities

	Standardized Coefficient Beta	t	significance
(Constant)		9.790	0.000
Age	0.300	3.122	0.002
s-cholesterol	0.225	2.361	0.02
s-Mg	-0.194	-2.122	0.03

Adjusted R^2= 23%

4. Discussion

There is increasing evidence that low intakes of Mg are associated with various metabolic diseases, including hypertension, cardiac arrhythmia, cardiovascular disease and diabetes mellitus [4, 8, 19]. The extensive reviews of epidemiological studies on drinking water composition and CVD supported the hypothesis that soft drinking water with the low supply of Mg from drinking water increased the risk of CVD mortality and possibly played a role in developing CVD [20, 21].

The present study shows an inverse association between the diastolic blood pressure and hardness of drinking water, and serum level of Mg is independent predictor of diastolic blood pressure in normotensive healthy subjects. It supports the previously presented hypothesis that water hardness and particularly Mg content may have a role in the etiology of hypertension [13, 14].

Large retrospective study which assessed Mg and Ca content in drinking water in subjects who died from hypertension compared with those who died from other causes demonstrated that magnesium levels in drinking water were inversely related to the risk of death from hypertension [22]. Many clinical studies have shown some forms of hypomagnesemia (serum and/or tissue) in hypertensive patients, with significant inverse correlations between magnesium concentration and blood pressure. A number of factors influence circulating concentrations of Mg. Age, gender, educational level, obesity, smoking habits, alcohol consumption and physical exercise are known to affect the intake of Mg, Ca, and Na. Data from the Vanguard study demonstrated that independently of weight reduction, diet-induced changes in systolic blood pressure were significantly related to changes in urinary excretion of magnesium, potassium and calcium relative to sodium [6]. In agreement with previous studies, after adjustment for age, gender, and BMI, the serum concentration of Mg in our subjects is still negatively associated with systolic and diastolic blood pressure, with the last stronger. Magnesium depletion may be due to dys-regulation of factors controlling magnesium status: intestinal hypoabsorption of magnesium, reduced uptake and mobilization of bone magnesium, urinary leakage, or hyperadrenoglucocorticism by decreased adaptability to stress. Long-term magnesium deficiency in experimental animal's potentates responses to vasoconstrictor agents, attenuates responses to vasodilator agents, increases vascular tone and elevates blood pressure. However, our subjects are healthy people of younger age, not obese, normotensive, with mean serum level of Mg within referent range, but they differ between the groups according to the hardness of drinking water, from the areas separately, as well as according to serum levels of Mg. The subjects from Pozarevac, the area with the hardest drinking water, show the highest level of serum Mg, the lowest level of serum ratio Ca^{2+}/Mg, lower level of serum Na, and the lowest level of diastolic blood pressure. Systolic blood pressure did not differ between the groups. There was the difference for the ratio of serum Ca^{2+}/Mg between the subjects from the hardest drinking water, Pozarevac, and the softener drinking water, Grocka. Since a major part of the magnesium and calcium intake is known to be dietary, the chief limitation of our study is omitted diet recordings from our participants. However, we supposed that the dietary habits of participants did not differ between the investigated areas.

Exact molecular mechanisms of Mg vascular actions are unclear, but Mg probably influences intracellular free Ca concentration, which is fundamental in myocardial regulation, endocrine and renal secretion, and smooth muscle contraction. In vascular smooth muscle cells, Mg antagonizes Ca by inhibiting transmembrane calcium transport and calcium entry. It also acts intracellularly as a Ca antagonist thereby modulating the vasoconstrictor actions of intracellular Ca, a major determinant of vascular contraction [2, 13, 25]. Low magnesium causes an increase in intracellular Ca, with associated vascular contraction and increased tone. Intracellular Mg depletion has been demonstrated in many tissues (heart, lungs, kidney, bone, and muscle) and cell types (vascular smooth muscle cells, erythrocytes, platelets, and lymphocytes) in both human and experimental hypertension [1, 13].

Magnesium is important as a co-factor in lipid metabolism. The rate limiting step in cholesterol synthesis, at HMG-CoA reductase, can be activated through magnesium requiring enzymes [28]. It has been suggested that low magnesium may impair HMG-CoA reductase inactivation via phosphorylation. Magnesium is also important for the activity of the extracellular enzymes Lecithin-Cholesterol acyl transferase and Lipoprotein lipase [28]. Our results agreed with the statement that low Mg might have the impact on the physiological processes that affect serum lipid levels, because we showed significant negative correlation between serum triglycerides and Mg level, and positive correlation between serum cholesterol and Ca^{2+}/Mg level. Magnesium also influences glucose and insulin homeostasis, and hypomagnesemia is associated with metabolic syndrome, with hypertension as a part of that cluster [29, 30].

In conclusion, data from clinical trials of magnesium therapy in hypertension have been disappointing. However, increasing evidence indicates that low magnesium may play a pathophysiological role in the development of hypertension. Our study results demonstrate that areas with hard drinking water and adequate supply of Mg from drinking water, may prevent hypertension.

Acknowledgments

This work was supported by grant from the Serbian Ministry of Science and Environmental Protection No1352. The authors are thankful to all healthy volunteers who participated in the study as well as to the medical staff of Zemun Clinical Hospital and Institute of Blood Transfusion in Belgrade.

References

[1] Kousa A, Moltchanova E, Viik-Kajander M, Rytkonen M, Tuomilehto J: Geochemistry of ground water and the incidence of acute myocardial infarction in Finland. *J Epidemiol Community Health*, 58 (2004) 136

[2] Rylander R. Environmental magnesium deficiency as a cardiovascular risk factor. *J Cardiovasc Risk*, 3 (1996) 4

[3] Nerbrand C, Agréus L, Arvidsson Lenner R, Nyberg P, Svärdsudd K: The influence of calcium and magnesium in drinking water and diet on cardiovascular risk factors in individuals living in hard and soft water areas with differences in cardiovascular mortality. *BMC Public Health*, 3 (2003) 21

[4] He K, Liu, K., Daviglus, ML., Morris, SJ, Loria, CM, Van Horn, L, Jacobs Jr, D.R., Savage, P.J. Magnesium intake and incidence of metabolic syndrome among young adults. *Circulation*, 113 (2006) 1675

[5] Kousa A, Havulinna S.A, Moltchanova E, Taskinen O, Nikkarinen M, Salomaa V, Karvonen M. Magnesium in well water and the spatial variation of acute myocardial infarction incidence in rural Finland. *Applied Geochemistry*, 23 (2008) 632

[6] Resnick, L.M., Oparil, S., Chait, A., Haynes, R.B., Kris-Etherton, P., Stern, J.S., Clark, S., Holcomb, S., Hatton, D.C., Metz, J.A., McMahon, M., Pi-Sunyer, F.X., McCarron, D.A. Factors affecting blood pressure responses to diet: the Vanguard study. *Am J Hypertens*, 13 (2000) 956

[7] Jee SH, Miller ER, Guallar E, Singh VK, Appel LJ, Klag MJ: The effect of magnesium supplementation on blood pressure: A metaanalysis of randomized clinical trials. *Am J Hypertens*, 15 (2002) 691

[8] Ragnar Rylander and Maurice J Arnaud. Mineral water intake reduces blood pressure among subjects with low urinary magnesium and calcium levels *BMC Public Health*, 4 (2004) 56

[9] Bucher HC, Cook RJ, Guyatt GH, Lang JD, Cook DJ, Hatala R, Hunt DL: Effects of dietary calcium supplementation on blood pressure. A meta-analysis of randomized controlled trials. *JAMA*, 275 (1996) 1016

[10] Rubenowitz E, Axelsson G, Rylander R. Magnesium in drinking water and body magnesium status measured using an oral loading test. *Scand J Clin Lab Invest*, 58 (1998) 423

[11] Cowna, J.A. The biological chemistry of magnesium. VCH Publishers, New York, 2000

[12] Altura, B.M. and Altura, B.T. Magnesium in cardiovascular biology. *Sci Amer* (Science and Medicine), 1995, p 28-37

[13] Touyz MR. Role of magnesium in the pathogenesis of hypertension. *Molecular Aspects of Medicine*, 24 (2003) 107

[14] Sontia B and Touyz MR. Role of magnesium in hypertension. *Archives of Biochemistry and Biophysics*, 458 (2007) 33

[15] Eaton, Andrew D. et al. 3500-Ca EDTA. Standard Methods for the Examination of Water and Wastewater, 19th ed. Washington, DC, American Public Health Association, 1995

[16] Eaton, Andrew D. et al. 2340-C EDTA. Standard Methods for the Examination of Water and Wastewater, 19th ed. Washington, DC, American Public Health Association, 1995

[17] Eaton, Andrew D. et al. 2520-B Conductivity Method. Standard Methods for the Examination of Water and Wastewater, 19th ed. Washington, DC, American Public Health Association, 1995

[18] Eaton, Andrew D. et al. 2540-B Total Solids Dried. Standard Methods for the Examination of Water and Wastewater, 19th ed. Washington, DC, American Public Health Association, 1995

[19] Rasic-Milutinovic Z, Perunicic-Pekovic G, Pljesa S, Dangic A, Libek V, Bokan LJ, Cankovic-Kadijevic M. Magnesium deficiency in type 2 diabetes. *Hippokratia,* 8 (2004) 179

[20] Sauvant M.P. and Pepin D. Drinking water and cardiovascular disease. *Food Chem Toxicol,* 40 (2002) 1311

[21] Monarca S, Donato, F, Zerbini I, Calderon R.L, Creau, GF. Review of epidemiological studied on drinking water hardness and cardiovascular diseases. *Eur J Cardiovasc Prev Rehabil,* 13 (2006) 495

[22] Yang CY and Chiu HF. Calcium and magnesium in drinking water and the risk of death from hypertension. *Am J Hypertens,* 12 (1999) 894

[23] Runyan, AL, Sun, Y, Bhattacharya, SK, Ahokas, RA, Chhokar, VS, Gerling, IC. Responses in extracellular and intracellular calcium and magnesium in aldosteronism. *J Lab Clin Med,* 146 (2005) 76

[24] R.M. Touyz, and E.L. Schiffrin. Activation of the Na^+-H^+ Exchanger Modulates Angiotensin II-Stimulated Na^+-Dependent Mg^{2+} Transport in Vascular Smooth Muscle Cells in Genetic Hypertension. *Hypertension,* 34 (1999) 442.

[25] W. Weglicki W, G. Quamme, K. Tucker, M. Haigney, L. Resnick. Potassium, Magnesium, and electrolyte imbalance and complications in disease management. *Clin Exp Hypertens,* 27 (2005) 95

[26] K. Kisters, F. Wessels, F. Tokmak, E.R. Krefting, B. Gremmler, M. Kosch, M. Hausberg. Early-onset increased calcium and decreased magnesium concentrations and an increased calcium/magnesium ratio in SHR versus WKY. *Magnes Res* , 17 (2004) 264

[27] Appel LJ, Moore T, Obarzanek E, Vollmer W, Svetkey L, Sacks F et al. A clinical trial of the effects of dietary patterns on blood pressure. Research Group. *N Engl J Med,* 336 (1997) 1117

[28] Rosanoff A and Seelig MS. Comparison of mechanism and functional effects of magnesium and statin pharmaceuticals. *J Am Coll Nutr,* 23 (2004) 501S

[29] Paolisso G, and Barbagallo M. Hypertension, diabetes mellitus, and insulin resistance: the role of intracellular magnesium. *Am J Hypertens,* 10 (1997) 346

[30] He K, Song Y, Belin RJ, Chen Y. Magnesium intake and the metabolic syndrome: epidemiologic evidence to date. *J Cardiometab Syndr,* 1 (2006) 351

Tap water quality regarding metal concentrations in Timisoara city, Romania

Gabriela Vasile[1], Liliana Cruceru[1], Jana Petre[1], Adriana Anghelus[2], Daniela Gheorghe[2], Diana Landi[2], Adriana Stefanescu[2]

[1]National Research and Development Institute for Industrial Ecology, Road Panduri no. 90-92, district 5, code 050663, Bucharest, Romania

[2]AQUATIM Company, Street Gheorghe Lazar no 11A, code 300081, Timisoara, Timis County, Romania

Corresponding author e-mail: ecoind@incdecoind.ro

Abstract

The aim of the study was to identify the risk prevalence of relevant metals in in-building installation systems in Timisoara City. In the study were collected more than 250 tap water samples in order to get an overview of the overall current contamination level of drinking water at the point of consumption. In the monitoring program were included three water plants, fifteen points from the control program of the company, thirty-three tap waters (first draw and fully flushed sampling procedure) and thirty-two tap water samples (random daytime sampling). The quality of drinking water produced by AQUATIM Company was in accordance with the European Directive 98/83/EC requirements. In samples collected from customer's tap the percent of non-compliance samples was around 50% in first draw, 10% in fully flushed and 25% in random daytime samples. The domestic distribution systems have an important influence to the quality of drinking water delivered by the AQUATIM Company.

1. Introduction

Access to safe drinking water is a basic concern for human health and health protection. According to the World Health Organization (WHO) and European Council Directives, a concentration of microorganisms, parasites or substances posing a possible risk to human health has to be prevented (WHO, 2008). The provision of safe drinking water is one of the main requirements of drinking water supply infrastructure. Therefore, the monitoring of the drinking water from source to tap is an essential step towards hygiene safety.

At the European Community level, Directive 98/83/EC (Council Directive, 1998) regulates water quality at the tap. The objective of this directive is to protect human health from adverse effects resulting from contamination of water intended for human consumption for drinking, cooking, food preparation or other domestic purposes (Roccaro et al, 2005).

As a result of the Council Directive 98/83/EC, water authorities around Europe are obliged to monitor water for public use, so that the consumer is provided with safe and substance - free water.

In some European countries such as Romania or Germany, the water distributors have to ensure that microbial and chemically clean water reaches water meters. After that, the owner of the building is responsible for the water quality. Up to the water meter the drinking water quality is very good (Volker at al, 2010), but the water provided by the water distributors may have a higher microbial and chemical quality than that available from taps in the customers households.

Household pipe can have a considerable impact on the water quality, which was already shown in several large scale studies addressing metal concentrations in tap water after overnight stagnation (Haider et al, 2002; Vasile et al, 2009; Ziez et al 2003; 2007). All these studies reported an increased concentration of lead, cadmium, copper, iron and nickel after stagnation in household tap water in Austria, Germany and Romania.

Numerous studies have highlighted and reviewed the influence of water quality parameters (e.g. pH, dissolved oxygen, temperature, alkalinity, chloride, sulfate, phosphate and organic matter) and operating conditions (stagnation time and pipe age) on copper release into drinking water from copper pipes (Darren, 2010).

Lead in drinking water is a major public health concern. It can create irreversible intellectual impairment in infants and young children, even at blood lead levels below 10 µ/L. (Jusko et al, 2008). Generally, source waters are free of lead, but significant amounts of lead may be present in the tap water due to dissolution of lead corrosion products, which are formed in water distribution network and domestic plumbing systems.

In the absence of lead pipe, Pb-based solder and brass fittings (materials containing up to 8% lead) are know to be dominant lead sources in public water supply system (Kimbroungh, 2001).

In a domestic plumbing system several types of materials could release lead into water. First source could be Pb-pipes as a predominant source (Cheng et Foland, 2004). Second, before 1987, solders used in plumbing systems contained significant amounts of lead are even today solders containing lead (low quantity, around maxim. 0.2%). The Pb-bearing solders could be in direct contact, and, therefore, release lead into water. In addition, some faucet assemblies and fixtures are also problematic sources of lead, as shown by Gulson et al (1994). Other materials, even those with low Pb contents may contribute to water Pb as well.

In Europe, random daytime (RDT) sampling (1st liter taken during office hours, without fixed stagnation) and sampling after 30 minutes of stagnation (30 Ms) (1st and 2nd liter) were identified as the best approaches for estimating exposure and detecting homes with elevated lead concentration in tap water (Deshommes, 2010; Hayes, 2009; Hayes et al, 2010).

Iron release from corroded iron pipes is the principal cause of "colored water" problems in drinking water distribution systems. These corrosion scale deposits reduce the hydraulic capacity of the pipes and can adversely affect water quality during distribution. Some consequences are colored water when iron is released from corrosion scales, high demand for chloride and dissolved oxygen, biofilm growth, adsorption, and accumulation of substances such as arsenic, which can be released on modification of water quality (Sarin et al., 2004).

Tap water from the municipal supply systems is the source of drinking water for a majority of homes in Romania. The aim of the study was to identify the risk prevalence of relevant metals in in-building installation systems in Timisoara City from Romania with the population more than 400,000 inhabitants. In the study were collected more than 250 tap water samples in order to get an overview of the overall current contamination levels of drinking water at the point of consumption.

2. Experimental Data

In February and June 2010 were collected more than 250 tap water samples delivered by AQUATIM COMPANY, an important Romanian Drinking Water Producer. In addition, were collected samples from Drinking Water Plants and 15 control points of the Company in order to establish a baseline for comparison of the data obtained in the customers monitoring plan.

AQUATIM Company delivers the drinking water in Timisoara town and in the surrounding area. Two sources of raw water are used: surface water and groundwater.

- Bega Water Plant - about 67% of the total quantity of water is produced using surface water from Bega River.
- Urseni Water Plant - approximately 30% of the total quantity of water is produced using groundwater from 18 drilling points at a depth of 60 to 80 m, and 40 drilling points at a depth of 110 to 160 m.
- Ronat Water Plant is used only in certain cases, only to ensure the high water consumption (especially in the evening), using groundwater from five drilling points.

AQUATIM Company checks daily the quality of drinking water in 30 monitoring points, situated in different locations in the aria, such as elementary schools, kindergartens, markets, fountains, public institutions

The samples were collected in accordance with the monitoring plan established between INCD-ECOIND and the specialists from AQUATIM Company. In figure 1 is presented the map of public network system in Timisoara City. The locations of control points are marked with green, the Water Plants with blue and customer's tap with red.

Figure 1. Timisoara tap water customer's monitoring plan

In order to obtain a large database, the samples were collected from customer's cold line-pipe with three different sampling techniques:

- first draw sampling (from kitchen, first in the morning, before using the tap) - 33 monitoring points from the customers with residence in different parts of the city (C1-C33);
- fully flushed sampling procedure after flushing five minutes the tap - same 33 points and other 15 points from the Drinking Water Producer, points situated in markets, schools, street fountains (P1-P15);
- random daytime procedure (within office hour, without previous flushing of the tap) - 32 sampling points situated in old buildings from the center of the city - medical centers, pharmacies, schools, private companies, public institutions (1P-32P).

The parameters Al, As, Cd, Cu, Cr, Fe, Mn, Ni, Pb, Se, Sb and Zn were analysed using inductively coupled plasma atomic emission spectroscopy ICP-EOS technique (OPTIMA 5300 DV Perkin Elmer with Flow Injection Hydride Generation System FIAS 400) after digestion of drinking water samples with nitric acid and concentration of the acid solutions from 150 mL to 25 mL.

For each set of samples was prepared a blank sample using the same procedure, blank obtained with ultra pure water and 5 mL of nitric acid suprapur. The WinLab 32 soft of OPTIMA 5300 DV equipment extract the blank value of the metal from the unknown concentration. Therefore, the obtained values of the parameters represent only the concentrations from the analyzed samples.

In the study were prepared samples on three different level of concentration, using Standard Reference Materials (Quality Control Standard Perkin Elmer 21, 100 mg/L As, Cd, Cr, Cu, Fe, Mn, Ni, Pb, Sb, Se, Zn si Quality Control Standard Perkin Elmer 7A, 100 mg/L Al), nitric acid and ultra pure water. The recovery percents were situated in the range 94.5% ÷ 114.5% and were used for calculation of the results.

In table 1 are presented the detection limits obtained with the equipment and analytical methods used in the study and also the maxim admissible value for the metal concentration according to Romanian Legislation (Law 458, 2002).

Table 1. Detection limits, maxim admissible value according to Romanian Legislation and analytical techniques applied in the study

Parameter	**Al**	**As**	**Cd**	**Cu**	**Cr**	**Fe**
Max. Admissible Value (µg/L)	200	10	5	100	50	200
LOD (µg/L)	1	0.3	0.4	0.6	0.5	0.3
Analytical technique	ICP-EOS	ICP-EOS-FIAS	ICP-EOS	ICP-EOS	ICP-EOS	ICP-EOS
Parameter	**Mn**	**Ni**	**Pb**	**Se**	**Sb**	**Zn**
Max. Admissible Value (µg/L)	50	20	10	10	5	5 000
LOD (µg/L)	0.1	1	1	0.4	0.1	0.5
Analytical technique	ICP-EOS	ICP-EOS	ICP-EOS	ICP-EOS-FIAS	ICP-EOS-FIAS	ICP-EOS

3. Results and Discussions

The quality of drinking water provided by AQUATIM Company was situated in the limits imposed by the Romanian Legislation. The results obtained in the monitoring program of metallic parameters in drinking water samples collected from the customer taps are compare with maximum admissible values for metal concentrations according to Law 458/2002 (with subsequent modifications) on water quality for human consumption.

The monitoring data show important influences on the tap water quality of the material used in the internal distribution system within the customer buildings. The data indicates real problems for Cu, Fe and Pb.

The materials used in drinking water domestic installations in the selected points for tap water survey were galvanized steel, lead, copper, steel, cast iron, polyvinyl chloride (PVC). In the local public network, in same points, the majority pipes consist of galvanized steel, cast iron, high-density polyethylene (PEHD) and polyethylene (PE) (table 2). The materials responsible for metals leaching are cast iron (Fe. Mn), copper (Cu), lead (Pb), galvanized sheet (Fe, Mn, Ni, Zn).

Table 2. Materials used in domestic distribution and public network systems

Type of material	Cast iron	PEHD	Copper	Pb	PVC	Steel	PE
Primary material of domestic distribution system	**7.2%**	-	**32.1%**	-	13.3%	**21.4%** Galvanized sheet	25% pexal
Primary material of public network	**41.9%**	3.2%	-	-	9.7%	12.9%	32.3%
Branch pipe	-	3.5%	-	**17.9%**	-	28.6%	50%
Control points AQUATIM	**33.3%**	60%			1%		

In the tables 3 to 7 are presented statistical data (minim, maxim, mean, median values, porcent of non-compliance samples) for first draw, fully flushed and random daytime results.

In thirty-two samples collected with random daytime procedure from the histhorical center of Timisoara City, 28.13 % are non-compliance samples (highest concentrations than admissible values for Cu, Fe, Mn and Pb) (Table 3). These data show that internal distribution systems affect drinking water quality. In order to use a better drinking water, the tap must be washed before the water is collected.

Table 3. Random Daytime Data, June 2010 (µg/L)

Parameter / Element	Al	Cu	Fe	Mn	Pb	Zn
Minimum value	13.3	0.9	5.9	1.2	< 1	2.1
Maximum value	67.7	**1064**	**602**	**314**	**16.5**	2404
Median value	28	11.8	38.8	5	< 1	124
Mean value	31.5	57.4	90.3	20.5	2	335
Standard deviation	11.9	184	131	56.7	3.2	565
Maximum admisible value	200	100	200	50	10	5000
% of Non-compliance samples /element	0	**9.38%**	**12.5%**	**6.35%**	**2.13%**	0
No. of Non-compliance samples /element	0	**3**	**4**	**2**	**1**	0
Total % of Non-compliance samples	**28,13 %**					
Total of Non-compliance samples	**9**					

Problems occur in thirty-three tap water samples collected with first draw sampling procedure, for which some of the obtained values of Cu, Fe, Ni, Pb are higher than admissible values indicating an influence of the local equipments (pipe, tap, fitting) to the drinking water quality (Tables 4 and 6).

Table 4. First Draw Data, February 2010(µg/L)

Parameter / Element	Al	Cu	Fe	Mn	Ni	Pb	Zn
Minim value	22.4	4.7	10.7	0.5	< 1	< 1	0.5
Maxim value	**735**	**397**	**1633**	49.7	**23.2**	**22.4**	1881
Median value	82	24.3	65.5	4.5	< 1	1.5	88.2
Mean value	101	80.7	186	7.2	5.2	4.2	321
Standard deviation	118	111	365	9.3	5	5.3	449
Maxim admisible value	200	100	200	50	20	10	5000
% of Non-compliance samples /element	**3.03%**	**27.27%**	**15.15%**	0	**3.03%**	**15.15%**	0
No. of Non-compliance samples /element	**1**	**9**	**5**	0	**1**	**5**	0
Total % of Non-compliance samples	**45,45 %**						
Total of Non-compliance samples	**15**						

It is not recommended to use first draw water for cooking or drinking purposes, because in this water can be leached high concentrations of metals depending on the retention time and material type. In the tap water collected from copper cold water pipes high concentrations of copper was recorded, much more than 100 µg/L, which is the maximum admissible value in Drinking Water Romanian Law. If the tap is washed more than 5 minutes, the copper content in tap water decreases, being situated under the limit.

Table 5. Fully Flushed Data, February 2010(µg/L)

Parameter / Element	Al	Cu	Fe	Mn	Ni	Pb	Zn
Minim value	22.7	2.5	15.2	3.9	< 1	< 1	0.5
Maxim value	**329**	52.8	**294**	12.5	3	**11.6**	873
Median value	82.9	6.3	36.1	4.3	< 1	< 1	3.6
Mean value	90.1	8.8	68.6	6.2	< 1	2.2	47.3
Standard deviation	53.7	9.2	68.2	4.6	0.6	3.0	152
Maxim admisible value	200	100	200	50	20	10	5000
% of Non-compliance samples /element	**3.03%**	0	**9.1%**	0	0	**6.06 %**	0
No. of Non-compliance samples /element	**1**	0	**3**	0	0	**2**	0
Total % of Non-compliance samples	**12,1%**						
Total of Non-compliance samples	**4**						

Table 6. First Draw Data, June 2010(µg/L)

Parameter / Element	Al	Cu	Fe	Mn	Ni	Pb	Zn
Minim value	17.8	5.4	7,5	1	< 1	< 1	0.5
Maxim value	199	**1029**	**589**	20.1	**32.7**	**36**	1570
Median value	28.4	35	60	3.9	< 1	< 1	340
Mean value	40.5	125	106	6.7	4.8	4.5	470
Standard deviation	37	210	130	5.3	8.6	9.5	398
Maxim admisible value	200	100	200	50	20	10	5000
% of Non-compliance samples /element	0	**39.1%**	**17.4%**	0	**8.7%**	**13%**	0
No. of Non-compliance samples /element	0	**9**	4	0	**2**	**3**	0
Total % of Non-compliance samples	**56,5%**						
Total of Non-compliance samples	**13**						

Another problem observed was related to iron content, possible leached by the cast iron or unprotected steel pipes. In addition, branch pipe and short piece of lead old pipe included in the internal distribution system has a negative influence on the tap water quality

Table 7. Fully Flushed Data, June 2010 (µg/L)

Parameter / Element	Al	Cu	Fe	Mn	Ni	Pb	Zn
Minim value	15	1.8	7.1	1.3	< 1	< 1	0.5
Maxim value	63.5	69.5	159	20.8	< 1	**12.1**	151
Median value	31.7	5.6	26.1	3.1	< 1	< 1	15.2
Mean value	34.9	11.2	41.8	4.2	< 1	1.6	32.8
Standard deviation	13.5	15.1	45.2	4	0	3.2	38.1
Maxim admisible value	200	100	200	50	20	10	5000
% of Non-compliance samples /element	0	0	0	0	0	**8.7%**	0
No. of Non-compliance samples /element	0	0	0	0	0	**2**	0
Total % of Non-compliance samples	**8.7%**						
Total of Non-compliance samples	**2**						

The metal concentrations recorded in tap water collected with tap flushing procedure were situated, almost in all cases, in the limit values. The values of Pb in the flushed tap waters were situated in most cases below the detection limit of the method used (1 µg/L) or were recorded very low values, close to the limit of detection. Only 2 samples (same in both compagnes) had Pb concentrations higher than the admissible limit (tables 5 and 7).

Table 8. Fully Flushed Data - Control Points, February 2010 (µg/L)

Parameter / Element	Al	Cu	Fe	Mn	Ni	Pb	Zn
Minim value	22.7	2.5	15.2	3.9	< 1	< 1	0.5
Maxim value	**329**	52.8	**294**	12.5	3	**11,6**	873
Median value	82.9	6.3	36.1	4.3	< 1	< 1	3.6
Mean value	90.1	8.8	68.6	6.2	< 1	2.2	47.3
Standard deviation	53.7	9.2	68.2	4.6	0.6	3.0	152
Maxim admisible value	200	100	200	50	20	10	5000
% of Non-compliance samples /element	**6.66%**	0	**20%**	0	0	**13.33%**	0
No. of Non-compliance samples /element	**1**	0	**3**	0	0	**2**	0
Total % of Non-compliance samples	**26.66%**						
Total of Non-compliance samples	**4**						

In tables 8 and 9 are presented results from control points samples recorded in February and June 2010. Samples were collected with fully flushed procedure. The data shows high contents for Al, Fe and Pb in some points (27% - winter champagne; 14% - summer period; non-compliance samples).

Table 9. Fully Flushed Data - Control Points, June 2010 (μg/L)

Parameter / Element	Al	Cu	Fe	Mn	Ni	Pb	Zn
Minim value	17.8	1.8	19.5	2	< 1	< 1	5.8
Maxim value	65.5	80.4	**290**	22.8	2.2	**12.9**	2681
Median value	26	7.4	47.5	5.1	< 1	< 1	83
Mean value	29.3	15.5	81.4	7.3	< 1	2.5	338
Standard deviation	12.8	19.7	75.9	6.5	0.5	3.7	672
Maxim admisible value	200	100	200	50	20	10	5000
% of Non-compliance samples /element	0	0	**14.3%**	0	0	**7.1%**	0
No. of Non-compliance samples /element	0	0	**2**	0	0	**1**	0
Total % of Non-compliance samples	**14,3%**						
Total of Non-compliance samples	**2**						

4. Conclusions

The quality of drinking water provided by AQUATIM Company was situated in the limits imposed by the Romanian Legislation.

High concentrations of Cu, Fe, Ni and Pb in first draw samples were recorded in apartments where the majority of the material for installation was copper, cast iron, galvanized sheet and the branch pipe was made from Pb.

In samples collected from customer's tap the percent of non-compliance samples was around 50% in first draw, 10% in fully flushed and 25% in random daytime samples.

The customers were advised to don't use the first draw water for cooking and drinking purpose, because in this water can be leached high concentrations of metals depending on the retention time and material type. Water volume and time of stationary are the most important parameters that determine the concentration of metallic elements released from the materials of consumer distribution installations.

Materials used in water supply domestic installations have a major contribution in deterioration of water quality provided by the local distribution network, due to the processes of water stagnation and lack of maintenance of the internal distribution materials.

If the limit values of metals in drinking water are exceeded the recommendations are either flushing the tap more than five minutes and then use water for household consumption or replacement of the pipes and fittings in both, local or domestic distribution systems.

This research demonstrates that materials used in water distribution systems are part of the overall treatment process that affect the water quality which consumers drink at their tap. The interaction between water and the infrastructure used for its supply are fundamental in producing safe drinking water. Subtle reactions between water and different materials used for its transport can result in alterations that affect the finale quality delivered to consumers.

References

Cheng, Z., Foland, K., (2004), Laed isotopes in tap water: implications for Pb sources within a municipal water supply system, *Appl. Geochem.*, **20**, 353-365.

Council Directive 98/83/EC, **1998**, on the quality of water intended for human consumption.

Darren, L.A., Mallikarjuna, N.N., (2010), A comprehensive investigation of copper pitting corrosion in a drinking water distribution system, *Corros. Sci.*, **52**, 1927-1938.

Deshommes, E., Laroche, L., Nour, S., Cartier, C., Prevost, M., (2010), Source and occurrence of particulate lead in tap water, *Water Res.*, **44**, 3734-3744.

Gulden, B., Law, A., Korsch, M., (1994), Effect of plumbing systems on lead content of drinking water and contributions to lead body burden, *Sci. Total. Environ.*, **144**, 279-284.

Haider, T., Haider, M., Wruss, W., (2002), Lead in drinking water of Vienna in comparison to other European countries and accordance with recent guidelines, *Int. J. Hyg. Environ. Health*, **205**, 399-403.

Hayes, R.C., (2009), Computational modeling to investigate the sampling of lead in drinking water, *Water Res.*, **43**, 2647-2656.

Hayes, R.C., Aertgeerts, R., Barrott, L., Becker, A., Benoliel, M. J., Croll, B., (2010), *Best practice guide on the control of lead in drinking water*, Hayes, R.C. (Ed), IWA Publishing, London, 13-23.

Kimbrough, D., (2001), Brass corrosion and the LCR monitoring program, *J. Am. Water Works Assoc.*, **1**, 81-91.

Law 458, (2002), concerning drinking water quality, Official Monitor of Romania, no. 552, modified by Law 311, (2004), Official Monitor of Romania, Part 1, 382 (romanian).

Roccaro, P., Mancini, G., Vagliasindi, F., (2005), Water intended for human consumption - Part I. Compliance with European water quality standards, *Desalination*, **176**, 1-11.

Sarin, P., Snoeyink, U.L., Bebec, J., Jim, K. K., (2004), Iron release from corroded iron pipes in drinking water distribution systems: effect of dissolved oxygen, *Water Res.*, **38**, 1259-1269.

Vasile, G.G., Dinu, C., Chiru E., (2009), Monitoring of metal concentrations in tap waters in Bucharest supply system, 3rd International Conference COST ACTION 637 "Metals and related substances in drinking water", Ioannina, Greece, 50.

Volker, S., Schreiber, C., (2010), Drinking water quality in household supply infrastructure. A survey of the current situation in Germany, *Int. J. Hyg. Environ. Health*, **213**, 204-209.

Zietz, B.P., de Vergara J.D., (2003), Copper concentrations in tap water and possible effects on infant's health-results of a study in Lower Saxony, Germany. *Environ. Res.*, **92**, 129-138.

Zietz, B.P., Lass, J., (2007), Assessment and management of tap water lead contamination in Lower Saxony, Germany, *Int. J. Environ. Health. Res.*, **17**, 407-418.

WHO, (2008), *Guidelines for drinking water quality. 3rd ed. Recommendations. Incorporating 1st and 2nd Addenda*, volume 1, World Health Organization, Geneva.

The need for an integrated approach to control metal and metalloid contamination of drinking Water

Colin Hayes
School of Engineering, Swansea University, Singleton Park, Swansea, SA2 8PP, UK
Corresponding author e-mail: c.r.hayes@swansea.ac.uk

Abstract
The metal and metalloid contaminations that can arise in a water supply system, from source to tap, are reviewed and the control options identified. Generally, point source problems are easier to monitor and rectify. The dependency on sampling and monitoring is outlined in the context of both regulatory and operational needs. Problems have been experienced with the monitoring of diffuse metal contamination (Cu, Ni, Pb) at consumers' taps and in consequence, public health protection is poor or absent in some countries. The potential extent of problems with lead in drinking water reveals the need for a more integrated approach to control. The main conclusions are that: (i) Health Authorities need to determine the extent of problems with metals and metalloids in drinking water in their area and pursue appropriate improvements; (ii) a European Drinking Water Inspectorate could considerably strengthen the enforcement of quality standards and take a leading role in the implementation of risk assessment and risk management; (iii) the public reporting of drinking water quality could be much improved; and (iv) risk assessment is not a perfect science and should endeavour to look broadly at all the relevant information that is available.

1. Introduction

The metals and metalloids most commonly associated with drinking water are listed in Table 1 together with the EU standards that apply [(1)], their main significance and the principal control options.

Table 1. Metals and metalloids in drinking water

Metal or metalloid	EU standard (µg/l)	H	A	MB	Control
Aluminium	200		✔		Source treatment and process control
Antimony	5	✔			Source treatment (rare)
Arsenic	10	✔			Source treatment (common)
Cadmium	5	✔			Source protection (industry)
Calcium	-			✔	Source and point-of-use treatment
Chromium	50	✔			Source protection (industry)
Copper	2000		✔		Restrict use & corrosion control
Iron	200		✔		Source treatment & pipe rehabilitation
Lead	25 (10)	✔			Pipe removal & corrosion control
Magnesium	-			✔	Source and point-of-use treatment
Manganese	50		✔		Source treatment
Mercury	1	✔			Source protection (industry)
Nickel	20	✔			Restrict use & corrosion control
Selenium	10	✔			Source treatment (rare)
Sodium	200,000	✔			Source treatment or blending
Zinc	-		✔		Restrict use & corrosion control

H = health A = aesthetic MB = mineral balance

The major issues are summarised in Figure 1 from "source to tap" and span several circumstances: (i) natural source contamination, (ii) contamination from treatment chemicals, (iii) pick-up from corroding water mains and (iv) pick-up from domestic pipe-work systems.

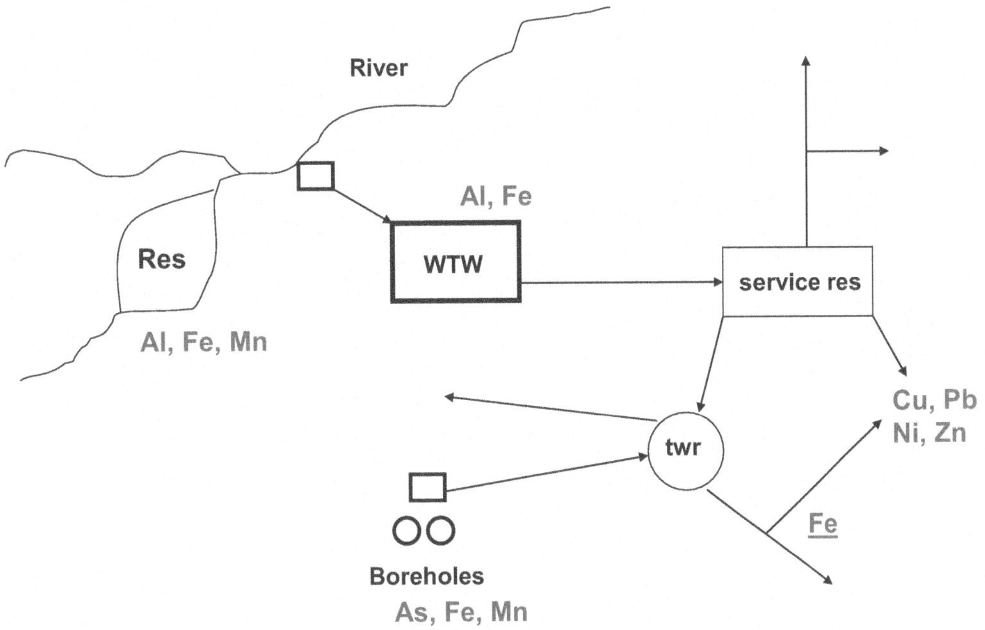

Figure 1. The major issues from source to tap

2. Control problems

At the municipal scale, point sources are easier to identify, monitor and control through source protection or treatment, whereas diffuse sources are much more difficult. Examples of the latter are (i) Fe from old iron water mains which can be difficult to pin-point and (ii) Cu, Pb, Ni from domestic pipe-work systems due to the variable contact time between water and metal components. It can be noted here that the EU Directive [(1)] has mostly failed to tackle Cu, Pb and Ni at the tap because of sampling problems. A further complication is that there are between 2 and 10 million small/very small supplies in the EU [(2)], serving at least 10% of the population. Monitoring and control are very much limited by a lack of resources and knowledge and the main concerns are As (feasibility of corrective treatment?) and Pb (pipe removal?). In essence, these point source problems are diffuse in control terms. The expected worsening impact of water stress as a consequence of climate change will compound these control problems (eg: deterioration in resource quality).

3. Problems with metals at the tap

Limited random daytime (RDT) sampling data from Europe [(3)] suggests that Cu is not a major issue. Problems with Cu pipes will likely be localised (pitting corrosion, influence of natural organic matter, acidic waters). However, there is scope for developing rapid corrosion testing methods. Limited RDT sampling data from Europe [(3)] indicates that Ni non-compliance is significant in some areas. A relaxation of the current standard from 20 µg/l to 70 µg/l (new WHO Guideline Value) would solve the problem, as would ortho-phosphate dosing. 95% of the UK's public water supplies are dosed with ortho-phosphate for plumbosolvency control and it appears that nickel-solvency is also much reduced, based on the high level of compliance (99.79%) observed in 2009 for Ni in the UK [(4)].

A range of data indicates widespread problems with Pb and up to 25% of homes in the EU [(5)] could be at risk of exceeding the WHO GV of 10 µg/l. The evidence comes from: (i) plumbosolvency testing - all types of drinking water in contact with Pb pipes can exceed the WHO Guideline Value, unless specifically treated to reduce plumbosolvency [(6)]; (ii) limited estimates from 12 EU countries of the percentage of homes that are supplied by Pb pipes (7); (iii) survey data from across Europe assembled by COST Action 637 via conferences and a research project, as summarised by Hayes and Skubala (5).

4. Risk assessment for lead in drinking water

The extent of risk associated with lead in drinking water is not necessarily recognised by all EU Member States because of the continuation of inadequate sampling practices (5, 8). Failure to comply with the WHO Guideline Value (and EU standard from December 2013) is mainly determined by the plumbosolvency of the drinking water and the extent of occurrence of lead pipes in the water supply system (6). Computational modelling can predict (9) the extent of non-compliance for a range of these factors, as illustrated by Table 2.

Table 2. Predicted extent of Pb problems (from 9)

Plumbosolvency			Percentage houses in zone > 10 µg/l based on RDT samples, for each %Pb occurrence				
Category	**M**	**E**	**10% Pb**	**30% Pb**	**50% Pb**	**70% Pb**	**90% Pb**
Very high	0.3	450	6.5	18.9	31.6	45.1	56.6
High	0.2	300	5.2	16.7	28.0	38.7	49.0
Moderate	0.1	150	3.9	12.1	20.2	28.9	37.0
Low	0.06	90	2.5	7.7	13.5	18.4	23.5
Phosphate dosed	0.02	30	0.4	1.1	2.1	2.7	3.2

[M is the initial mass transfer rate (µg/m^2/s) and E is the equilibrium concentration for lead associated with each plumbosolvency condition.]

At first glance, the predicted levels of failure appear very high. However, data from 55 case studies, based on actual RDT sampling, validates these predictions in general terms, as summarised in Table 3.

Computational modelling also enables the severity of risk from lead in drinking water to be predicted, when coupled with data from epidemiological studies. An example is given in Figure 2, which has the following features: (i) the curvi-linear relationship (10) between water Pb and blood Pb concentrations is generalised - there is much scatter around the curve shown; (ii) the reductions of 1 to 4.6 in IQ in children for increases in blood Pb between 10 and 20 µg/dl derive from three studies (11, 12, 13), whereas the reduction of 7.4 in IQ in children for increases in blood Pb between 0 and 10 µg/dl derives from only one study (13) - it can be noted here that further epidemiological studies have demonstrated a link between IQ reduction (14) and neurobehavioural outcomes (15) with elevated blood lead concentrations in the range 0 to 25 µg/dl and 0 to 20 µg/dl, respectively (iii) the trigger for action in the US to prevent Pb poisoning in children is 10 µg/dl (16).

Table 3. Zonal failure rates for lead in drinking water in 55 case studies (BE, FR, ND, PT, UK), based on real RDT sampling and 10 µg/l

Percentage RDT samples > 10 µg/l	Number and percentage of zones in each category
0 to 9.9	18 (32.7%)
10 to 19.9	18 (32.7%)
20 to 29.9	11 (20.0%)
30 to 39.9	3 (5.5%)
40 to 49.9	4 (7.3%)
50 to 59.9	1 (1.8%)

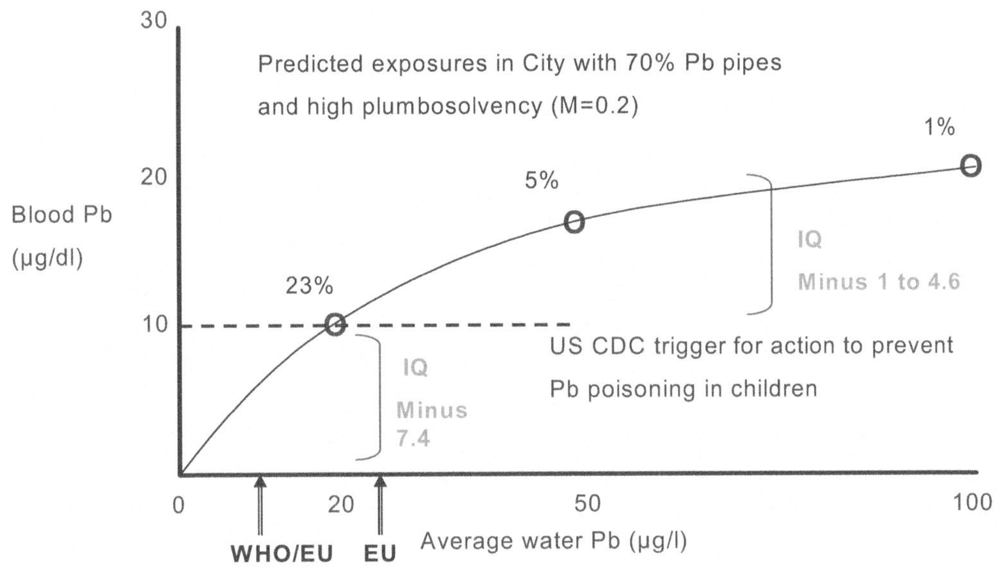

Figure 2. Pb in drinking water and reductions in the IQ of children- a risk assessment

For the water supply system circumstances in this example (high plumbosolvency and 70% houses with Pb pipes), it can be predicted that 1% of houses (equivalent to the same percentage of children) are exposed to an average water Pb concentration of 100 µg/l or higher, associated with an IQ reduction of up to 12, that 5% of houses (children) are exposed to an average water Pb concentration of 50 µg/l or higher, and that 23% of houses (children) are exposed to an average water Pb concentration of 20 µg/l or higher. The IQ reductions associated with these exposures are tentative as they rely in part on a single epidemiological study [(13)]. However, whether the IQ reductions associated with this range of average water Pb concentrations are 12, 6 or 3 is somewhat immaterial; the epidemiological studies all demonstrate that IQ reductions in children occur when blood Pb is elevated, which in the 21st Century must simply be regarded as unacceptable. The value of predicting the severity of risk as a function of population is that it can help alert health authorities of the potential magnitude of health effects in their area and help to set priorities. It is interesting to note from the general relationship shown in Figure 2 between water Pb and blood Pb that: (i) the current EU standard for Pb of 25 µg/l is border-line in relation to the US trigger for preventative action (ie: no safety margin), and (ii) the WHO Guideline Value and future EU standard of 10 µg/l appears to afford a moderate measure of protection when benchmarked by the US trigger for preventative action.

5. The case for a more integrated approach to control

The risk assessment (above) indicates that problems with lead in drinking water (at the zonal level) are likely to be significant wherever there are lead pipes in sufficiently numbers, unless corrective water treatment has been initiated, which outside the UK and the Netherlands is very limited in the EU [(6)]. However, there have been some positive developments:

- 2008 Recommendations [(17)] to the EC to adopt, in the next Drinking Water Directive, risk management strategies, highlighting metal leaching from domestic systems (particularly Pb), operational monitoring to supplement compliance monitoring, and the adoption of random daytime sampling
- 2009 Adoption of Pb as a core parameter in the UN/WHO Protocol on Water and Health [(18)] and guidelines [(19)] for Pb based on random daytime sampling (As and Fe were also adopted)
- 2010 Technical Digest on Pb published by the EC [(20)]
- 2010 IWA Best Practice Guide on the Control of Lead in Drinking Water [(6)]
- 2010 IWA Guide on Pb for Small Communities [(21)]

Presently, the revision of the Drinking Water Directive awaits political progress and the implementation of the Protocol on Water and Health is at an early stage. The exact manner and timescale of the resolution of problems with lead in drinking water, by these regulatory means, is therefore uncertain. The recommendations by the WHO [22] that drinking water safety plans should be implemented to strengthen regulatory control creates a fascinating challenge: any water supplier who implements a drinking water safety plan must assess the risks from lead (and all the other metals and metalloids) which means getting to grips with sampling metals at the tap. Unfortunately, the implementation of drinking water safety planning in Europe is at an early stage. The probable answer to these problems, at least in the short term, lies with health authorities, from national to local level. The relevance of metals and metalloids to human health, particularly arsenic and lead, demands that health authorities: (i) are proactive about drinking water in their area; (ii) are involved with the water suppliers in drinking water safety planning; (iii) maintain consumer awareness programmes; (iv) have a strategy for dealing with small/very small supplies. They should also consider undertaking blood surveillance for Pb and urine surveillance for As.

Reports on drinking water quality (by Member States to the EC and by the Parties to the Protocol) need to be improved: presently there are missing reports, different formats are used, or there is an unclear basis for the data submitted [23, 24]. It should be possible for: (i) consumers to easily check the quality of the drinking water supplies they receive, and (ii) for national/regional compliance to be published in an informative manner (a good example is www.dwi.gov.uk). It can also be noted here that safe drinking water is now a human right [25].

It also appears that enforcement of the Drinking Water Directive could be improved. Enforcement at the EU level is slow, inconspicuous and appears more concerned with legal transposition, whereas enforcement at national/regional level appears to range from none to extensive (as judged by the numerous discussions at the International Conferences of COST Action 637). Enforcement of the standards for Cu, Pb and Ni seems unlikely under the current Directive in the absence of an agreed approach to sampling metals at the tap. This raises the question: how about a European Drinking Water Inspectorate? Its roles could include: (i) providing guidance; (ii) certifying drinking water safety plans; (iii) prosecuting non-compliance; and (iv) providing reassurance through meaningful reports.

An integrated approach to controlling metals and metalloids in drinking water is summarised in Figure 3, and embraces source to tap safety planning. The approach highlights adequate monitoring, regulatory enforcement, awareness and health surveillance as necessary and complimentary functions alongside water system operational control. This integrated approach would deliver better health protection and facilitate the optimisation of water supply operation.

6. Conclusions

1. Problems have been experienced with the monitoring of diffuse metal contamination (Cu, Ni, Pb) at consumers' taps and in consequence, public health protection is poor or absent in some countries.
2. The potential extent of problems with lead in drinking water reveals the need for a more integrated approach to control.
3. Health Authorities need to determine the extent of problems with metals and metalloids in drinking water in their area and pursue appropriate improvements.
4. A European Drinking Water Inspectorate could considerably strengthen the enforcement of quality standards and take a leading role in the implementation of risk assessment and risk management.
5. The public reporting of drinking water quality could be much improved.
6. Risk assessment is not a perfect science and should look broadly at all the relevant information that is available.

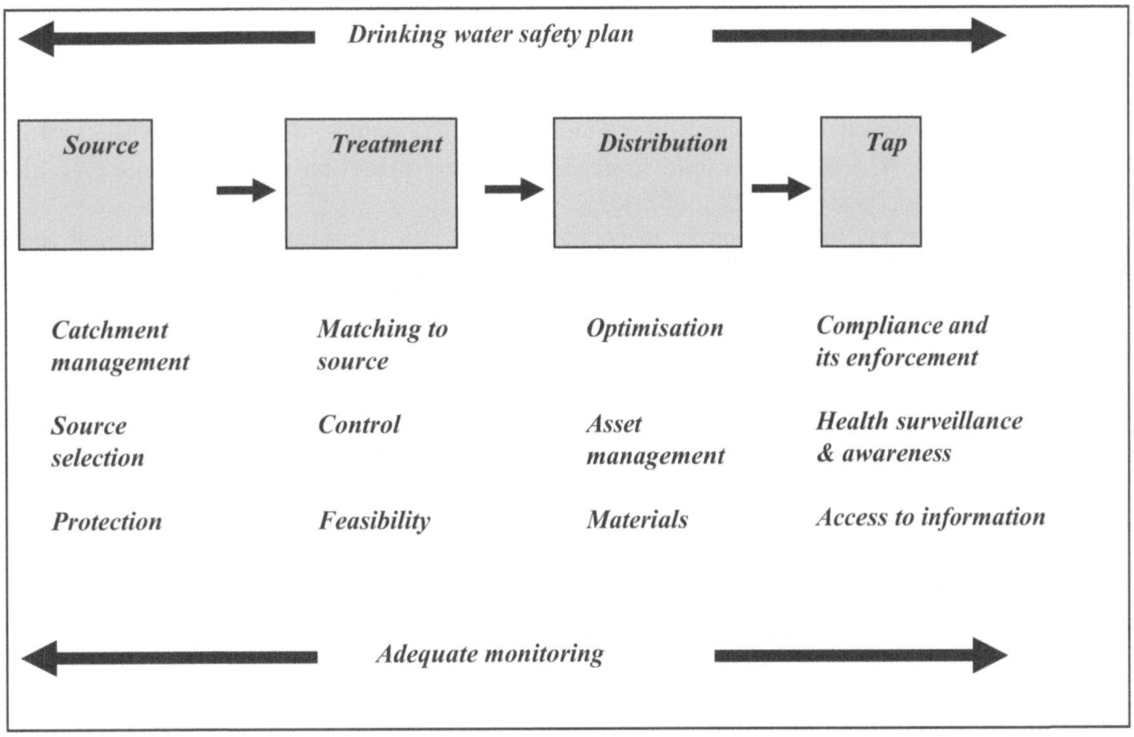

Figure 3. An integrated approach to control

References

1. European Commission (1998). Council Directive (98/83/EC) of 3 November 1998 on the quality of water intended for human consumption. Official Journal, L330/32, 5 December 1998.
2. Hulsmann, A.(2005). Small systems large problems. A European inventory of small water systems and associated problems. WEKNOW/ENDWARE
3. Skubala, N D and Hayes, C R (2009). A review of lead, copper and nickel in European drinking water. Proceedings of the 2nd International Conference on Metals and Related Substances in Drinking Water. October 2008, Lisbon, COST Action 637.
4. Drinking Water Inspectorate. www.dwi.gov.uk
5. Hayes, C.R. and Skubala, N.D. (2009b). Is there still a problem with lead in drinking water in the European Union? Journal of Water and Health, 07.4, 569-580.
6. International Water Association. Best Practice Guide on the Control of Lead in Drinking Water. Editor, Dr C R Hayes. ISBN 13: 9781843393697
7. Van den Hoven, T. J. L., Buijs, P. J., Jackson, P. J., Gardner, M., Leroy, P., Baron, J., Boireau, A., Cordonnier, J., Wagner, I., do Mone, H. M., Benoliel, M. J., Papadopoulos, I. and Quevauviller, P. (1999). Developing a new protocol for the monitoring of lead in drinking water. EUR 19087.
8. Hulsmann A. and Cortvriend J. (2006). Water 21. Magazine of the International Water Association.
9. Hayes, C. R. (2010). Computational modelling methods for assessing the risks from lead in drinking water. Journal of Water and Health, 08.3, 532-542.
10. Quinn, M.J. and Sherlock, J.C. 1990 The correspondence between U.K. "action levels" for lead in blood and in water. Food Additives and Contaminates, **7**, 387-424.
11. Tong, S.L., Baghursr, P., McMichael, A., Sawyer, M. and Mudge, J., 1996 Lifetime exposure to environmental lead and children's intelligence at 11-13 years: The Port Pirie cohort study. British Medical Journal, **312**, 1569-1575.
12. Pocock, S.J., Smith, M., Baghurst, P. 1994 Environmental lead and children's intelligence- a systematic review of the epidemiologic evidence. British Medical Journal, 309, 1189-1197

13. Canfield, R.L., Kreher, D.A., Cornwell, C. and Henderson, C.R. Jr. (2003). Low-level lead exposure, executive functioning, and learning in early childhood. Child Neuropsychology, 9, 35-53.
14. Bellinger, D.C. 2008 Very low lead exposures and children's neurodevelopment. Current opinion in Paediatrics, 20, 172-177.
15. Chiodo, L. M., Covington, C., Sokol, R. J., Hannigan, J. H., Jannise, J., Ager, J., Greenwald, M. and Delaney-Black, V. 2007 Blood lead levels and specific attention effects in young children. Neurotoxicology and Teratology, **29**, 538-546.
16. CDC (1991). Preventing lead poisoning in young children. US Department of Health and Human Services, Public Health Service, Atlanta, Georgia.
17. Hoekstra, E J, Aertgeerts, R, Bonadonna, L, Cortvriend, J, Drury, D, Goossens, R, Jiggins P, Lucentini, L, Mendel, B, Rasmussen, S, Tsvetanova, Z, Versteegh, A and Weil, M, 2008. The advice of the Ad-Hoc Working Group on Sampling and Monitoring to the Standing Committee on Drinking Water concerning sampling and monitoring for the revision of the Council Directive 98/83/EC. Office for Official Publications of the European Communities, Luxembourg, EUR 23374 EN - 2008.
18. United Nations Economic Commission for Europe/WHO.(2007) Protocol on Water and Health. GE.06-26870- January 2007-4.290. Geneva.
19. Hoekstra, E J, Hayes, C R, Aertgeerts, R, Becker, A, Jung, M, Postawa, A, Russell, L and Witczak, S (2009). Guidance on sampling and monitoring for lead in drinking water. Office for Official Publications of the European Communities, Luxembourg, EUR 23812 EN - 2009.
20. Hayes, C R and Hoekstra, E J (2010). Technical Digest on Lead in Drinking Water. Office for Official Publications of the European Communities, Luxembourg, EUR 24265 EN - 2010.
21. International Water Association (2010). Guide for Small Community Water Suppliers and Local Health Officials on Lead in Drinking Water. Editor, Dr C R Hayes. IWA Publishing. ISBN 13 9781843393801.
22. World Health Organization. 2004 Guidelines for Drinking-water Quality: Third Edition, Vol. 1, Recommendations, WHO Geneva.
23. European Commission 2008. The quality of drinking water in the European Union. Synthesis report on the quality of drinking water in the Member States of the European Union in the period 1999-2001 Directive 80/778/EEC, 14 April 2008
24. Juszczak, T (2010). Pilot reporting under the Protocol on Water and Health - preliminary results.
25. http://www.unece.org/env/water/Protocol_implementation_reports.html
26. International Water Association (2010). Water 21, October edition.

Uranium in drinking water

P. Andrew Karam

Bureau of Environmental Emergency Preparedness and Response
New York City Department of Health and Mental Hygiene
New York, USA

Corresponding author e-mail: pkaram@health.nyc.gov

Uranium is a toxic, radioactive heavy metal that is ubiquitous in geologic materials. Uranium's concentration in geologic materials is determined by its geochemical properties and its eventual dissolution into groundwater is controlled by its chemical properties and those of the water with which it comes in contact (Murphy and Shock 1999; Faure and Mensing (2004). As a large ion uranium tends to be concentrated in granites and similar igneous rocks. And because uranium is soluble in oxidizing waters (and insoluble in anoxic waters), uranium will dissolve into surface waters that come in contact with granitic rocks - when these waters become anoxic the uranium precipitates from solution causing elevated uranium concentrations in organic-rich sedimentary rocks such as coal and black shale. Uranium concentrations also tend to be elevated in phosphate rocks and in minerals containing rare earth elements (such as monazite). Oxygenated waters flowing through uranium-bearing rocks can thus become enriched in dissolved uranium, reaching concentrations as high as 500 µg/l in surface waters and higher in some groundwaters (Kim 1986).

As noted above, uranium is a radioactive heavy metal that, when ingested, can be retained for decades in the bone. In sufficiently large quantities uranium is toxic to the kidneys and has been associated with kidney damage; uranium seems to have little significant effect on other organ systems (National Research Council 2008 and 1988). Although radioactive, uranium has so long a half-life that its radiotoxicity is not normally important in either the short or the long term; health effects are typically determined by uranium's chemical toxicity (National Research Council 2008). Uranium concentrations in Swedish waters tend to be higher than those of many other nations (Rosborg and Surbeck 2004) because of the elevated concentrations of uranium in Swedish rocks. There is evidence that, at kidney concentrations in excess of 1-3 µg/g, uranium begins to damage the kidneys but at present there is no consensus on the concentrations of uranium in drinking water required to cause harm. In considering this topic the National Research Council (1988) concluded that "exposure to natural uranium is unlikely to be a significant health risk in the population and may well have no measureable effect." Nonetheless it is prudent to take steps to reduce uranium concentrations when they exceed recommended limits of 15 µg/L; this can be accomplished by chemical treatment or by passing the uraniferous waters through ion exchange or filtration media (Rosborg and Surbeck 2008).

References

Faure G and Mensing TM. *Isotopes: Principles and Applications*, 3rd Edition. Wiley, New York, 2004

Kim JI. Chemical behavior of transuranic elements in natural aquatic systems, in *Handbook on the Physics and Chemistry of the Actinides.* (Freeman AJ and Keller C eds). Elsevier Science Publishers, Amsterdam

Murphy WM and Shock EL. Environmental Aqueous Geochemistry of Actinides, in *Uranium: Mineralogy, Geochemistry, and the Environment* (Burns PC and Finch R eds). Mineralogical Society of America, Washington DC 1998

National Research Council. *Review of the Toxicologic and Radiologic Risks to Military Personnel from Exposures to Depleted Uranium During and After Combat.* National Academies Press, Washington DC, 2008

National Research Council. *Health Risks of Radon and Other Internally Deposited Alpha-Emitters: BEIR IV.* National Academies Press, Washington DC, 1988

Rosborg I and Surbeck H. Uranium. In *Water Treatment for Metals Control* (2008)

Arsenic in drinking water and non-insulin-dependent diabetes in Zrenjanin Municipality, Serbia

Dragana Jovanovic[1], Zorica Rasic-Milutinovic[2], Gordana Perunicic-Pekovic[2], Katarina Paunovic[3], Tanja Knezevic[1], Miroslav Radosavljevic[1], Snezana Plavsic[1], Melita Dimitric[4], Radivoje Filipov[4]

[1] Institute of Public Health of Serbia "Dr Milan Jovanovic Batut", Belgrade, Serbia
[2] Departments of Endocrinology, University Hospital Zemun, Belgrade, Serbia
[3] Institute of Hygiene and Medical Ecology, School of Medicine, Belgrade, Serbia
[4] Institute of Public Health of Zrenjanin, Zrenjanin, Serbia

Corresponding author e-mail: dragana_jovanovic@batut.org.rs

Abstract

Introduction: Arsenic from drinking water has been shown to be associated with increased rates of diabetes incidence, prevalence and mortality, when occurring in concentrations above 200 µg/L.

Methods: The aim of this cross sectional study was to evaluate the association between drinking water arsenic exposure and the incidence of non-insulin-dependent diabetes in Zrenjanin municipality in 2006-2008. The exposed population in Zrenjanin consumes arsenic in drinking water (mean = 70 µg/L, range 0.5 to 256 µg/L). The unexposed population comprised Central Serbian population where arsenic is not present in drinking water. The data on the incidence of type 2 diabetes were obtained from Population-based Diabetes Registry and from the Institute of Public Health of Zrenjanin. This registry also included data on the family history of diabetes, overweight (defined as body mass index over 25 kg/m^2) and central obesity (defined as waist circumference greater than 102 cm for man and greater than 88 cm for women) for exposed and unexposed population. Standardized incidence rates (SIR) were calculated by direct standardization, using the World (ASR-W) standard population.

Results: The two populations were comparable by family history of diabetes and prevalence of overweight persons ($p>0.05$). The unexposed population in Central Serbia was shown to have higher prevalence of central obesity ($p<0.001$). Standardized incidence rate of non-insulin-dependent diabetes in 2008 was two times higher among the exposed population in Zrenjanin (226.7 per 100.000), compared to the unexposed population in Central Serbia (107.9 per 100.000). Standardized incidence rate in Zrenjanin was also higher in the previous years: in 2007, SIR in exposed population = 150.7 versus SIR in unexposed population = 114.0 per 100.000; and in 2006, SIR in exposed population = 224.8 versus SIR in unexposed population = 136.4 per 100.000. Relative risk for the occurrence of non-insulin-dependent diabetes was significantly higher in exposed population, RR = 2.01 (95% Confidence Interval = 1.83 - 2.20).

Conclusion: This cross sectional study showed that study population exposed to arsenic in drinking water had higher incidence rate and was at double risk for the occurrence of non-insulin-dependent diabetes.

1. Introduction

Type 2 diabetes accounts for 90-95% of all cases of diabetes and is a major public health problem worldwide [1].Well known risks factors of type 2 diabetes include older age, obesity, physical inactivity, family history, and genetic polymorphisms. In addition, environmental toxicants, including arsenic, have been suggested to play an etiologic role in diabetes development [2].High chronic exposure to inorganic arsenic in drinking water has been related to diabetes development, but the effect of exposure to low to moderate levels of inorganic arsenic on diabetes risk is unknown. Arsenic from drinking water has been shown to be associated with increased rates of diabetes incidence, prevalence and mortality, when occurring in concentrations above 200 µg/L [3, 4]. Arsenic has been proposed to induce insulin-dependent and non-insulin-dependent diabetes, probably through increased oxidative stress by inducing the development of insulin resistance and endothelial dysfunction [2]. However, biological mechanisms for an association between chronic arsenic exposure and increased diabetes risk remain unknown [5, 6, 7].

2. Materials and Methods

The aim of this cross sectional study was to evaluate the association between drinking water arsenic exposure and the incidence of non-insulin-dependent diabetes in Zrenjanin municipality in 2006-2008.

The exposed population in Zrenjanin consumes arsenic in drinking water (mean = 70 µg/L, range 0.5 to 256 µg/L). The unexposed population comprised Central Serbian population where arsenic is not present in drinking water.

Arsenic concentrations in drinking water were obtained from National water quality monitoring programs from 2006 to 2008. Incidence and mortality data for bladder cancer were obtained from National Cancer Register, supported by the CanReg3 programme package (Department of Descriptive Epidemiology, IARC, Lyon, France, 2002-2005) [8]. These data were available for the same three-year period (2006 to 2008).

This registry also included data on: family history of diabetes, prevalence of overweight persons (defined as body mass index over 25 kg/m^2), and prevalence of central obesity (defined as waist circumference greater than 102 cm for man and greater than 88 cm for women). Standardized incidence rates (SIR) were calculated by direct standardization, using the World (ASR-W) standard population.

3. Results

The two populations were comparable by family history of diabetes (Table 1). The prevalence of overweight persons was higher among men in exposed area and among women in unexposed area (Table 2). The unexposed population in Central Serbia was shown to have higher prevalence of central obesity (Table 3).

Table 1. Family history of diabetes in two investigated areas by gender

Family history of diabetes	Exposed population in Zrenjanin	Unexposed population in Central Serbia	p value
Men	88 (36.4%)	438 (33.3%)	0.375
Women	103 (34.4%)	475 (33.2%)	0.637
Total	191 (35.3%)	913 (33.2%)	0.370

Table 2. Prevalence of overweight persons in two investigated areas by gender

Overweight persons	Exposed population in Zrenjanin	Unexposed population in Central Serbia	p value
Men	163 (67.4%)	795 (60.4%)	0.044
Women	104 (34.8%)	625 (43.7%)	0.005
Total	267 (49.4%)	1420 (51.7%)	0.323

Table 3. Prevalence of central obesity in two investigated areas by gender

Obese persons	Exposed population in Zrenjanin	Unexposed population in Central Serbia	p value
Men	65 (26.9%)	468 (35.6%)	0.009
Women	52 (17.4%)	412 (28.8%)	<0.001
Total	117 (21.6%)	880 (32.0%)	<0.001

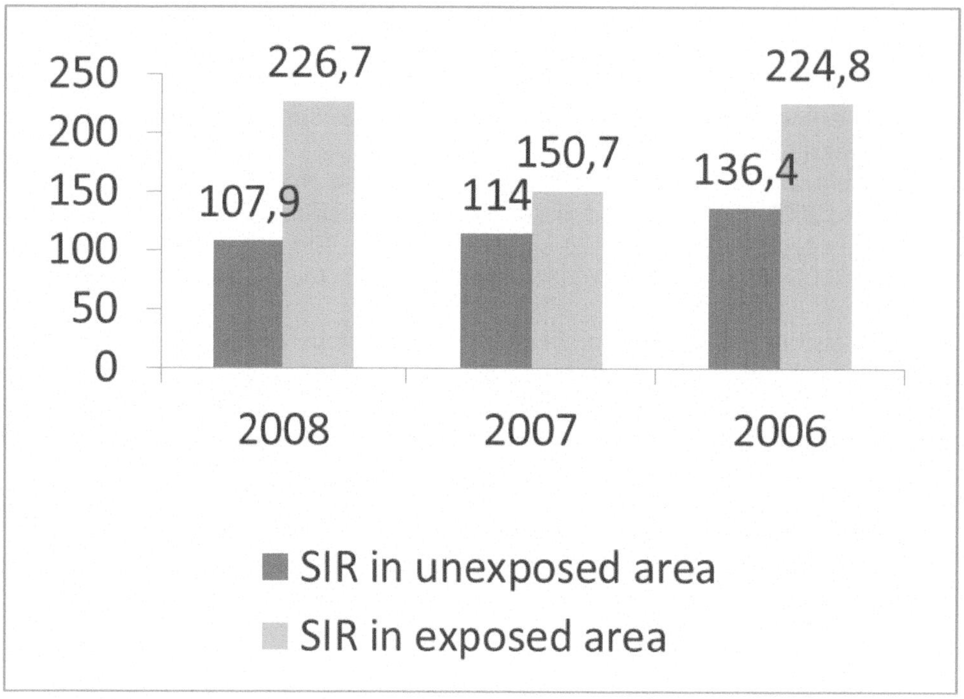

Figure 1 Standardized Incidence Rates per 100.000 (SIR) in exposed and unexposed areas in Serbia in 2008, 2007 and 2006

Relative risk for the occurrence of non-insulin-dependent diabetes was significantly higher in exposed population, RR = 2.01 (95% Confidence Interval = 1.83 - 2.20).

4. Discussion

We found higher standardized incidence rates for diabetes type 2 in the exposed population compared to unexposed population in the observed three-year period. Also, significantly higher relative risk for non-insulin-dependent diabetes was observed in exposed population. Unexposed population had significantly higher prevalence of central obesity for both genders and prevalence of overweight for female suggested that arsenic exposure may have a role in the higher prevalence of diabetes type 2.

The evidence on the association of arsenic exposure with diabetes risk is inconclusive. Our findings are supported by the cross-sectional study conducted in United States [9]. Study conducted in southwestern Taiwan confirmed that the subjects in the arseniasis-endemic area had an elevated prevalence of diabetes compared with the nonendemic area (odds ratio = 2.7 after adjustment for age and sex) [10]. Evidence of diabetogenic effect of inorganic arsenic was also provided by the Mexican case-control study with two and three-fold higher risk of having diabetes in subjects in intermediate and highest total arsenic concentration in urine [11].

Studies in Taiwan and Bangladesh consistently identified an increased risk of diabetes with increased arsenic exposure, with relative risks ranging from 1.46 to 10.1 (median, 2.40) and with a pooled relative risk estimate using and inverse variance weighted random-effects model of 2.52 (95% CI, 1.69-3.75; p heterogeneity < 0.001) [12]. In contrast cross-sectional study, Health Effects of Arsenic Longitudinal Study in Bangladesh, did not observe association between arsenic exposure and significantly increased risk for diabetes mellitus type 2.

Our study shares limitations common to similar ecological studies, including the lack of data on individual arsenic exposure, lack of biomarkers data and the lack of data on the presence of other risk factors for non-insuline dependent diabetes such as nutrition, stress, hypertention, socioeconomic status, phisical activity.

5. Conclusion

This cross-sectional study showed that population exposed to arsenic in drinking water had higher incidence rate and was at double risk for the occurrence of non-insulin-dependent diabetes. Our finding supports the hypothesis that low levels of exposure to inorganic arsenic in drinking water, a widespread exposure worldwide, may play a role in diabetes prevalence.Prospective studies in populations exposed to a range of inorganic arsenic levels are needed to establish whether this association is causal.

References

[1] S. Wild, G. Roglic, A. Green, R. Sicree, H. King. Global prevalence of diabetes: estimates for the year 2000 and projections for 2030. Diabetes Care 27 (2004) 1047-53.

[2] M.P. Longnecker, J.L. Daniels. Environmental contaminants as etiologic factors for diabetes. Environ Health Perspect 109(suppl 6) (2001) 871-876.

[3] M.S. Lai, Y.M. Hsueh, C.J. Chen, M.P. Shyu, S.Y. Chen, T.L. Kuo, et al. Ingested inorganic arsenic and prevalence of diabetes mellitus. Am J Epidemiol.139 (1994) 484-492.

[4] M. Rahman, M. Tondel, S.A. Ahmad, O. Axelson. Diabetes mellitus associated with arsenic exposure in Bangladesh. Am J Epidemiol 148 (1998) 198-203.

[5]NRC (National Research Council). 1999. Arsenic in Drinking Water. Washington, DC:National Academy Press.

[6]NRC (National Research Council). 2001. Arsenic in Drinking Water. 2001 Update. Washington DC:National Academy Press.

[7] C.H. Tseng. The potential biological mechanisms of arsenic-induced diabetes mellitus. Toxicol Appl Pharmacol 197 (2004) 67-83.

[8]CanReg3 programme package. Department of Descriptive Epidemiology, IARC, Lyon, France, 2002-2005.

[9] A. Navas-Acien, E.K. Silbergeld, R. Pastor-Barriuso, E. Guallar. Arsenic Exposure and Prevalence of Type 2 Diabetes in US Adults. JAMA. 300 (2008) 814-822.

[10] S.L. Wang, J.M. Chiou, C.J. Chen, C.H. Tseng, W.L. Chou, C.C. Wang et al. Prevalence of non-insulin-dependent diabetes mellitus and related vascular diseases in southwestern arseniasis-endemic and nonendemic areas in Taiwan. Environ Health Perspect. 111 (2003) 155-159.

[11] J.A. Coronado-González, L.M. Del Razo, G. García-Vargas, F. Sanmiguel-Salazar, J. Escobedo-de la Peña. Inorganic arsenic exposure and type 2 diabetes mellitus in Mexico. Environ Research. 104 (2007) 383-392.

[12] A. Navas-Acien, K. Ellen, E.K. Silbergeld, A. Robin, R.A. Streeter, M. Jeanne et al. Arsenic Exposure and Type 2 Diabetes: A Systematic Review of the Experimental and Epidemiologic Evidence. Environ Health Perspect. 114 (2006) 641-648.

[13] Y. Chen, H. Ahsan, V. Slavkovich, G.L. Pettier, R.T. Gluskin, F. Parvez et al. No association between arsenic exposure from drinking water and diabetes mellitus: a cross-sectional study in Bangladesh. Environ Health Perspect. 119 (2010) 1299-305.

Does water softening improve eczema in children? Results of a clinical trial - the softened water eczema trial (swet)

Ian H Pallett[1], Kim S Thomas[2], Tara Dean[3], Tracey H Sach[4], Karin Koller[2], Anthony Frost[5], Hywel C Williams[2].

[1] British Water, 1 Queen Anne's Gate, London SW1H 9BT, UK
[2] Centre of Evidence Based Dermatology, University of Nottingham, Nottingham, UK
[3] School of Health Sciences and Social Work, University of Portsmouth, Portsmouth, UK
[4] School of Pharmacy, University of East Anglia, Norwich, UK
[5] UK Water Treatment Association, Loughborough, UK

Corresponding author e-mail: ian.pallett@britishwater.co.uk

Abstract
Anecdotal reports have suggested that water softening may be beneficial to eczema sufferers; consequently a clinical trial was commissioned to investigate whether water softening is beneficial to children with eczema. The aims and design of the study are outlined but the results cannot be discussed until they have been published formally.

1. Introduction

Eczema is an extremely itchy and painful skin condition that affects 1 in 5 school children,This can lead to scratching, bleeding, secondary infection, sleep loss, poor concentration and psychological distress to the child and the entire family.The cost of treating eczema is substantial both for the health treatment provider and for families.

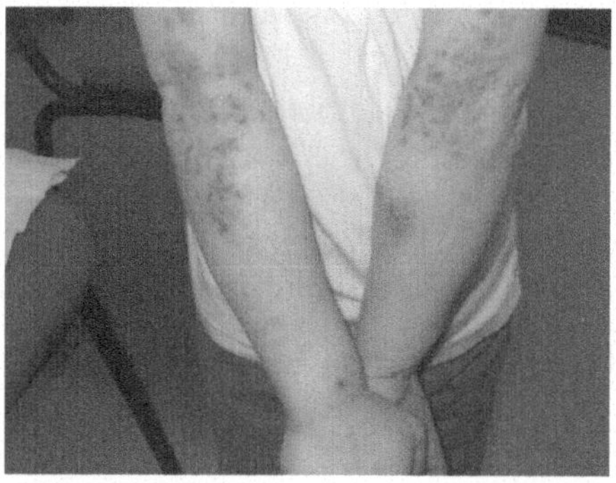

Figure 1 Typical childhood eczema

Hard water has been linked with increased incidence of eczema in children in the UK [1], Japan [2] and Spain [3]. Doctors and water companies often receive anecdotal reports of the benefits of softened water. As a result a trial was commissioned to investigate whether water softening improves the severity of eczema in children - the Softened Water Eczema Trial (SWET).
The aims of the trial were:

- To assess whether ion-exchange water softening reduces the severity of eczema in children with moderate to severe eczema
- If so, to establish the likely cost and cost-effectiveness of the intervention.

2. Materials and methods

2.1 Study design.

The SWET trial was an observer-blind, randomised controlled trial of 12-weeks duration, followed by a 4-week observational cross-over period [4], Table 1.

Table 1. Study design

	STUDY PERIOD = 16 weeks		
	0 to 12 weeks	12 to 16 weeks	
Group A	Usual eczema care + water softener installed (n = 155)	Unit removed	Option to purchase unit at reduced cost
Group B	Usual eczema care + delayed installation (n = 155)	Unit installed	

2.2 Recruitment.

Participants were recruited in 8 UK centres: Nottingham, Cambridge, London (x 2), Isle of Wight, Portsmouth, Lincoln, and Leicester. All participants lived in hard water areas (≥ 200 mg L-1 of calcium carbonate) and had a home suitable for straight forward installation of a water softener. In total 336 children aged 6 months to 16 years, with moderate or severe eczema were enrolled into the trial.

2.3 Interventions.

The effect of softened water from an ion-exchange water softener was compared with the usual eczema care. A generic water softening unit was produced for the trial (Figure 2) and water hardness was checked weekly. All water entering the home was softened, with the exception of a drinking water tap at the kitchen sink.

Figure 2. Generic water softener fitted under kitchen sink

3. Results and discussion

The study is now complete and a paper has been submitted for publication. Unfortunately, the trial's results are under embargo until the paper has been published. Full details will be available in due course from the SWET website www.swet-trial.co.uk

3.1 Main Outcomes

The primary outcome was the difference between Groups A and B in mean change in disease severity at 12 weeks compared with baseline. Secondary outcomes included night-time movement due to scratching, use of eczema medication, eczema symptoms, eczema control, and quality of life.

3.2 Academic/industry partnership

This trial would not have been possible without the expertise and support of both the academic and industry partners. A consortium of water softener supply companies provided generic water softeners, technical support, salt supplies and water testing. The independent Trade Associations British Water and the UK Water Treatment Association were vital in this collaboration.

Acknowledgments

Funding - primary. This trial was funded by the National Institute for Health Research, Health Technology Assessment Programme (NIHR HTA) - project number HTA 05/16/01. The views and opinions expressed in this article are those of the authors and do not necessarily reflect those of the NIHR Health Technology Assessment Programme.

Funding - supplementary. A consortium of water industry representatives led by their trade associations British Water and the UK Water Treatment Association contributed technical and financial support to this trial.

References

[1] McNally N.J.W.H., et al., Lancet, 1998, 352, 527-531.
[2] Miyake Y., et al., Environmental Research, 2004, 94(1), 33-37.
[3] Arned-Pena A., et al., Salud Publica Mex., 2007, 49(4), 295-301
[4] Thomas K.S., et al., British Journal of Dermatology, 2008, 159(3), 561-6

Preliminary assessment of metal concentrations in drinking water in the city of Szczecin (Poland): human health aspects

J. Górski [1], M. Siepak [1], S. Garboś [2], D. Święcicka [2]

[1] Adam Mickiewicz University; Department of Hydrogeology and Water Protection; 16 Maków Polnych Str., 61-606 Poznań, Poland

[2] National Institute of Public Health, National Institute of Hygiene; Department of Environmental Hygiene; 24 Chocimska Str., 00-791 Warsaw, Poland

Corresponding author e-mail: gorski@amu.edu.pl

Abstract

The paper presents the results of determination of aluminum (Al), arsenic (As), cadmium (Cd), copper (Cu), lead (Pb), zinc (Zn), nickel (Ni), iron (Fe), manganese (Mn), calcium (Ca), magnesium (Mg), sodium (Na), chlorides (Cl^-) and sulphates (SO_4^{2-}) in water samples collected directly at consumers from the water pipe network of the city of Szczecin (Poland). Samples of tap water for chemical analysis were taken in June 2010 using the random daytime sampling (RDT) method. A total of 100 sites were sampled. The sampling covered 16 km^2 of the city area divided into a grid of 400 x 400 m squares, with a sampling point located in the centre of a square. The determinations of metals were performed using inductively coupled plasma mass spectrometry (ICP-MS) and inductively coupled plasma optical emission spectrometry (ICP-OES). The determinations of Cl^- and SO_4^{2-} were performed using the high performance ion chromatography technique (HPIC). The study revealed increased concentrations of metals eluted from the water pipe network. This concerns mainly Fe (19% of samples above drinking water quality standards) and Pb (5%). In several cases the maximum admissible concentration levels (MACLs) for Mn, Cu and Ni were also exceeded. The MACL for Al was not observed to be exceeded, despite the use of aluminum compounds in the process of water treatment. Significant influence of the type of material used in domestic plumbing systems on the increased concentrations of metals in water (especially Fe, Mn, Pb, Cu and Ni) was observed.

1. Introduction

The increased concentrations of metals in water at consumers may result from their presence in water sources and their ineffective removal during the water treatment process, secondary introduction of metals during the water treatment process (e.g. Al) and the elution of metals from water pipe networks. Many authors stress the great influence of the latter process, especially in the conditions of increased water corrosiveness (1-7). This process may be the cause of the increased lead concentrations related to the use of lead pipes in old water pipe networks (8).

The above issues were analysed based on the study of metals contents in water at consumers in Szczecin. The study was conducted under the project entitled "Metals and accompanying substances in drinking water in Poland", carried out within the COST Action 637. The city of Szczecin was selected as one of ten problem areas, for which detailed studies of metal concentrations at consumers were conducted.

The study aim in the area of Szczecin was to identify processes of releasing metals from the old water pipe network, which prevails in the central part of the city and includes lead pipes in apartments, in the conditions of supplying the water pipe network with surface water of relatively low mineralisation and hardness. When selecting the study object, the following factors were taken into consideration: long-lasting transfer of surface water taken from Lake Miedwie (about 30 km from the city centre) and the fact of using aluminum sulphate in the water treatment process. It should be mentioned that, during the selection process, the occurrence of above normal concentrations of such metals as Fe and Mn, which were recorded by Chief Sanitary Inspectorate at consumers despite very low concentrations of these metals in raw water (much below the drinking water norm with reference to the Minister of Health

Regulation of 20 April 2010 or the European Community Council Directive 98/83/EC (10)), was also considered.

2. Materials and Methods

2.1 Study area

Szczecin is one of the oldest and largest Polish cities. It is located in north-western Poland, in Zachodniopomorskie Province, close to the Polish-German border. According to the data of 31 December 2009, the city had a population of 406 307 citizens (11). The rivers flowing through the city include the Odra, the Regalica, the Parnica (joining the two), as well as many smaller ones. The city area is 301 km^2 (12). The city is divided into four quarters: Północ (North), Prawobrzeże (Right Bank), Śródmieście (Centre) and Zachód (West) (Figure 1).

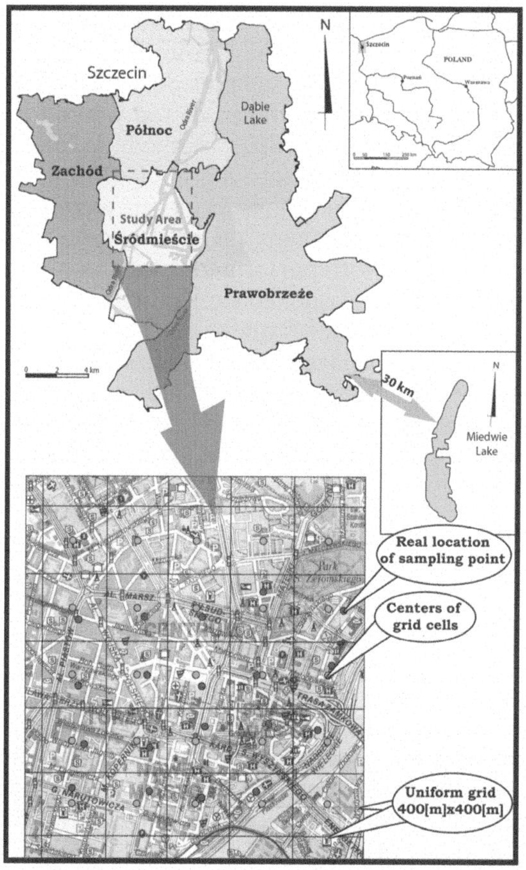

Figure 1. Study area with marked sampling points (RDT method).

2.2 Waterworks for the city of Szczecin

At present Szczecin is supplied with drinking water from the Lake Miedwie surface water capture and from the Pilchowo ground water well field. The Miedwie waterworks produces from 82.000 to 85.000 m^3 of water daily, which constitutes about 90% of water used in the city during the day. It supplies the eastern and central part of the city. Lake Miedwie is situated about 30 km south-east from Szczecin (Figure 1). The water is taken from the lake by the intake located at the height of 6 m above the lake bottom and about 16 m under the water table. The water flows gravitationally to the pumping station situated about 425 m from the water capture. The water is then directed to the water treatment station, where it undergoes coagulation with aluminum sulphate, filtration, and disinfection with chlorine (Figure 2). Next, it is pumped into the pipe network and distributed to consumers. In 2008 water from the Lake Miedwie was characterised by the following average concentrations: temperature 16.6°C; pH 8.6; EC 536 µS/cm; As <10 µg/L; Cd, Cu, Pb, Ni <1 µg/L; Zn<30 µg/L; Fe 7 µg/L; Mn <20 µg/L; Cl^- 46.2 mg/L and SO_4^{2-} 106 mg/L. In the part of the city were water goes from Miedwie waterworks samples for chemical analysis were collected at consumers.

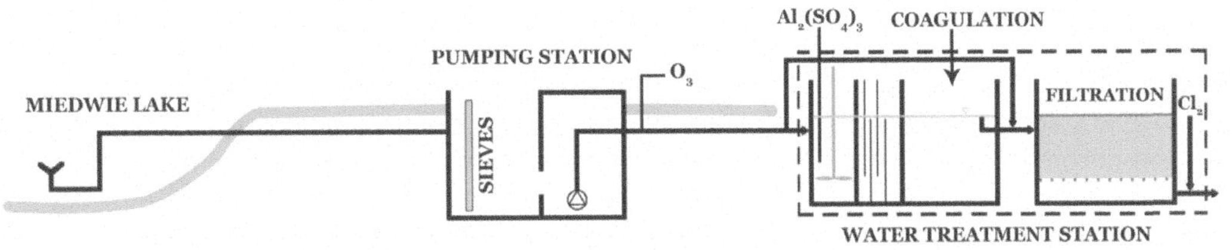

Figure 2. Miedwie water works - a cross-section.

2.3 Distribution network

The earliest information about a water pipe network (made of wood) in Szczecin was recorded in 1577 when, on the order of duke Jan Fryderyk, Pomeranian Dukes' Castle was being rebuilt and supplied with running water. In the following years, more water pipe networks were built, and the population took water from public wells: their number and location changed with the development of the city. On 01.10.1865 the first water distribution company named Pomorzany was founded in Szczecin. As the city developed, more water pipe networks were constructed. At present the total length of water pipe network amounts to about 1163 km. 39.2% of pipes in the distribution system are made of cast iron. The remaining pipes are made of: steel (21.9%), PE (16.9%), PVC (11.1%), asbestos cement (3.1%), spheroid cast iron (2.24%) and other materials (5.5%) - Table 1. It is also possible that some fragments of the piping are made of lead.

Table 1. Materials used in the construction of water pipe network in Szczecin.

Material type	Total length of pipes [km]	Share in the total network length [%]
Cast iron	455.6	39.2
Spheroid cast iron	26.0	2.24
Steel	254.5	21.9
Asbestos cement	36.0	3.1
PVC	129.0	11.1
PE	196.4	16.9
Other	63.9	5.5
Total	**1162.2**	**100**

2.4 Sample collection

Samples of tap water for chemical analysis were taken in June 2010 using the random daytime sampling (RDT) method (Figure 1). A total of 100 sites were sampled. 10% doubled samples and 10% on-field blank samples were also collected. The sampling covered 16 km^2 of the city area divided into a grid of 400 x 400 m squares, with a sampling points located in the centre of a squares. The samples with volume of 100 ml were collected in Nalgene® bottles (HDPE) and preserved with 0.5 ml of 60% HNO_3 Ultrapur® (Merck; Darmstadt, Germany). The water pH, electrolytic conductivity and temperature were determined in the samples using a Multi 350i/SET (WTW, Germany) meter.

2.5 Chemical analysis

For the determination of Al, As, Cd, Ni and Pb, inductively coupled plasma mass spectrometry (ICP-MS) was applied (XSeries II CCT spectrometer, Thermo Electron Corporation, UK). For the determination of Cu, Fe, Mn and Zn (and occasionally Ni and Pb when they occurred at high concentrations) inductively coupled plasma optical emission spectrometry (ICP-OES) with a CID detector was used (IRIS Advantage Duo ER/S spectrometer, Thermo Jarrell Ash, USA). Additionally, in 10 selected samples the determinations of Ca, Mg and Na were performed using inductively coupled plasma optical emission spectrometry (ICP-OES) (IRIS Advantage Duo ER/S spectrometer, Thermo Jarrell Ash, USA). Determinations of Cl^- and SO_4^{2-} using high performance ion chromatography (HPIC) with a Metrohm apparatus, model 881 Compact IC Pro (Metrohm, Switzerland) were also performed. Total alkalinity was determined by titration of a water sample against methyl orange indicator. Operating conditions for ICP-MS, ICP-OES and HPIC determinations were listed in Table 2, Table 3 and Table 4.

Table 2. Operating conditions for ICP-MS determinations.

SAMPLE INTRODUCTION SYSTEM / PARAMETER	TYPE / VALUE
PLASMA TORCH	QUARTZ, EQUIPPED WITH SILVER SCREEN
SPRAY CHAMBER	"IMPACT-BEAD" TYPE (COOLED TO 2°C WITH PELTIER SYSTEM)
NEBULIZER	GLASS CONCENTRIC
R.F. FREQUENCY	27.12 MHz
FORWARD POWER	1400 W
ARGON FLOW RATES:	
- COOL	13 l/min
- AUXILIARY	0.72 l/min
- NEBULIZER	0.95 l/min
TARGET ANALYTE ISOTOPES MONITORED	^{27}Al, ^{75}As, ^{114}Cd, ^{60}Ni, ^{208}Pb
INTERNAL STANDARD	^{89}Y
NUMBER OF POINTS PER PEAK (CHANNELS PER MASS)	1
DWELL TIME PER ISOTOPE	10 ms
SWEEPS PER RUN	230
ACQUISITION TIME - MAIN RUN	30 s - PEAK JUMPING
NO. OF RUNS PER SAMPLE	3
SAMPLE PUMPING FLOW RATE	approx. 0.8 ml/min
UPTAKE AND WASH TIMES	60 s

LOD (µg/L)				
Al	As	Cd	Ni	Pb
0.15	0.15	0.012	0.08	0.12

2.6 Reagents

For preparation of mixed calibration solutions for calibration of ICP-MS and ICP-OES spectrometers multi-element stock solution "CertiPUR ICP multi-element standard solution IV" (Merck; Darmstadt, Germany) with concentrations of elements at the level of approx. 1000 mg/l and arsenic stock solution "CertiPUR arsenic ICP standard" (Merck; Darmstadt, Germany) with concentration of As at the level of 1000 mg/l, ULTRAPUR concentrated nitric acid (60 %; Merck; Darmstadt, Germany) and deionized water prepared with the use of Millipore Simplicity 185 system were applied. In the case of ICP-MS measurements yttrium additions at the level of 10 µg/l for calibration solutions, blank solutions and samples were applied as internal standard. Stock solution of yttrium "CertiPUR yttrium ICP standard" (Merck; Darmstadt, Germany) with concentration of Y at the level of 1000 mg/l was used for that purpose.

For checking the trueness of metal determinations certified reference materials: CRM TMDA-51.3 "*A high level fortified standard for trace elements*" (Environment

Table 3. Operating conditions for ICP-OES determinations.

SAMPLE INTRODUCTION SYSTEM / PARAMETER	TYPE / VALUE
PLASMA TORCH	QUARTZ, HORIZONTAL DUO
SPRAY CHAMBER	CYCLONE
NEBULIZER	GLASS CONCENTRIC
R.F. FREQUENCY	27.12 MHz
FORWARD POWER	1150 W
ARGON FLOW RATES:	
- PLASMA	15 l/min
- INTERMEDIATE	1 l/min
- OPTICS INTERFACE	4 l/min
- PURGING OPTICS	4 l/min
- PURGING CID DETECTOR	80 units
- NEBULIZER PRESSURE	26 psi
SAMPLE PUMPING FLOW RATE	110 rpm (approx. 2 ml/min)
WASTE PUMPING FLOW RATE	110 rpm
RINSING TIME	60 s
NO. REPLICATES/SAMPLE	4
INTEGRATION TIME IN THE RANGE OF WAVELENGTHS: 175 - 275 nm INCLUDING: Ni - 231.604 nm; Cu - 224.700 nm; Pb - 220.353 nm; Fe - 238.204 nm; Zn - 206.200 nm; Mn - 257.610 nm; Ca - 317.933 nm; Mg - 279.079 nm; Na - 589.592 nm;	50 s (AXIAL OBSERVATION SYSTEM)

LOD (μg/L)

Cu	Zn	Ni	Pb	Fe	Mn	Ca	Mg	Na
1.5	0.49	1.0	8.0	0.71	0.19	70	64	56

Table 4. General conditions and parameters of the analytical technique (IC)

Analyte	Analytical parameters	Element	LOD [mg/L]
Anions	Metrosep A Supp 5 - 150/4.0 column Metrosep A SUPP 4/5 Guard/4.0 Sequential suppression system: chemical suppressor MSM II and MCS suppressor CO_2 Conductivity detection Eluent 3.2 mmol/L Na_2CO_3/1.0 mmol/L $NaHCO_3$, flow rate 0.7 mL/min	Cl^- SO_4^{2-}	0.01 0.02

Canada) and SRM 1643e "*Trace Elements in Water*" (National Institute of Standards & Technology, USA) were applied. In order to acidify samples ULTRAPUR concentrated nitric acid (60 %; Merck; Darmstadt, Germany) was applied (0.5 ml of 60 % HNO_3/100 ml of sample). In the case of observation of high turbidity for samples collected, they have been filtered by PTFE 0.45 µm Millex-LCR syringe filters. During the determinations using high performance ion chromatography (HPIC) standard solutions produced by Merck (Merck, Darmstadt, Germany) and CPAchem (C.P.A. Ltd. Stara Zagora, Bulgaria) were used. The mobile phase was prepared using the reagents produced by Fluka (Sigma-Aldrich, Steinheim, Switzerland).

3. Results and Discussion

The study revealed that water at consumers in Szczecin is characterised by the pH ranging from 7.46 to 8.07 and electrolytic conductivity from 600 to 672 µS/cm. Total alkalinity ranged from 3.0 to 3.5 mval/L (Table 5). Distributions of metal concentrations presented in Figure 3 has shown the widest ranges of variability changes for Cu and Zn, and then for Pb, Fe, Cd, Ni and Mn. Taking into consideration much lower and stabilised concentrations of these metals in raw water, the above data indicate the elution of metals from the water pipe network. This process does not refer to, or is of little significance for As and Al, which are characterised by a narrow range of variability of concentrations. The concentrations of metals which exceed the Polish MACLs for drinking water (based on the European Community Council Directive 98/83 EC (EU)) were observed for Fe (19% of samples), Pb (5%), Mn (2%), Cu (1%) and Ni (1%) (Table 5). The concentrations of As, Al and Cd were below the MACLs. When analysing the above date, a significant share of samples which exceeded the MACL for Pb should be emphasised. This fact should be linked to the use of lead pipes, which may be part of the network in the city centre.

Table 5. Statistical values for determinations of physicochemical parameters of water samples collected at consumers.

Parameters		Minimum	Average	Maximum	SD	Limits for drinking water in EU (98/83/EC)/ number of exceedances %
Temperature	[°C]	14.3	18.0	24.0	1.85	-*
pH	-	7.46	7.63	8.07	0.08	6.5-9.5/0
EC	[µS/cmm	600	623	672	7.32	2 500/0
Alkalinity	[mval/L]	3.0	3.2	3.5	0.2	-
Al	[µg/L]	3.77	15.8	95.6	9.73	200/0
As		0.15	0.55	0.91	0.12	10/0
Cd		0.01	0.05	1.75	0.24	5/0
Cu		1.50	22.4	2240	271.2	2 000/1
Pb		0.12	0.84	22.1	3.52	10/5
Zn		11.9	135.2	2790	384.1	-*
Ni		0.79	1.49	65.3	6.63	20/1
Fe		22.1	105.5	2870	302.8	200/19
Mn		0.96	5.47	98.0	11.6	50/2
Ca^{2+}	[mg/L]	70.0	73.7	87.0	7.02	-
Mg^{2+}		12.8	16.1	17.3	1.82	-
Na^{+}		21.7	26.3	27.9	2.56	200/0
Cl^{-}		47.6	61.0	66.6	7.29	250/0
SO_4^{2-}		117.9	119.1	121.4	0.94	250/0

-* - unlimited

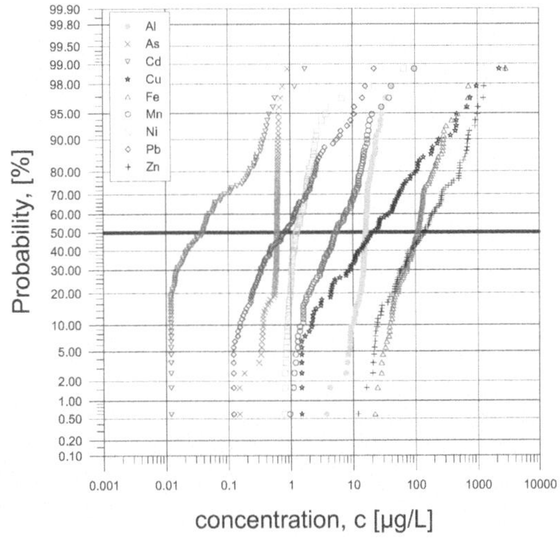

Figure 3. Metal concentrations in piped water in Szczecin obtained using the RDT method.

In spite of using raw water coagulation with $Al_2(SO_4)_3$, Al concentrations were relatively low, ranging from 3.77 to 95.6 µg/L (average 15.8 µg/L). However, they were much higher when compared with similar studies conducted in Poznań (Poland), where coagulation by aluminum compounds is not used (Figure 4).

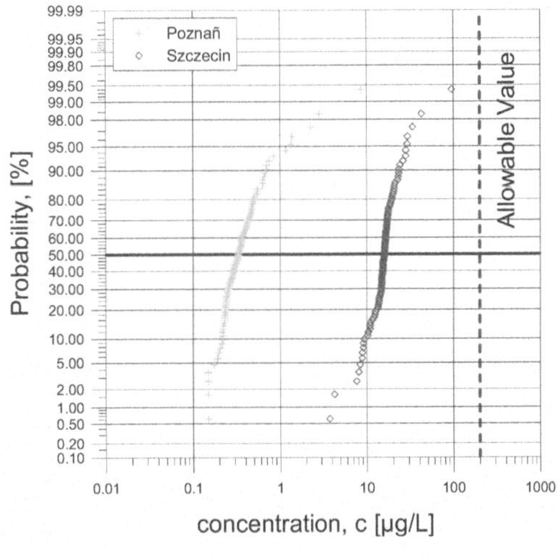

Figure 4. Comparison of aluminum concentration in water samples collected in Szczecin and Poznań at consumers.

The fact of eluting metals from water pipe networks is also confirmed by the data presented in Figure 5. The data indicate that the type of material used in domestic plumbing is of a special importance. This is especially noticeable for Fe, Pb, Mn and, to a lesser extent, for Ni and Cd, which show the highest concentrations in copper installations. In the copper installations also higher concentrations of Fe, Pb and Mn are recorded in comparison with PVC installations. Summing up, it is noteworthy that the main problem in Szczecin lays in the occurrence of lead in water at consumers in 5% samples, which, according to Hayes (8) requires detailed studies, and even remedial actions.

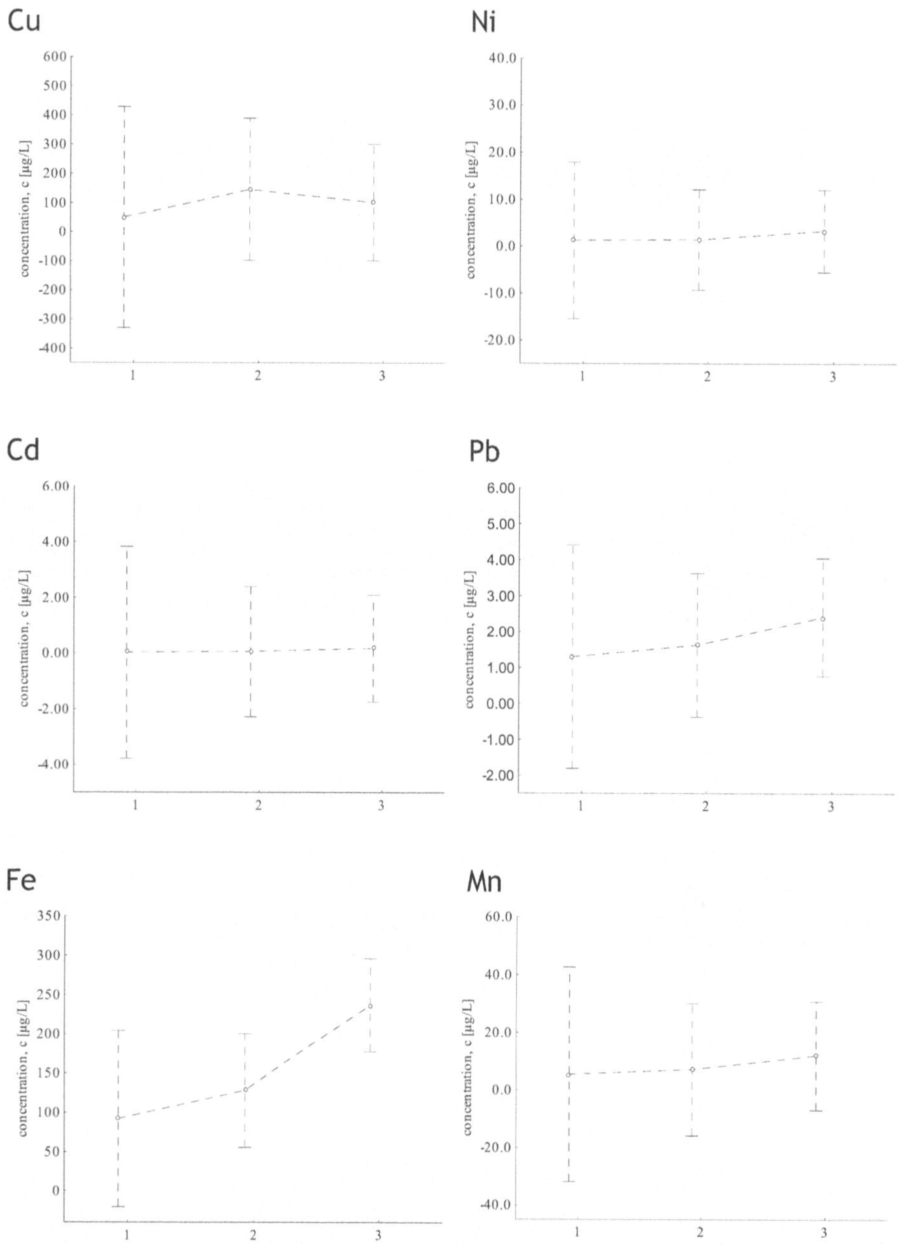

Figure 5. Metal concentrations in piped water depending on the type of material used in domestic plumbing (1- PVC; 2- copper pipes; 3- galvanized pipes).

4. Conclusions

The study of water from domestic plumbing at consumers in Szczecin using the RDT sampling method revealed the occurrence of increased metal concentrations. The concentrations of most metals were much higher compared to their content in raw water taken from Lake Miedwie, which indicates the process of eluting metals from water pipe networks. This process is also confirmed by a clear correlation between the type of material used in domestic plumbing and the metal concentration in water. The highest concentrations of such metals as Fe, Pb, Mn and, to a lesser extent, of Ni and Cd were observed in the installations built from galvanized steel pipes, and the concentrations of Cu in copper installations.

The maximum admissible concentration levels in drinking water were exceeded according to Polish regulations (based on the European Community Council Directive 98/83/EC) for Fe (19% of samples), Pb (5%), Mn (2%), Cu (1%), and Ni (1%). The concentrations of the remaining metals (As, Al and Cd) were much below the MACLs. It should be emphasized that the Al standard was not exceeded in spite of the applied technique of coagulation with aluminum sulphate. However, concentrations of Al were considerably higher than in Poznań tap water, where coagulation with aluminum sulphate is not used. In turn, there was a fairly substantial number of samples exceeding the Pb standard, which can be due to the presence of pipe sections made of lead. In this situation according to Hayes C.R. (8) system-wide measures may be required in addition to resolving any localized clusters.

Acknowledgments

The research was financed from the 2009-2010 research fund as project No. 398/N COST/2009/0 of the Ministry of Science and Higher Education.

References

[1] Schock M.R., *Water quality and treatment*, 4th Edition. McGraw-Hill Inc. New York, **1990**, 997-1111.
[2] Smith D.W., *Municipal and rural water supply and water quality*, Sozański M., (ed), **1994**, 3-16.
[3] Sobesto J., *Ibidem*, **1994**, 997-1002.
[4] Tamasi G., Cini R., *The Science of the Total Environment*, **2004**, *327*, 41-51.
[5] Al-Malack M.H., *Journal of Hazardous Materials B82*, **2001**, 263-274.
[6] Toczyłowska B., *Municipal and rural water supply and water quality*, Sozański M., (ed), **1994**, 45-54
[7] S. Karavoltsos, et. al., *Desalination*, **2008**, 224, *317-329*
[8] Hayes C. R., *Metals and related substances in drinking water*, Ioannina Greece, **2009**, 60-65.
[9] Minister of Health Regulation of 20 April 2010, amending the regulation on quality of water intended for consumption by people (Dziennik Ustaw - Polish Journal of Laws No. 72, item 466)
[10] Council Directive of 3 November 1998 on the quality of water intended for human consumption, 98/83/EC
[11] *Demographic Yearbook of Poland*, Dmochowska H., (ed), **2010,** 530 pp.
[12] *Concise Statistical Yearbook of Poland*, Dmochowska H., (ed), **2010,** 724 pp.

Section 3

Mineral Balance in drinking water

Influence of mineral composition of drinking water on acid-base balance of human body

Frantisek Kozisek[1,2], Hana Jeligova[1], Vladimira Nemcova[3], Ivana Pomykacova[1]

[1] National Institute of Public Health, Department of Environmental Health, CZ-10042 Prague, Czech Republic

[2] Department of General Hygiene, Third Faculty of Medicine, Charles University, CZ-10000 Prague, Czech Republic

[3] Institute of Public Health, CZ-70200 Ostrava, Czech Republic
Corresponding author e-mail: water@szu.cz

Abstract

Recent systematic review and meta-analysis found significant evidence of an inverse association between magnesium levels in drinking water and cardiovascular mortality. Other studies suggest also beneficial effect of water calcium and magnesium on other diseases. As drinking water, comparing to diet, usually provides only minority of total daily intake of essential nutrients, several mechanisms have been suggested to explain this effects. The most recent hypothesis indicates that negative health effects related to drinking low mineral (acidic) water may be caused by an increased urinary excretion of minerals induced by acidosis. We tried to verify this hypothesis with pilot experiment with healthy volunteers (4 men and 4 women, age 20-47). The first week, they drank medium hard tap water and the other week demineralised water treated by reverse osmosis. They collected 24-hour urine samples twice a week to measure net endogenous acid production and excretion of essential elements in laboratory. This pilot study did not find any significant impact of low mineral water on acid-base balance and mineral excretion, but provided important experience regarding the design of future full scale studies.

1. Introduction

The issue of health effects of drinking soft water (or generally water with low mineral content) has been discussed in scientific literature for almost 100 years [1] and in more systematic way since 1957 [2]. Results of number of epidemiologic studies done in the 1960s - 1970s were summarized in compelling dictum "soft water, hard arteries", widely accepted by both water and public health experts. Recent extensive and systematic review, summarizing data from several thousands of papers on water hardness and health published in English, found significant evidence of an inverse association between magnesium levels in drinking water and cardiovascular mortality [3]. Following meta-analysis of case control and cohort studies [4] calculated a pooled odds ratio 0.75 (95%CI 0.68, 0.82; $p < 0.001$) which means that people consuming drinking water with magnesium 8.3 - 19.4 mg/l had the risk of cardiovascular mortality lower by 25 %, in comparison with people using water with Mg content of 2.5 - 8.2 mg/l. A number of other papers suggest also beneficial effect of water calcium (Ca), magnesium (Mg) and some other minerals on other diseases.

As drinking water, comparing to diet, usually provides only minority of total daily intake of essential nutrients, several mechanisms have been suggested to explain how waters with various hardness, but representing relatively tiny difference in contribution to recommended daily intake of magnesium (3 % vs. 10 %), may cause 25 % difference in cardiovascular:

- essential elements are present in water in free ionic forms which are better absorbable (bioavailable) than elements in complexes in food;
- cooking vegetables, rice, pasta, potatoes etc. in soft water supports higher leaching (losses) of minerals from such food which means not only lower intake of elements (Ca, Mg) from water, but also from food in soft water areas;

- due to processing and refinement of modern food, substantial part of population (more than 50 % of adults) in western countries has daily intake of Mg lower than recommended daily intake [5]; in case of borderline deficiency maybe also relatively small contribution from drinking water plays key role in disease development.

The most recent hypothesis [6], based on analysis of previous epidemiological studies, indicates that negative health effects related to drinking soft (low mineral) water may be caused by an increased urinary excretion of minerals induced by acid conditions in the body. Water with low mineral content, including low bicarbonate content, has lower buffer capacity and is more acidic, and its regular consumption supports induction of acidosis. Dietary intervention studies have shown that acid-base conditions influence the homeostasis of minerals and metabolic acidosis is linked to significant morbidity of several diseases, including osteoporosis [7, 8].

While opposite hypothesis - drinking water high in bicarbonates decreases net acid excretion (i.e. has alkalising effect), lowers bone resorption and lowers calcium excretion - was experimentally proved by several intervention studies [9, 10], acidifying effect of drinking water with low mineral content and low bicarbonates has not been experimentally tested up to know. We conducted pilot study with human volunteers to explore necessary design of the study and try to find if there is any such acidifying effect and if it leads to increase of excretion of essential minerals with urine.

2. Materials and Methods

The group of eight healthy volunteers (4 men and 4 women, age 20 - 47) used two different types of drinking water in two consecutive weeks for drinking and cooking purposes. Participants were asked to consume usual amount of water (and drinks prepared from it) and reduce intake of other drinks. Participants were also asked to keep usual nutritional habits, but to reduce intake of some foodstuff known to be extremely acidifying or alkalising. After the experiment, they provided information on their usual diet composition. The first week, they used medium hard tap water and the other week demineralised water treated by reverse osmosis (produced centrally by one RO unit and distributed to participants in bottles). Basic chemical characteristics of waters used are provided in the Table 1.

Table 1. Selected chemical characteristics of tap water and osmotic water used in the study.

Parameter	Unit	Tap water	Osmotic water
Conductivity	mS/m	31 - 41	1 - 2
Ca	mg/l	32 - 54	0,5 - 1,0
Mg	mg/l	8 - 11	1 - 2
Alkalinity	mmol/l	1 - 2	0,2
Bicarbonates	mg/l	55 - 110	12

The participants measured every day pH value of their morning urine sample and twice a week (on the 3rd and 7th days), they collected 24-hour urine samples to measure net endogenous acid production and excretion of essential elements in laboratory. The 24-hour urine samples were immediately transported to laboratory and analysed for pH, urea, creatinine and seven essential elements (chloride, phosphate, sulphate, sodium, potassium, calcium and magnesium). Personal measurements of pH by participants with commercially available urine test strips proved not to be sensitive enough and therefore data are not presented here. Concentrations of elements in urine were adjusted to creatinine excretion.

Design of the study was approved by the ethical committee of the Institute of Public Health in Ostrava. All participants were informed beforehand about the aim and design of the study and asked to immediately contact two physicians leading and supervising the study in case of any health problems. However, no such problems of volunteers were noticed throughout our experiment.

3. Results and Discussion

All findings from the urine analysis were in physiological range in case of all participants (for parameters where reference ranges for urine tests exist). Statistical evaluation of data showed that there were no statistically significant differences between urine pH, urea or elements excretion during the first and the second weeks, i.e. between tap water and osmotic water consumption. It means we did not observe any trends in element excretion, neither increase or decrease. The results for pH, urea, calcium and magnesium are showed in Figures 1 to 4.

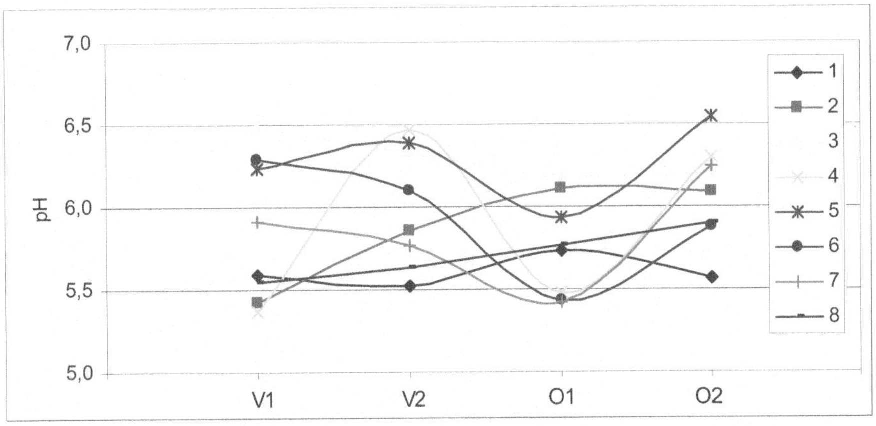

Figure 1. Results of pH of urine of 24-hour samples taken on the 3rd and 7th days of the first week (tap water consumed - V1 and V2) and of the second week (osmotic water consumed - O1 and O2).

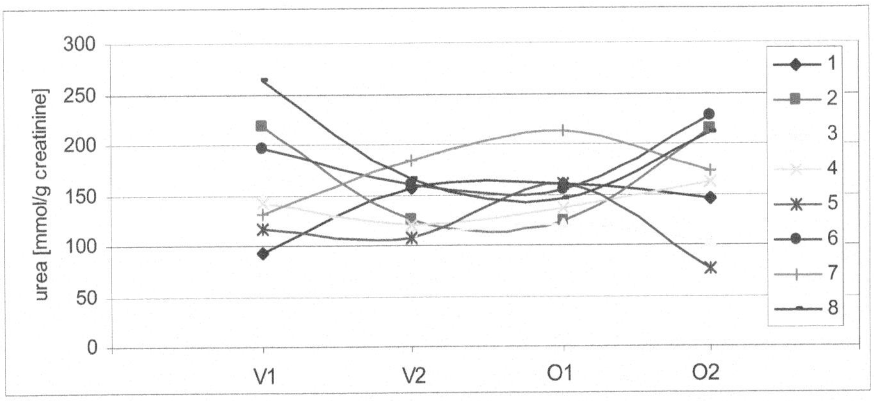

Figure 2. Results of urea content in urine (adjusted to creatinine excretion) of 24-hour samples taken on the 3rd and 7th days of the first week (tap water consumed - V1 and V2) and of the second week (osmotic water consumed - O1 and O2).

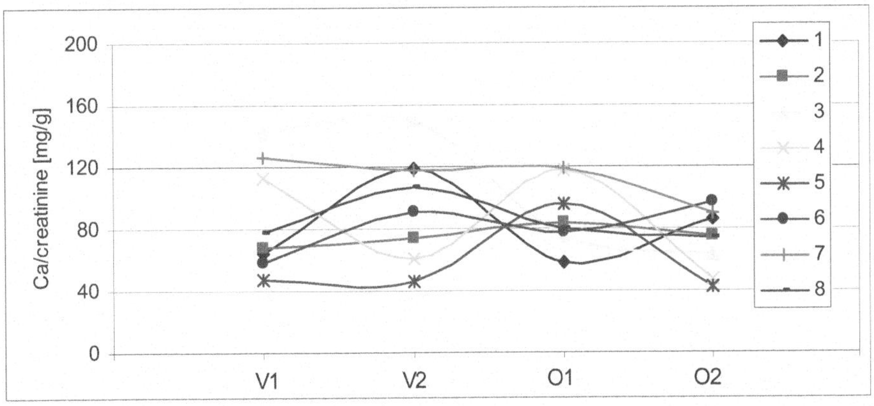

Figure 3. Results of calcium concentration in urine (adjusted to creatinine excretion) of 24-hour samples taken on the 3rd and 7th days of the first week (tap water consumed - V1 and V2) and of the second week (osmotic water consumed - O1 and O2).

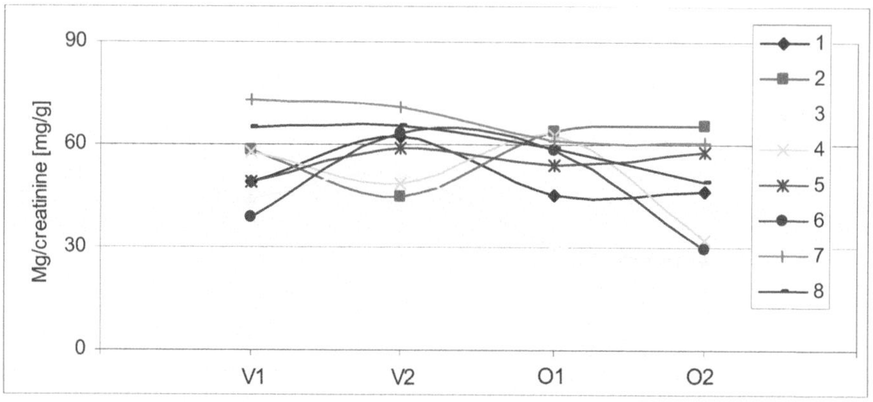

Figure 4. Results of magnesium concentration in urine (adjusted to creatinine excretion) of 24-hour samples taken on the 3rd and 7th days of the first week (tap water consumed - V1 and V2) and of the second week (osmotic water consumed - O1 and O2).

If we did not find any clear impact of consumption of low mineral water (including low bicarbonate content) on acid-base balance of the organism, as measured through changes in urine pH and urea or some essential elements excretion, what does it mean or how it can be interpreted? Before we suggest possible explanation, we have to realize the nature of this study. It was only a pilot study trying at first to verify the design of the experiment. We did not try to control all possible influencing factors, but preferred for the beginning to observe the situation of usual daily life and practice. Bearing it in mind we suggest the following possible explanations of our findings:

The hypothesis tested is wrong, i.e. drinking water of low mineral and bicarbonate contents cannot substantially influence acid-base balance of human body.

It was too short period of drinking low mineral water to show the effect, because adaptation mechanisms were still not overcome. It is known from anecdotal evidence that the first symptoms of negative health effects of drinking demineralised water occur usually after several weeks of the exposition. Nevertheless, we thought that some changes on biochemical level may be seen much earlier than any clinical signs of health problems appear.

"Too healthy" volunteers with good adaptation mechanisms and good nutritional status were recruited for the study and they could more easily compete with this burden than more sensitive/vulnerable groups of population which recruitment would be questionable from an ethical point of view.

Influence of food on acid-base balance is generally much more important than influence of water quality. If we did not simultaneously control acid renal load of food consumed, it may be hard to detect minor influence of acid renal load of water consumed. Especially if most of our volunteers consumed enough vegetable and meat not every day, and so their food acid load was lower than in average population.

Tap water used might not be very different (because of the effect of acidification) from osmotic water to show any effect and difference in renal acid load. Wynn et al. [11] estimated acid load (expressed as potential renal acid load - PRAL) of 150 mineral waters and found that 30 % had a positive PRAL and were therefore acidifying. 95 % of PRAL positivity could be explained by sulphate content. If we use the formula used by Wynn to compare the PRAL of our tap and osmotic waters, tap water had higher PRAL (i.e. on theory had more acidifying effect) than osmotic water just because of higher sulphate content. However, the main shortcoming of this approach is not taking bicarbonate content into account and no experimental verification of the assessment.

Contrary to our first reasoning, we cannot exclude the possibility that the results support the hypothesis. What if "no change" in ions excretion in fact means increase of excretion because of lower total intake of the minerals? For example, because of differences in calcium contents between the waters used, drinking osmotic water entailed lower total intake of calcium by 50 - 100 mg (if we presuppose the same intake from food).

4. Conclusions

Our experiment provides several important experiences and lessons for future studies, testing the hypothesis of effect of low mineral water. The trial should last longer, probably at least 3 to 4 weeks - while renal acid load of any food may be observed immediately (in 24-hour urine sample), renal acid load caused by water used may be seen after longer period when base pool is depleted (?). Other option is to keep the volunteers on uniform diet. While recruiting some more vulnerable group like people with some specific relevant illness would be unacceptable for ethical reasons, some studies searching for effect of alkali load of mineral waters took "advantage" of e.g. calcium deficient female to show the effect in an easier way [9]. The volunteers on more "unhealthy" food not providing enough essential elements and alkali may be the option. Using two waters of more different PRAL as calculated theoretically [11] may also help to find possible effects.

Systematic analysis of information from "field" experiments could help to understand and verify the hypothesis. It may be e.g. case histories (anecdotal evidence) from people using tap water treatment in their households, e.g. reverse osmosis or AEW (alkaline electrolysed water) units. Both negative and positive health effects after using such waters for cooking and drinking purposes have been reported.

Acknowledgments

Prepared within the research project "Metals and relating elements in drinking water" (Ministry of Education of the Czech Republic, program COST No. 1715/2007-32).

References

[1] Thresh, J.C., *The Lancet*, **1913**, 182(4702): 1057 - 1058.
[2] Kobayashi, J., *Berichte des Ohara Instituts für Landwirtschaftliche Biologie*, **1957**, 11: 12-21.
[3] University of East Anglia + Drinking Water Inspectorate. Review of evidence for relationship between incidence of cardiovascular disease and water hardness. Final report. Norwich - London, **2005**. Available at: www.dwi.gov.uk.
[4] Catling, L.A., Abubakar, I., Lake, I.R., Swift, L., Hunter, P.R. *Journal of Water and Health*, **2008**, 6(4): 433-442.
[5] Atkinson, S.A., Costello, R., Donohue, J.M., In: Cotruvo J., Bartram J. (eds.) Calcium and Magnesium in Drinking-water: Public health significance. Geneva, World Health Organization, **2009**, Chapter 2 (p.17-36).
[6] Rylander, R., *Journal of Nutrition*, **2008**, 138: 423S-425S.
[7] Remer, T., *Seminars in Dialysis*, **2000**, 13(4): 221-226.
[8] Remer, T., Dimitriou, T., Manz, F., *American Journal of Clinical Nutrition*, **2003**, 77(5): 1255-60.
[9] Burckhardt, P., *Journal of Nutrition*, **2008**, 138(2): 435S-437S. Erratum in: J. Nutr., 2008, 138(9):1730.
[10] Burckhardt, P., In: Burckhardt P., Dawson-Hughes B., Weaver C. (Eds.). Nutritional Influences on Bone Health. Springer-Verlag, London **2010**; Chapter 26.1 (p. 181-185).
[11] Wynn, E., Raetz, E., Burckhardt, P., *British Journal of Nutrition*, **2009**, 101(8): 1195-9.

Magnesium and calcium in drinking water and mortality due to cardiovascular disease in the Netherlands

Cindy de Jongh[1], Margeet Mons[2], Annemarie van Wezel[1]

[1]KWR Watercycle Research Institute, Nieuwegein, the Netherlands
[2]Prorail, Utrecht, the Netherlands

Corresponding author e-mail: Cindy.de.Jongh@kwrwater.nl

Abstract

Epidemiological studies showed conflicting results on the possible relation between the calcium and magnesium content of drinking water and the protective effect against cardiovascular disease. To obtain more clarity on this possible association, a large prospective cohort study was performed in the Netherlands. More than 120,000 cohort members aged 55-69 years were followed during 10 years. The study design allowed us to correct for a broad spectrum of potential confounders, including diet. This cohort study gave no evidence for an overall protective effect of a higher tap water hardness or a higher magnesium or calcium concentration against risk of dying from cardiovascular disease. However, a higher magnesium content of tap water (> 4 mg/L) was associated with a lower risk of dying from stroke in men with a low magnesium intake through their diet. In women with a low dietary magnesium intake an opposite results were found, but the association was not significant.

1. Introduction

During the past decades several studies have reported a possible protective effect of water hardness, or minerals contributing to water hardness, on cardiovascular mortality. However, other studies showed no effect or even opposite effects [1, 4, 6]. These inconclusive results of previous studies prompted us to investigate the association between tap water calcium and magnesium concentration or total hardness and risk of dying from ischemic heart disease or stroke [3]. More understanding on possible (beneficial) public health effects may be of importance in policy making regarding drinking water softening.

2. Materials and Methods

This study was performed in the framework of the Netherlands Cohort Study on Diet and Cancer [5]. In 1986, a cohort of 120,852 men and women aged between 55 and 69 years provided detailed information on their dietary and lifestyle habits. This large group was followed for mortality due to ischemic heart disease or stroke during a period of ten years. Data on tap water hardness and content of calcium and magnesium was obtained from all pumping stations in the Netherlands based on postal codes. The multivariate case-cohort analysis was based on 1944 ischemic heart disease and 779 stroke-mortality cases and 4114 subcohort members. For each gender hazard ratio's and corresponding 95% confidence intervals were calculated by Cox proportional hazard models. With this large, prospective cohort study it was possible to correct for a broad spectrum of potential confounders, including established risk factors for cardiovascular disease [3].

3. Results

In both men and women, we observed no relationship between tap water hardness or calcium and magnesium content and dying from ischemic heart disease or stroke (Table 1, [3]). When restricting the analysis to subjects with the 20% lowest dietary magnesium intake, we observed a statistically significant association between a higher magnesium content of tap water (> 4 mg/L) and a lower risk of mortality due to stroke in men (hazard ratio = 0.38 [0.15-0.94]; Table 1). Also a lower risk was found for ischemic heart disease, but this was not significant (hazard ratio = 0.69 [0.38-1.28]). In contrast, in women with a low dietary magnesium intake we found a non-significant increased risk for mortality due to ischemic heart disease or stroke (hazard ratios = 1.13 [0.48-2.63] and 1.47 [0.56-3.87], respectively) for a tap water magnesium concentration of higher than 4 mg/L (Table 1).

Table 1. Hazard ratio's (95% confidence intervals) for mortality due to ischemic heart disease and stroke in relation to calcium and magnesium concentration in tap water and total water hardness. Table derived from [3].

	Hazard ratio [95%-confidence interval]			
	Men		Women	
	Ischemic heart disease	Stroke	Ischemic heart disease	Stroke
Hard vs soft water[a]	1.04 [0.85-1.28)	0.90 [0.66-1.21]	0.93 [0.71-1.21]	0.86 [0.62-1.20]
High vs low calcium conc[b]	0.91 [0.60-1.38]	1.18 [0.62-2.22]	1.11 [0.59-2.07]	1.01 [0.47-2.19]
High vs low magnesium conc[c]	1.23 [0.82-1.86]	0.69 [0.37-1.31]	0.89 [0.50-1.59]	0.77 [0.38-1.57]
	Men with low dietary magnesium intake (<285 mg/day)		Women with low dietary magnesium intake (<255 mg/day)	
Magnesium concentration > 4 mg/L vs < 4 mg/L	0.69 [0.38-1.28]	0.38 [0.15-0.94]*	1.13 [0.48-2.63]	1.47 [0.56-3.87]

[a] hardness >2 mmol/L vs <1.5 mmol/L

[b] highest quintile vs lowest quintile (>82 mg/L vs <40 mg/L)

[c] highest quintile vs lowest quintile (>8.5 mg/L vs <3.8 mg/L)

* $P < 0.05$

4. Conclusion and Discussion

This study gave no evidence for an overall protective effect of a higher tap water hardness or a higher magnesium or calcium concentration against risk of dying from cardiovascular disease [3]. However, a higher magnesium content of tap water (> 4 mg/L) was associated with a lower risk of dying from stroke in men with a low magnesium intake through their diet. In women with a low dietary magnesium intake an opposite results were found, but the association was not significant. We found no biological explanation for these findings in the literature. Further studies may focus on the possible effect of tap water magnesium in subjects with a marginal magnesium intake through their food.

In the Netherlands, the magnesium concentration in tap water ranges between 0.9 and 16.1 mg/L. About 15% of the Dutch population has drinking water with a magnesium concentration lower than 4 mg/L. The total hardness of drinking water is predominantly determined by the calcium and magnesium content of the water. On average, 50% of the drinking water in the Netherlands is softened using pellet reactors or by nanofiltration or reverse osmosis techniques. The latter two techniques not only remove calcium ions but also magnesium ions from the drinking water.

Based on the results of this study, we derived the theoretical public health gain if all drinking water would contain at least 4 mg magnesium per litre [2]. We estimated a yearly theoretical avoidance of 26 premature deaths (range 3 to 36) among men aged 55-70 years with a low dietary magnesium intake. This number is low compared with the number of premature deaths due to smoking (20,000), overweight (8,000) or traffic accidents (1,200).

Acknowledgments

This project was conducted within the Joint Research Program of the Dutch water companies by researchers from KWR Watercycle Research Institute and Maastricht University.

References

[1] Catling LA, Abubakar I, Lake IR, Swift L, Hunter PR. A systematic review of analytical observational studies investigating the association between cardiovascular disease and drinking water hardness. *J Water Health* **2008**;6:433-42.

[2] de Jongh CM, Mons MN, van Wezel AP. Resultaat onderzoek relatie calcium en magnesium in drinkwater en hart- en vaatziekten. *H2O* **2010**;43:43-5.

[3] Leurs LJ, Schouten LJ, Mons MN, Goldbohm RA, van den Brandt PA. Relationship between tap water hardness, magnesium and calcium concentration and mortality due to ischemic heart disease or stroke in the Netherlands. *Environ Health Perspect* **2010**;118:414-20.

[4] Monarca S, Donato F, Zerbini I, Calderon RL, Craun GF. Review of epidemiological studies on drinking water hardness and cardiovascular diseases. *Eur J Cardiovasc Prev Rehabil* **2006**;13:495-506.

[5] van den Brandt PA, Goldbohm RA, van 't Veer P, Volovics A, Hermus RJ, Sturmans F. A large-scale prospective cohort study on diet and cancer in The Netherlands. *J Clin Epidemiol* **1990**;43:285-95.

[6] WHO. Calcium and Magnesium in Drinking-water - Public Health Significance. Geneva; **2009**.

Mineral balance and quality standards for desalinated water: the Israeli experience

Asher Brenner[1] and Abraham Tenne[2]

[1]*Dept. of Environmental Engineering, Ben-Gurion University, Beer-Sheva 84105, Israel*
[2]*Head of Desalination Division, Israel Water Authority, Tel-Aviv 61203, Israel*

Corresponding author e-mail: brenner@bgu.ac.il

Abstract
Due to the increasing problems of water shortage across the world, there is a growing trend of producing new water sources by desalination of seawater and brackish water. In desalinated water, the levels of alkalinity and essential ions, such as calcium and magnesium, are very low. Therefore, desalinated water may be associated with inferior taste, and corrosion problems that may result in the release of metals into water distribution pipes. In addition, the total daily dietary intake of such minerals may be reduced in some populations consuming tap water. For these considerations, new quality standards to reduce boron content and to stabilize the desalinated water have been recently enforced by legislation for all desalination plants in Israel. These quality standards actually take into account the dietary need for the nutritional supply of calcium through tap water consumption. This can be obtained mainly through dissolution of calcium carbonate. The need to add to desalinated water other minerals essential for human health, or required to enable efficient wastewater treatment and fulfill agronomic needs, should also be resolved in a sustainable manner.

1. Introduction

Growing problems of water scarcity and environmental pollution have motivated Israel to develop a careful water resources management system, based on effective integration of natural water sources, new water supplies, and reclaimed wastewater. The national policy has promoted and enforced water conservation in the urban, industrial, and agricultural sectors. However, there is a growing need for production of new water sources by desalination of seawater and brackish groundwater, as well as by reclamation and reuse of an increasing proportion of municipal wastewater.

Future water management in Israel is liable to become more complex, due to mutual effects caused by the mixing of multiple-quality water sources. While agricultural irrigation is the main target of reclaimed water, salt accumulation in soils and groundwater cannot be eliminated by gradual increase of desalinated water supplies. In addition, the comprehensive reuse of treated wastewater may ultimately cause a long-term buildup of toxic chemicals in the closed cycle of water supply and wastewater treatment and reuse.

2. Water balance in Israel

Basic data regarding water demand forecast for 2020 in Israel are given in Table 1. Two aquifers in Israel are the main sources of fresh water, the coastal and the mountain aquifers. Their annual production potential is approximately 300 and 350 Mm^3/Y, respectively. Other small local aquifers can add another 250 Mm^3/Y. The Sea of Galilee is a surface water source that can supply approximately 300 Mm^3/Y. There are also various local small aquifers of brackish water, especially in the southern part of Israel (The Negev Desert). This water is partly used in agriculture and industry, and its maximum production potential is approximately 300 Mm^3/Y. As can be seen in Table 1, most of the water is destined for the agricultural sector, which is gradually converting to the use of marginalwater, especially treated effluents. The 2020 forecast of the specific municipal water consumption is 105 m^3-capita/Y, based on a population forecast of 8.5 M.

Population growth, frequent drought incidents and environmental pollution have exerted increasing pressure on available natural resources and have mandated the development of a sustainable water resource management concept. The implementation of such a concept could supply a major part of agricultural water demand and should enable disposal of effluents without any health hazard or environmental nuisance. It has become a national policy to gradually substitute higher quality water supplies by reclaimed wastewater for direct non-potable applications.

In order to fulfill existing and projected water shortages, several seawater desalination plants have been designed and are gradually being erected and implemented along the Mediterranean Sea shore (see Table 2). Thus, by the year 2020, according to the figures presented in Tables 1 and 2 (and taking into account additional local desalination plants for brackish water, currently producing 30 Mm^3/Y and in 2020 planned to supply 80 Mm^3/Y), approximately 35% of the total fresh water supplies and 72% of the urban water supply will consist of desalinated water. This change of raw water supplies will dramatically alter the mineral composition of tap water and may also affect the composition of the reclaimed wastewater. This situation requires careful future management.

For several reasons, future management is not as simple as may be reflected in Table 1. In Israel, as in many dry regions, most of the precipitation occurs during a short season of 4-5 months. Furthermore, there is a steep precipitation gradient from north (600-800 mm annual rainfall) to south (less than 100 mm annual rainfall) along a distance of approximately 500 km. This situation requires careful design of water conduits (from north to south) and storage reservoirs (from winter to summer). There is also uneven distribution of population (consuming water and consequently producing wastewater).

Table 1. Year 2020 water demand forecast in Israel (Mm^3/Y).

	Fresh Natural	Desalinated	Effluents	Brackish & Runoff	Total	% of grand total	% of fresh*
Agriculture	450	/	550	200	1,200	44%	24%
Urban	250	650	/	/	900	33%	49%
Industry	100	/	/	50	150	6%	5%
Others**	400	/	50	/	450	17%	22%
Total	**1,200**	**650**	**600**	**250**	**2,700**	100%	
% of grand total	45%	24%	22%	9%	100%		

** % of fresh water consumption of the total fresh water consumption including desalinated water (1,850 Mm^3/Y)*
***Agreements with state neighbors, aquifer rehabilitation, and nature conservation*

Table 2. Planned desalination plants in Israel.

Plant	Year	Capacity (Mm^3/Y)
Ashkelon	2006	120
Palmachim	2007	45
Hadera	2010	130
Sorek	2013	150
Ashdod	2014	100
West Galil	2016	50

The coastal plain is densely populated while the southern Negev Desert is much less so, but has the highest reserves of land for agriculture. Therefore, wastewater conveyance systems (from center to south) are required. Storage systems are also necessary for reclaimed wastewater, because it is continuously produced during the entire year, while agricultural demand is highest during the summer. Storage can be provided in open reservoirs (the most common practice in Israel) or by aquifer recharge. Both strategies affect water quality, due to chemical and biological processes occurring during long storage periods.

3. Desalination

Desalination converts water with high dissolved solids content into water with a very low dissolved solids content. Reverse osmosis (RO) is the most common process applied today in the desalination of seawater or brackish water for the purpose of drinking water supplies. While osmosis is a natural phenomenon of water diffusion through a semi-permeable membrane, due to a concentration gradient (the motion is from the low solute concentration to the high solute concentration), reverse osmosis requires input of an external pressure to drive a flow in the opposite direction.

Nanofiltration (NF) is a more moderate high pressure driven process in which monovalent ions pass freely through the membrane, while highly charged multivalent salts are rejected due to the special structure of the NF membrane surface, which is negatively charged at neutral and alkaline media. For these two processes, the separation is based not only on physical mechanism related to membrane pore size (relatively small) that serves as a barrier, but also on the chemical structure of membrane material that can dissolve, attract, or reject various substances. In addition to the production of potable water, there are other alternative uses of desalination technologies that include: softening, natural organic matter removal for disinfection by-products control, and specific contaminant removal such as heavy metals, radio-nuclides, or emerging micro-pollutants.

In Israel, there is another problem related to the quality of desalinated water, based on its conversion after primary use to municipal wastewater destined for reclamation and reuse. Boron toxicity to plants may limit the application of reclaimed wastewater originating from desalinated seawater, because of the high content of boron (approximately 5 mg/L) and its limited rejection in conventional reverse osmosis processes. The new desalination plants planned in Israel are therefore required by the Israel Water Authority to upgrade their processes to reduce boron levels to 0.3-0.4 mg/L. Therefore, modification of conventional RO technology is required to answer this quality standard demand.

4. Mineral balance and quality standards for desalinated water

In desalinated water, the levels of alkalinity and essential minerals, such as calcium and magnesium, are very low. Therefore, desalinated water may be associated with inferior taste, and corrosion problems that may result in the release of metal colloids (including heavy metals) into water distribution pipes. Therefore, desalinated water is usually stabilized before distribution in order to avoid the problems of corrosion and "red water" incidents in water distribution pipes. Stabilization practices typically involve mixing the desalinated water with un-desalinated source water, or adding the needed minerals and alkalinity, using, for example, limestone.

The total daily dietary intake of ions such as calcium and magnesium might be reduced in some populations consuming desalinated water. Table 3 presents figures based on the Dietary Reference Intakes (DRIs) data established by the United States Institute of Medicine [1, 2, 3], regarding the Recommended Dietary Allowance (RDA). This figure is the average daily dietary intake level that is deemed sufficient to meet the nutrient requirements of nearly all (97 to 98 percent) individuals in a life stage and gender group, taking into account total intake from food, water, and supplements. The data are related to the most essential minerals and trace elements needed for human health. These substances are divided into three groups, according to the RDA levels (in mg per day). It is obvious that trace elements should not be added to any source of water, due to their low RDA that can be obtained from other sources, and because of technological limitations regarding dose adjustment. Of the micro-nutrients, fluoride is the only one that can be contributed mainly through water consumption. Therefore, to reduce tooth decay, it is commonly added to public water supplies in many regions of the world. The required level is relatively low (0.5-1 mg/L) and depends on climate (affecting water consumption). This treatment is still a controversial issue, since excessive fluoridation can cause skeletal and dental fluorosis, and possibly increased bone fracture risk. Furthermore, the use of fluoridated toothpaste and mouthwash has achieved similar protective results in many places. In Israel, desalinated water as other water sources (when fluoride concentration is below 0.6 mg/L) are still fluoridated, according to specific standards issued by the Ministry of Health (0.8-1.0 mg/L, depending on climate conditions).

Table 3. RDA ranges according to the U.S. National Academies Press Nutrition Dietary Reference Intake (US-IOM, 1997, 2000, 2004).

Nutrient group	RDA range, mg/d	Minerals
Macro-nutrients	Hundreds to thousands	Calcium Chloride Magnesium Phosphorus Potassium Sodium
Micro-nutrients	Less than 15	Fluoride Iron Manganese Zinc
Trace elements	Less than 1	Chromium Copper Iodine Molybdenum Selenium

Of the six macro-nutrients detailed in Table 3, the RDA of calcium and magnesium is not always obtained by regular consumption of foods and drinks, especially in industrialized countries. This is due to low consumption of vegetables and dairy products and due to drinking liquids poor in minerals. The 2005 *Dietary Guidelines for Americans* (United States Department of Agriculture and United States Department of Health and Human Services 2005) reported that less than 60% of adult men and women in the United States met the *Adequate Intake* values for calcium and magnesium, and both calcium and magnesium were listed as nutrients consumed in amounts low enough to be of concern [4]. A deficiency of these two essential nutrients may be a major factor in many chronic disorders and common health problems. One of their major contributions has been widely reported to be related to the cardiovascular system

Some of the dietary intake can be obtained by drinking water. However, in many places the water sources supplied are surface water (dam water) or desalinated water, which are very poor in these essential minerals. In addition, in many developed countries, water is softened (in central treatment plants or by domestic installations). Thus the consumed water contains neither calcium nor magnesium in significant quantities. In some cases, water treatment processes can affect mineral concentrations and contribute to the total intake of essential minerals for some individuals. Water stabilisation, for instance, based on dissolution of calcium carbonate can supply the required levels of alkalinity and calcium to prevent corrosion. It can thus account for the portion of calcium that can be obtained from drinking water. However, this technology does not supply sufficient magnesium, and the ratio of calcium to magnesium, following desalination/calcification, becomes very high and may not enable efficient absorption of magnesium [5].

The World Health Organization (WHO) recently assembled a diverse group of nutrition, medical, epidemiological and other scientific experts, together with water technologists, to address the possible role of drinking-water containing calcium and/or magnesium as a contribution to the daily intake of those minerals. Among the numerous issues addressed were the desirability and feasibility of re-mineralization for stabilization and potential health benefits [4]. Thus the issue of calcium and magnesium addition to drinking water (motivated by health related reasons) has been considered seriously by the WHO. It was also addressed by other national committees, such as the Israeli committee for the update of Israel drinking water standards.

Based on the recommendations of this committee, the Israeli Ministry of Health has proposed new quality standards for desalinated water, requiring the application of post-treatment for the conditioning of desalinated water, mainly through dissolution of calcium carbonate. This process can supply the proposed quality standards, detailed in Table 4. These quality standards actually cover the dietary need for nutritional supply of calcium in drinking water (>50 mg/L as $CaCO_3$), as required independently by the Israel Ministry of Health. These requirements can be obtained simply through dissolution of calcium carbonate. The Israeli committee for the update of Israel drinking water standards also recommended addition of magnesium to desalinated water (>42 mg/L as $CaCO_3$). However, the resulting new version of the drinking water quality standards does not include a requirement for magnesium addition, on the assumption that this mineral's intake-requirements can be obtained from food products more easily than calcium as well as for economic considerations. Typical design requirements and performance data for the three operating desalination plants in Israel are given in Table 5. It can be concluded that this technology has proved to be a reliable means of supplying the required quantity and quality of water. According to the experience gained so far regarding the build/operate/turn (BOT) tenders, the cost per cubic meter of desalinated water is relatively low.

Regarding the magnesium issue, the WHO [4] details many published studies regarding the beneficial health effects of calcium and magnesium. Despite the relatively low potential contribution from drinking water supplies, compared to food sources, obtaining the potential portion present in water supplies should be seriously considered, as can be concluded from many of the studies, some of which (not reviewed before by the WHO [4]) are cited herein.

Table 4. Proposed quality standards for desalinated water in Israel (Drinking water standards, issued by the Israeli Ministry of Health, 2010).

Monitoring	Monitoring Location	Parameter	Units	Maximum Concentration
Continuous	Exit of desalination process	Conductivity	□S/cm	95% of daily values < OV and not higher than 1.3(OV)
	Exit of calcification process	Turbidity	NTU	95% of daily values < 0.5 and not higher than 1.0
		pH	/	7.5-8.3 in 95% of daily values and not higher than 8.5
Batch	Exit of calcification process	Dissolved Ca^{++}	$ppmC_aCO_3$	80 - 120
		Alkalinity	$ppmC_aCO_3$	> 80
		CCCP**	$ppmC_aCO_3$	3 - 10
		LSI***	/	> 0

* OV=Operational value, approved by Health Authorities (specific to each plant).
** Calcium carbonate precipitation potential.
*** Langelier saturation index.

Table 5. Design requirements and performance data of desalination plants in Israel (Data obtained from the Israel Water Authority).

Parameter	Units	Design requirements			Performance		
		Ashklon	Palmachim	Hadera	Ashklon	Palmachim	Hadera
Chloride	ppm	<20	<80	<20	10-15	30-50	10-15
Boron	ppm	<0.4	<0.4	<0.3	0.2-0.3	0.3-0.38	0.2-0.3
pH	/	7.5-8.5	7.5-8.5	7.5-8.5	8-8.5	8-8.5	8-8.5
LSI	/	-0.2-0.5	-0.5-0.5	0-0.5	0-0.5	0-0.5	0-0.5
Alkalinity	$ppmC_aCO_3$	/	/	>80	45-50	40-45	80-100
Hardness	$ppmC_aCO_3$	>60	>75	80-120	90-110	85-95	80-100
Turbidity	NTU	<0.5	<0.8	<0.5	0.15-0.2	0.15-0.3	0.4-0.5

The crucial contribution of magnesium to human health has been demonstrated by Shechter [6] who found that oral magnesium supplementation in patients with coronary artery disease (CAD) for 6 months resulted in a significant improvement in exercise tolerance, reduction in exercise-induced chest pain, and better quality of life. The study suggests a potential mechanism whereby magnesium could beneficially alter outcomes in patients with CAD. Ferrándiz *et al.* [7] presented a study that provides statistical evidence of the relationship between mortality from cardiovascular diseases and hardness of drinking water. This relationship is stronger in cerebrovascular disease than in ischemic heart disease, is more pronounced for women than for men, and is more apparent with magnesium than with calcium concentration levels. Catling *et al.* [8] systematically reviewed and critically assessed 2,906 papers dealing with analytical observational epidemiology studies investigating the association between levels of drinking water hardness and cardiovascular disease. They found significant evidence of an inverse association between magnesium levels in drinking water and cardiovascular mortality following a meta-analysis of case control studies. However, evidence for calcium remained unclear.

On the other hand, there are also several studies with opposite results. Leurs et al. [9] investigated the possible association between tap water calcium or magnesium concentrations, total hardness and ischemic heart disease (IHD) mortality or stroke mortality and found no evidence for an overall significant association between tap water hardness, magnesium or calcium concentrations and IHD or stroke mortality. A study with 7735 men aged 40-59 years, including estimates of town-level water hardness and estimation of individual calcium and magnesium intakes, was carried out to follow up incidents of major coronary heart disease (CHD) and stroke and CHD mortality for 25 years [10]. This study suggests that neither high water hardness, nor high calcium or magnesium intake appreciably protect against CHD or cardiovascular disease (CVD), and therefore, initiatives to add calcium and magnesium to desalinated water cannot be justified.

It may be therefore concluded that more research is needed to investigate the effect of tap water hardness or specific calcium and/or magnesium levels on cardiovascular and heart disease and mortality. The WHO [4] also recommends research priorities, in order to build an evidence data base to inform decisions on managing "processed" drinking-water such as softened or desalinated water that significantly alter its mineral composition. This may result in future recommendations for a careful control of calcium and magnesium in drinking water supplies due to their proven contribution to health benefits in populations, as well as recommendations for the control of other elements that may also have health relevance. It should be noted, however, that unlike fluoride, the therapeutic window of magnesium is wide, and in the absence of renal failure, severe side-effects are extremely rare [11].

There are also other considerations regarding mineral balance, for desalinated water that may be converted to domestic wastewater to be further treated and reused in agriculture. Bi-carbonate (alkalinity) balance is crucial to sustain stable biological wastewater treatment, especially for nitrogen removal systems. In addition, massive municipal use of sodium-containing chemicals might deteriorate the sodium adsorption ratio (SAR) of the wastewater and thus affect soil properties.

5. Conclusions

Future management of water resources in Israel is indeed a complex issue, incorporating several measures such as modification of traditional water treatment schemes and quality standards, upgrading of wastewater treatment processes, and source control. New water sources based on desalinated water and reclaimed wastewater should be integrated with natural water supplies, in order to enable sustainable water management. This concept is crucial in water-scarce countries, where reclaimed wastewater should be regarded as a resource and not as a waste product requiring disposal. Management of complex water systems composed of multi-quality sources requires a revolution in the traditional treatment and regulation concepts applied so far.

In Israel, it is planned that by 2020, approximately 35% of the total fresh water supplies and 72% of the urban water supply will consist of desalinated water. This change of raw water supplies will dramatically change the mineral composition of tap water and may also affect the composition of the reclaimed wastewater. In desalinated water, the levels of alkalinity and essential minerals, such as calcium and magnesium, are very low. Therefore, the total daily dietary intake of such minerals might be reduced in some populations consuming desalinated water. New quality standards to reduce boron content and to stabilize the desalinated water have been recently enforced by legislation for all desalination plants. These measures actually cover the dietary need to account for nutritional supply of calcium (>50 mg/L as $CaCO_3$), and can be obtained mainly through dissolution of calcium carbonate. On the other hand, lack of other essential minerals in desalinated water, which are essential for human health, or are required to enable efficient wastewater treatment and fulfill agronomic needs, should also be resolved in a sustainable manner.

References

[1] US-IOM, *Dietary Reference Intakes for Calcium, Phosphorus, Magnesium, Vitamin D, and Fluoride.* Prepared by the Standing Committee on the Scientific Evaluation of Dietary Reference Intakes, Food and Nutrition Board, Institute of Medicine. National Academy Press, **1977**, Washington, DC.

[2] US-IOM, *Dietary Reference Intakes for Vitamin A, Vitamin K, Arsenic, Boron, Chromium, Copper, Iodine, Iron, Manganese, Molybdenum, Nickel, Silicon, Vanadium, and Zinc.* Prepared by the Standing Committee on the Scientific Evaluation of Dietary Reference Intakes, Food and Nutrition Board, Institute of Medicine. National Academy Press, **2000**, Washington, DC.

[3] US-IOM, *Dietary Reference Intakes for Water, Potassium, Sodium, Chloride, and Sulfate.* Prepared by the Standing Committee on the Scientific Evaluation of Dietary Reference Intakes, Food and Nutrition Board, Institute of Medicine. National Academies Press, **2004**, Washington, DC.

[4] WHO, *Calcium and Magnesium in Drinking-water: Public health significance,* **2009**, World Health Organization, Geneva, 180 pp.

[5] Whiting,S.J., Wood, R.J., Adverse effects of high-calcium diets in humans, *Nutrition Reviews,* **1997**, 55(1): 1-9.

[6] Shechter M., Effects of oral magnesium therapy on exercise tolerance, exercise-induced chest pain, and quality of life in patients with coronary artery disease, *Am J Cardiol,* **2003**, 91:517-521.

[7] Juan, F., Abellán J-J., Gómez-Rubio, V., López-Quílez, A., Sanmartín, P., Abellán, C., Martínez-Beneito, M-A., Melchor, I., Vanaclocha, H., Zurriaga, O., Ballester, F., Gil, J-M., Pérez-Hoyos, S., Ocaña, R., Spatial analysis of the relationship between mortality from cardiovascular and cerebrovascular disease and drinking water hardness, *Environmental Health Perspectives,* **2004**, 112(9): 1037-1044.

[8] Catling, L.A, Abubakar, I., Lake, I.R., Swift, L., Hunter, P.R., A systematic review of analytical observational studies investigating the association between cardiovascular disease and drinking water hardness, *J. Water Health,* **2008**, 6(4):433-442.

[9] Leurs, L.J., Schouten, L.J., Mons, M.N., R. Alexandra Goldbohm, R.A., van den Brandt, P.A., Relationship between tap water hardness, magnesium and calcium concentration, and mortality due to ischemic heart disease or stroke in the Netherlands, *Environmental Health Perspectives,* **2010**, 118(3): 414-420.

[10] Morris, R.W., Walker, M., Lennon, L.T., Shaper, A.G., Whincup, P.H., Hard drinking water does not protect against cardiovascular disease: new evidence from the British Regional Heart Study, *European J. of Cardiovascular Prevention & Rehabilitation,* **2008**, 15(2): 185-189.

[11] Saris N.E., Mervaala E., Karppanen H., Khawaja J.L., Lewenstam A., Magnesium: an update on physiological, clinical and analytical aspects, *Clin Chim Acta,* **2000**, 294:1-26.

Mineral balance in water: before and after treatment

Ingegerd Rosborg[1], Prosun Bhattacharya[1], Jimmie Parkes[2]

[1]Royal Institute of Technology, KTH, Stockholm, Sweden.
[2]Inter-Euro Technology Ltd., Carlow, Ireland.

Corresponding author e-mail: rosborg@spray.se

Abstract

The importance of minerals from drinking water is a question of increasing interest. Different treatment methods performed to remove undesirable substances of the water may completely alter the mineral balance of the water. Thus, the measure taken may eliminate one health problem for consumers of the water, but cause new. Softening filters substantially decrease a number of elements and ions, especially the important metal Ca, which is included in the building of bones and teeth, and irreplaceable in the heart and nerve function. In addition, a number of other elements in limestone decrease in concentration, some very important to the human health. The change in element concentrations in three different Swedish municipal water plants, with hard and mineral rich raw water, are reflected in this paper; one without- and two with softening filter. RO (Reverse Osmosis) filters completely de-mineralize water. This may cause de-mineralization of the whole body. No scientific studies on health effects from drinking RO treated drinking water were to be found, even though the method is rapidly increasing in use among the public and on water plants around the world. Thus, the concentrations of metals and ions in one sample of well water with RO filter installed at the kitchen tap is compared to mean levels of mineral elements in acid and alkaline well waters in a study from 2002.

Introduction

Background: The composition and concentration of minerals and salts play an important role in the quality of drinking water. The relative concentrations of various mineral elements and ratios of elements vary between waters of different hardness. By treatment a reduction or addition of specific minerals can lead to undesirable or even hazardous changes in the concentration of elements or ratios between elements. This has in general been overlooked. Careful monitoring becomes necessary when water is treated for consumption.

Water hardness: The water hardness is expressed in German degrees, °dH, and reflects the sum of all divalent metal ions, as e.g. Ca^{2+} and Mg^{2+}, where Ca is dominating. 10 mg CaO/L is equal to 1 °dH, Hardness due to Mg is for simplicity reasons converted to an equivalent amount of CaO. In addition to mentioned metals a large number of other metal ions as well as complex negative ions, e.g. CO_3, HCO_3, and SO_4, are present in higher or lower concentrations depending on the composition of the bedrock. As water is heated Ca ions precipitate as carbonates in e.g. pipes, coffee machines and different installations. In addition, more detergents are required in hard water.

Hardness is generally classified in accordance to this list:

Very soft	0-2 ° dH
Soft	2-5 ° dH
Medium hard	5-10 ° dH
Hard	10-20 °dH
Very hard	>20 ° dH

In accordance with the Swedish drinking water directive the following values are recommended for technical reasons:

Ca < 100 mg/L, desirably 20-60 mg/L
Mg < 30 mg/L, levels exceeding this level may cause taste and aesthetical problems
Alkalinity (bicarbonate, HCO3) > 70 mg/L, where too high levels may also cause Cu corrosion

Precipitation of $CaCO_3$ in pipes and installations take place as follows:

$$Ca^{2+} + 2\ HCO_3^- \Leftrightarrow CaCO_3 + CO_2 + H_2O$$

$$Ca^{2+} + HCO_3^- + NaOH \longrightarrow CaCO_3 + Na+ + H_2O$$

Swedish natural water outside calcareous areas are soft waters, and may contain detectable but low concentrations of elements like Al, Fe, Ca, K, Na, Si, Mn, Mg and heavy metals (Scheffer 1989, Lundegårdh 1995). Limestone, slate and sandstone, on the other hand, are easily weathered rocks that give higher concentrations of a number of elements and ions, e.g. Ca, Mg, HCO_3, K, Na, Fe, Mn, P and S in ground water originating from limestone (Scheffer 1989). In addition the elements Al, Cd, Co, Cr, Cu, Ni, Pb, V and Zn are generally present. The concentrations of Ca and Mg may vary widely in limestone areas from approximately 55 % CaO and less than 1 % MgO in calcite, which is dominating in the bedrock of Southern Sweden, to about 30 % CaO and 12 % MgO in Central European dolomite (FitzPatrick 1980).
Thus, all these elements can originate from drinking water, to a higher or a lower extent.

Ca, Mg and HCO_3 in the body: Calcium is needed for teeth, bone tissue, heart function, nerve impulses, pH regulation and contraction of muscles. Inadequate intakes of calcium have been associated with increased risks of osteoporosis, nephrolithiasis (kidney stones), colorectal cancer, hypertension and stroke, coronary artery disease, insulin resistance and obesity (Bowman & Russell, 2006). Magnesium is the fourth most abundant cation in the body and the second most abundant cation in intracellular fluid. It is a cofactor for some 350 cellular enzymes, many of which are involved in energy metabolism. It is also involved in protein and nucleic acid synthesis and is needed for normal vascular tone and insulin sensitivity. Low magnesium levels are associated with endothelial dysfunction, increased vascular reactions, and decreased insulin sensitivity. Alcoholism and intestinal mal-absorption are conditions associated with magnesium deficiency. Certain drugs, such as diuretics, some antibiotics and some chemotherapy treatments, increase the loss of magnesium through the kidney (Bowman & Russell, 2006).

Bicarbonate is the most important buffering agent in nature as well as in humans. Bicarbonate from water may decrease dissolution of bone tissue and raise the stomach and body pH (Frassetto et al. 2001). Sulphur, mainly present in drinking water as SO_4, is antagonistic against heavy metals, and is regarded as decreasing the health risks connected with intake of heavy metals. Sulphate (SO_4) is also active against constipation (Bergmark 1959).

Hard water as potential protection against diseases: A large number of scientific studies clearly show that intake of hard water for decades, with elevated levels of elements like e.g. calcium (Ca) and magnesium (Mg) protects against heart diseases (e.g. Rubenowitz et al 1999, 200, Rylander et al 1991, Yang et al 1998a). There are also some studies indicating that hard water may be protective against diabetes (e.g. Yang et al 1999a), osteoporosis (Frassetto et al 2001), and some forms of cancer (Yang et al 2000). In an American study Shroeder et al (1966) the death rates due to high blood pressure and arteriosclerosis were higher in cities where the drinking water had low conductivity, water hardness, concentration of magnesium, sodium, potassium, sulphate and barium, as well as low concentrations of bicarbonate, chlorine, silicon, lithium, strontium and vanadium, but high concentrations of copper. Most of these elements and ions are limestone related and appear at higher concentrations in hard water (Scheffer & Lundegård, 1995).

In addition, nutrient minerals may act protectively against toxic metals and ions, as e.g. Ca against Pb on a cell level, and hard water form a protective layer on pipes, preventing further dissolution of Pb. Certain ratios, as e.g. Ca/Mg or Na/K also are important, since there are intervals in which the ratios can be regarded as health bringing. Any treatment alters the mineral balance of water.

Results of the WHO conference on Ca and Mg from drinking water in Baltimore, 2006:
WHO, 2009, states; "Food is the principal source of both calcium and magnesium. Dairy products are the richest sources of dietary calcium, contributing over 50% of the total calcium in many diets. Some plant foods, including legumes, green leafy vegetables and broccoli, can also contribute to dietary calcium, but the content is lower than in dairy products, and the bioavailability of calcium in plant foods can be low if the concentration of oxalate or phytate is high. Dietary sources of magnesium are more varied; dairy products, vegetables, grain, fruits and nuts are important contributors.

Many of the ecological epidemiological studies conducted since the mid-1950s have supported the hypothesis that extra magnesium and/or calcium in drinking-water can contribute to reduced cardiovascular disease and other health benefits in populations. However, most of those studies did not cover total dietary intake and other important factors."

Thus, no recommendations of lowest acceptable levels of Ca and Mg were established as a consequence of the Baltimore conference. However, the discussion is not over, yet, which also is stated: "It is hoped that this publication will advance knowledge and contribute to further discussions on these and related issues in this area", WHO 2009.

What has not been considered is the fact that metals like e.g. Ca and Mg are more readily absorbed from water, since they appear as ions in water. The uptake from the intestinal fluid may be as bonded to some minor molecule, such as an amino acid or other organic compound. However, such molecules are always present in the intestines, indicating that metals taken in as free ions are more readily absorbed.

Uranium: U has three isotopes, whereas 99.28 % of natural U is ^{238}U, 0.7% is ^{235}U, 0.005% is ^{234}U. ^{238}U, which is the most abundant isotope of U, has a long half life and thus a low specific activity. This means that uranium is present at measurable chemical quantities even at low activity concentrations. It is thus considered not so much a radioactive contaminant but rather a chemical contaminant.

The average human gastrointestinal absorption of uranium is 1-2% (Wrenn et al 1985). The absorption in female Sprague Dawley rats increased from 0.17 to 3.3% when iron (III) was administered simultaneously (Sullivan et al 1986). Bone may be a target of chemical toxicity of uranium in humans (Komulainen et al 2005). Wrenn et al (1989) state, that no carcinogenic effects of administered doses of uranium to animals have been demonstrated. However, chromosome aberrations have been observed in germ cells of male mice after administration of uranul fluoride. Raymond-Wish, 2007, states that uranium is an endochrine-disrupting chemical and populations exposed to environmental uranium should be followed for risk of fertility problems and reproductive cancers (Arnault et al 2008). In a study by Kurttio et al (2002), uranium (U) in urine was statistically significantly associated with increased fractional excretion of calcium (Ca), glucose and phosphate (PO_4), indicating a renal effect of U from drinking water. In addition, re-absorption of e.g. Ca and phosphate, low molecular weight proteins and enzymes may be reduced. This may affect the calcium balance and increase blood pressure. The median U-concentration in drinking water in this study was 28 µg/L (range: 6-135 µg/L).

Common treatment processes

Softening filter: In order to reduce the hardness and avoid problems with precipitations on pipes and installations, softening filters are installed in houses or at drinking water plants. Softening filters are either ion exchange filters or precipitation filters, where preferably NaOH is added to increase the pH to a level where Ca precipitates as $CaCO_3$. A number of other limestone related substances then co-precipitate along with $CaCO_3$.

Reverse Osmosis (RO): "Pure water" or "clean water" has become a health issue, implicating that H_2O, water molecules, is the only desired content of drinking water. In the marketing of RO filters the degree to which the elimination of elements and ions is performed is often reflected. Thus, reduction of metals and ions to as close to 100% as possible, is an implication of a "good functioning "RO filter. In addition to desired reductions of different pollutants or heavy metals, also nutrient minerals like e.g. Ca, Mg, HCO_3 and SO_4 are almost completely eliminated. However, there are no scientific studies, so far, indicating health bringing properties of totally de-mineralized drinking water. There are in reality no studies at all of health effects from drinking RO water for weeks, months, years or decades. However, there are numerous studies clearly showing the importance of minerals from drinking water on the human body. RO has been used to eliminate e.g. As, since high concentrations of As in drinking water causes specific skin lesions and in the end may cause skin and other forms of cancer. By treating e.g. As contaminated drinking water with RO (Reverse Osmosis) all metals and ions are almost completely eliminated. Thus, by solving one health problem we cause another. Softening filter and RO filters probably have the greatest impact on the mineral balance of water of all methods that are used at water plants and in private households.

Aims

The aims with this study was to identify the treatment processes of some municipal drinking waters and evaluate the mineral balance in treated waters and compare it to untreated water.

Presentation of treatment processes, results and discussion

Kristianstad:

The Kristianstad drinking water is pumped up in glauconitic sand, below a 100 m thick limestone cliff. It has taken hundreds of years for the water to percolate through the cliff. Meanwhile mineral elements have been dissolved into the water. The wells, which are situated close to the town center, were originally artesian. Nowadays they are not. The water pressure has decreased, why there's a risk of leakage of pollutants from the city dump as well as increased U levels from nearby veins of water. In addition, NO_3 from intensive farming on the Kristianstad flatland is a threat to the water.

Kristianstad is surrounded by a large swamp, The Kingdom of Water, a UN biosphere, where a large number of rare birds, flowers and plants, as well as fish find their desired habitats. Thus, both the ground water and the surface water serve the city with nature's fortune.

Figure 1: The recently officially opened Naturum, in the center of the UN Biosphere, The Kingdom of Water, situated close to the center of Kristianstad (www.kristianstad.se).

Present situation: The only treatment of Kristianstad drinking water is aeration, in order to eliminate H_2S. Thus also Fe^{2+} is being oxidized to Fe^{3+}, which precipitates as Fe-hydroxides, reducing the Fe concentration to <0.01 mg/L. The Ca concentration is about 80 mg/L, Mg 6.5 mg/L, HCO_3 210 mg/L, F 0.35 mg/L. Another around 40 elements and ions did not change during the passage of the sand filter, except for Zn which increased and Sr that decreased (see Table 1).

Table 1: Average concentrations of some parameters in Kristianstad untreated and treated drinking water.

Parameter	Untreated	Trated	Unit	Parameter	Untreated	Trated	Unit
pH	8.0	8.0					
Ca	80	75	mg/L	NH_4	0.1	<det.	mg/L
Mg	6.5	6.5	mg/L	NO_3	<det.	<det.	mg/L
Na	12,7	12.4	mg/L	SO_4	21	15.9	mg/L
K	2.9	2.9	mg/L	Cl	22.2	19.2	mg/L
HCO_3	210	156	mg/L	F	0.53	0.43	mg/L
Fe	14.9	<det.	µg/L	PO_4	<det.	<det.	µg/L
Mn	2.7	0.6	µg/L	Zn	0.9	28.7	mg/L
Al	0.7	0.6	µg/L	Sr	174	149	µg/L

Malmö:

In Malmö the raw water is taken from a lake, and infiltrated in gravel and sand. The raw water is then pumped up as ground water. Elevated Cu levels in the sewage sludge became an increasing problem in the 1990's, and the sludge could not be used as a fertilizer in farming. High concentration of Natural Organic Matter (NOM), from algal bloom in the lake in combination with high water hardness is the main causes of the dissolution of Cu from pipes. Softening filter was installed. In the filter supplementation with NaOH was performed in order to decrease the hardness, one of the parameters causing the Cu dissolution. Thus, the Ca concentration decreases from around 65 mg/L to 30 mg/L, and HCO_3 from 190 mg/L to 140 mg/L by precipitation of especially $CaCO_3$, while the Mg concentration remains at 6 mg/L. In addition, Fe decreases from 30 µg/L to 20 µg/L, and Mn is almost eliminated. The Ca concentration in the treated water is decreased to half the original concentration. Ba and Sr are reduced by about 50%, respectively, due to co-precipitation of especially $BaCO_3$ and $SrCO_3$ along with the $CaCO_3$ precipitation. The SO_4 concentration is not reduced. Other parameters, about 40 analyzed, were not influenced by the treatment process.

Figure 2: Lake Vombsjön situated about 10 km from the center of Malmö (www.sjoboallehanda.se).

Uppsala:

Water from the local river, the river Fyris, is infiltrated into the Uppsala ridge and becomes artificial ground water of the same quality as the groundwater present in the ridge. Raw water is pumped up from four different wells and treated separately at two water plants, Baeckloesa and Graenby. Uranium in Uppsala drinking water has its origin in acid granite bedrock, which has a relatively high content of uranium, in gravel and sand material in the Uppsala ridge. In the 1990's the high U level was discussed, since it was supposed to be an indicator of high Radon. However, the correlation between high U and Rn concentrations was very weak. Toxicity of U in drinking water was not considered a health risk at that time. However, recent studies indicate that high U in drinking water may be harmful for especially the kidneys. In addition, some studies indicate that bone may be a target of chemical toxicity of uranium in humans (Komulainen et al 2005). Uranium intake may also negatively influence nerves and the brain, Bussy et al (2006), and cause oxidative stress (Taulan et al 2004).

The raw water in Uppsala is treated with addition of NaOH, as in Malmoe, in order to decrease the U concentration, which after this treatment is 35 µg/L. U co-precipitates with $CaCO_3$, and decreases from to about 18-21 µg/L in Graenby, and to 25-31 µg/L in Baecklösa, at present, tab. 2 (Oral comm. Soederstad, Uppsala Municipality). By the treatment HCO_3 is decreased from 325 mg/L to about 115 mg/L by precipitation of especially $CaCO_3$, Ca from above 90 mg/L to about 35 mg/L, while the Mg concentration is preserved at 15 mg/L. In addition to above mentioned elements and ions, pH, Mn, color, hardness, conductivity, and Cl are decreased by the softening treatment (see Table 1, below).

Table 2: Average concentrations of a number of parameters in Uppsala drinking water before and after treatment, Graenby / Baeckloesa together (in mg/L).

Parameter	Untreated	Treated	Parameter	Untreated	Treated
pH	7.4	8.2	Cu	<0.02	<0.02
Colour	8.15	0.45	F	0.99	0.87
Conductivity	62.1	41.2	Fe	0.03	0.03
COD_{Mn}	1.7	1.5	Mn	0.016	<0.005
Turbidity	0.12	0.12	Na	19.1	19.5
Ca	88.7	37.1	NH_4	<0.04	<0.04
Mg	12.1	12.1	NO_3	7.0	7.1
HCO_3	300	112	SO_4	38.8	38.0
Tot Hardness	15.2	8.0	U	27.7	22.8
Cl	29,7	43			

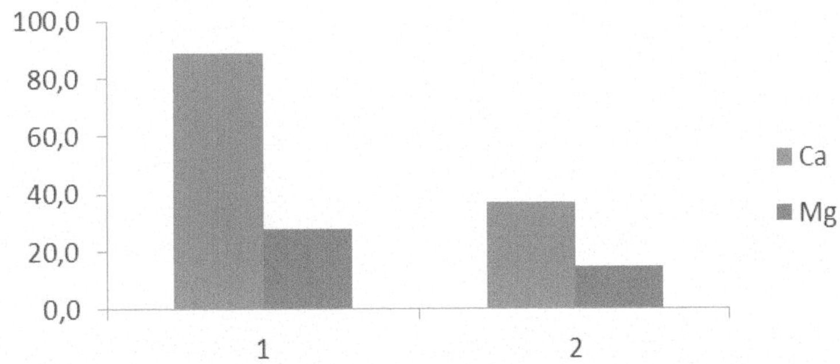

Figure 3: Ca and Mg concentrations in Uppsala drinking water before (1) and after (2) treatment

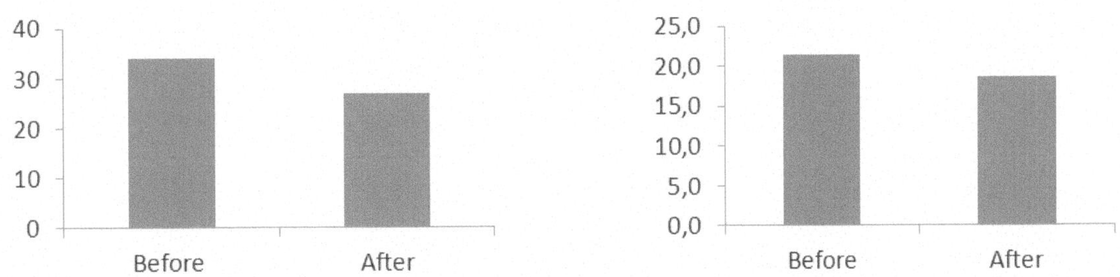

Figure 4: U concentrations before and after treatment at the Baeckloasa well (left) and Graenby well (right).

The treatment method was chosen in order to preserve the Mg concentration. Focus among many scientists has been on Mg from drinking water as the only element regarded as essential from drinking water. The RDI of Ca is 800 mg/day and for Mg 375 mg/day (www.slv.se), which gives a Ca/Mg ratio of about 2. The Ca/Mg ratio is decreased from 6-7 to about 2.5. The extra cellular Ca/Mg ratio is about 6-7 (Bowman & Russell, 2006). The question arises which ratio is most relevant in the case of those two mineral elements from drinking water. However, Ca acts antagonistic against U, since U is stored in bone tissues as U-phosphate. This antagonism inhibited by the decreased Ca level, while the U level still is higher than the provisional EU guide line. The U problem is not solved in Uppsala by taken measures. Maybe the most important element in this water is Ca, and should be preserved.

RO filter

Reverse osmosis is being used as a treatment method with increasingly frequency. The reason to that is that the method almost to 100% removes any contaminant aimed to remove. However, the treatment forms a completely de-mineralized drinking water, since all metals and ions are removed. One specific well water, among 47 acid well waters and 49 alkaline in a study from 2002, had RO filter installed in connection with the kitchen tap. The water analysis recorded the following data (Table 3). This water is almost completely de-mineralized. Hair element concentrations of the woman who had been drinking the water for more than 5 years were also extremely low compared to both mean acid and alkaline hairs. There is a big difference in elements and ions concentrations in acid compared to alkaline well waters, as alkaline had significantly much higher levels of a large number of elements and ions. However, RO treated water is even more de-mineralized.Women's hair was analyzed on about the same elements as water in the study from 2002, Table 4.

Table 3: Concentrations of metals and ions in a well water with RO treatment compared to mean concentrations in the other acid well waters in the study and alkaline well waters Rosborg 2002).

parameter	RO	mean acid	mean alkaline	unit	parameter	RO	mean acid	mean alkaline	unit
pH	5.47	5.91	7.7		Zn	0.026	0.012	0.153	mg/L
Na	3.51	7.48	24.8	mg/L	(NH_4-N)	<det.	0.008	0.13	mg/L
K	1.55	3.48	4.96	mg/L	(NO_3-N)	0.5	0.3	3.45	mg/L
Ca	1.66	10.9	57.1	mg/L	Cl	7.3	17.2	26.2	mg/L
Mg	0.9	2.1	2.32	mg/L	SO_4-S	2.3	4.1	15.8	mg/L
Fe	0.022	0.040	0.175	mg/L	F	<det.	0.36	0.039	mg/L
Mn	0.050	0.040	0.024	mg/L	HCO_3	1.6	14.2	169	mg/l
Cu	0.078	0.034	0.085	mg/L	As	0.07	0.2	1.0	µg/L
B	5.1	12.3	17.3	µg/L	Sr	15	50	165	µg/L
Ba	36	49	11.7	µg/L	Ti	1.6	12	42	µg/L
Co	0.1	0.06	0.2	µg/L	V	0.1	0.2	0.4	µg/L
Cr	0.05	0.15	3.6	µg/L	Pb	0.4	0.9	0.3	µg/L
Mo	0.1	0.1	3.5	µg/L	Cd	0.1	0.1	<det.	µg/L
Ni	0.1	1.0	2.3	µg/L	Hg	0.01	0.01	<det.	µg/L
Se	0.2	0.3	1.0	µg/L	Al	0.04	0.098	0.033	µg/L
Si	0.3	1.8	2.8	µg/L					

Table 4: Hair element concentrations of a woman drinking the RO treated water and the ratios "RO hair"/"alkaline hair" and "RO hair"/"acid hair".

	Al	Ca	Cu	K	Mg	Na	P	S	
	mikg/g	µg/g	µg/g	µg/g	µg/g	µg/g	µg/g	µg/g	
RO	2,72	126,00	80,92	37,83	41,95	81,91	205,49	45235,07	
acid/RO	2,29	**4,14**	1,22	5,18	1,16	4,37	0,68	0,89	
alk/RO	1,92	**12,91**	0,43	4,70	1,20	4,82	0,58	0,97	
	Zn	As	B	Ba	Cd	Co	Cr	Fe	Hg
	µg/g	ng/g	ng/g	ng/g	ng/g	ng/g	ng/g	ng/g	ng/g
RO	174,06	21,33	712,46	251,71	<det.	6,26	28,16	8056,03	674,06
acid/RO	0,95	0,82	0,77	5,94	<det.	7,44	7,15	1,28	0,81
alk/RO	0,97	1,38	0,75	2,06	<det.	4,97	11,08	4,42	0,56
	Mn	Mo	Ni	Pb	Rb	Se	Sr	Ti	V
	ng/g	ng/g	ng/g	ng/g	ng/g	ng/g	ng/g	ng/g	ng/g
RO	340,16	34,98	180,75	46,93	43,00	<det.	126,99	3962,32	29,00
acid/RO	3,29	0,57	2,99	**34,88**	7,06	<det.	**9,95**	0,42	0,54
alk/RO	4,45	1,05	2,72	**14,06**	4,73	<det.	**26,98**	0,38	0,92

The Ca concentrations of the woman drinking RO treated well water was 12 times lower than mean concentration among women drinking alkaline water and 4 times lower than women with acid well waters. This is a large difference, and health implications may be present. See further highlighted ratios.

Table 5: Suggested intervals of mineral elements and ions, and one ratio in drinking water.

parameter	interval	unit	parameter	interval	unit
pH	7.0 - 8.0		Zn	0.1 - 1.0	mg/L
Na	20 - 100	mg/L	(NH_4-N)	<0.1	mg/L
K	4.0 - 10.0	mg/L	(NO_3-N)	<1.0	mg/L
Ca	20 - 80	mg/L	Cl	20 - 100	mg/L
Mg	5.0 - 20	mg/L	SO_4-S	20 - 100	mg/L
Fe	0.05 - 0.15	mg/L	F	0.5 - 1.2	mg/L
Mn	0.02-0.10	mg/L	HCO_3	65 - 200	mg/l
Cu	0.01 - 0.10	mg/L	As	0.5 - 1.5	µg/L
B	50 - 500	µg/L	Sr	0.05 - 0.3	mg/L
Ba	<100	µg/L	Ti	< 50	µg/L
Co	0.2 - 1.2	µg/L	V	0.3 - 0.6	µg/L
Cr	2.0 - 10.0	µg/L	Pb	<0.01	µg/L
Mo	2.0 - 10.0	µg/L	Cd	<0.02	µg/L
Ni	1.0 - 5.0	µg/L	Hg	<0.001	µg/L
Se	0.5 - 5.0	µg/L	Al	<40	µg/L
Si	0.5 - 25	mg/L	Ca/Mg	2-7	

Conclusions

Calcium from drinking water as an important source to the daily intake is not as accepted as Mg. Treatment methods in order to decrease the hardness of water generally take into account that the method should restore the Mg level of the water. However, a more than 50% lower Ca concentration in the treated drinking water of Uppsala and Malmoe water will have an impact of the total daily intake, taken into account the high bioavailability of metals from water, since they almost to 100% appear as free ions in water, readily absorbed in the intestines as free ions or attached to organic molecules present in the intestinal fluid. Thus, treatment methods chosen in accordance with the general researcher's opinion of today, which in general regard only Mg from drinking water as important, may be considered too simplified in the future. RO osmosis water is almost completely de-mineralized. This may cause a de-mineralization of the whole body, as implicated by hair mineral analysis, after a few years of use, or even faster. Influence of RO treated or by any other treatment de-mineralized drinking water must indeed be studied in the near future.

Mineral composition of health bringing drinking water: The mineral composition of a health bringing drinking water is not completely identified, so far. A suggestion of intervals is presented below (Rosborg 2005). Future scientific studies on the subject will make it possible to revise the intervals and ratios.

Suggested future research

The most health bringing mineral element composition of a drinking water is not yet fully identified. In addition, there are probably different demands from different people on the mineral content of water. People having a low urine/body pH most probably need a more alkaline and mineral rich water than an alkaline individual, where minerals are not excreted through the urine to the same extent.

References

1. Arnault, E., Doussau, M., Pesty, A., Gouget, B., Van der Meeren, A., Fouchet, P., Lefevre, b. 2008. Natural uranium disturbs mouse folliculogenesis in vivo and oocyte meiosis in vitro. Toxicology, 247(2-3):80-87.
2. Bergmark, M. 1959. Bath and remedy. (Bad och bot, in Swedish). Natur och Kultur.
3. Bowman B, Russell R. 2006. Present knowledge in Nutrition, Ninth Edition, Volume 1. ILSI (Inetrnational Life Sciences Institute), pp 373-405.
4. Bussy, C., Lestaevel, P., Dhiex, B., Amourette, C., Paquet, F., Gourmelon, P., Houpert, P. 2006. Chronic ingestion of uranyl nitrate perturbs acetylcholinesterase activity and monoamine metabolism in male rat brain. NeuroToxicology, 27(2):245-252.
5. FitzPatrick, E.A.: 1980, Soils. Their formation, classification and distribution. Longman. London and New York.
6. Frassetto, L., Morris, R.C. Jr., Sellmeyer, D.E., Todd, K., Sebastian, A. 2001. Diet, evolution and aging - the pathophysiologic effects of the post-agricultural inversion of the potassium-to-sodium and base-to-chloride ratios in the human diet. Eur. J. Nutr. 40: 200-213.
7. Komulainen, H., Leino, A., Salonen, L., Auvinen, A., Saha, H. 2005. Bone as a Possible Target of chemical toxicity of Natural Uranium in Drinking Water. Environmental Perspectives, 113(1):68-72.
8. Kurttio, P., Auvinen, A., Salonen, L., Saha, H., Pekkanen, J., Makelainen, I., Vaisanen, S.B., Penttila, I.M., Komulainen, H.: 2002, Renal effects of uranium in drinking water. Environmental Health Perspectives 110, 337-342.
9. Raymond-Whish, S., Mayer, LP., O'Neil, T., Martinez, A., Sellers, MA., Christian, PJ., Marion, SL., Begay, C., Propper, CR., Hoyer, PB., Dyer, CA. 2007. Drinking Water with Uranium below the U.S. EPA Water Standard Causes Estrogen Receptor-Dependant Responses in Female Mice. Environmental Health Perspectives, 115(12):1711-1716.
10. Rosborg I, Nihlgard B, Gerhardsson L. 2003a Inorganic constituents of well water in one acid and one alkaline area of south Sweden. Water Air Soil Pollut 142, 261-277.
11. Rubenowitz, E., Axelsson, G., Rylander, R. 1998. Magnesium in drinking water and body magnesium status measured using an oral loading test. Scand J Clin Lab Invest, 58:423-428.
12. Rubenowitz, E., Axelsson, G. and Rylander, R. 1999. Magnesium and calcium in drinking water and death from acute myocardial infarction in women. Epidemiology 10:31-36.
13. Rubenowitz, E., Molin, I., Axelsson, G., Rylander, R. 2000. Magnesium in drinking water in relation to morbidity and mortality from acute myocardial infarction. Epidemiology 11:416-421.

14. Rylander, R., Bonevik, H. and Rubenowitz, E. 1991. Magnesium and calcium in drinking water and cardiovascular mortality. Scand. J. Work Environ. Health 17:91-94.
15. Sakamoto, N., Shimizu, M, Wakabayashi, I., Sakomoto, K. 1997. Relationship between mortality rate of stomach cancer and cerebrovascular disease and concentrations of magnesium and calcium in well water in Hyogo prefecture. Magnesium Research 10:215-223.
16. Scheffer, F.: 1989, Lehrbuch der Bodenkunde. Scheffer; Schachtschabel. -12., neu bearb. Aufl. Von P.Schachtschabel, H.-P. Blume, G. Brummer, K.-H. Hartge und U. Schwertmann. Ferdinand Enke Verlag, Stuttgart. 491pp.
17. Shroeder, HA. 1966. Municipal Drinking Water and Cardiovascular Death Rates. JAMA, 195(2):125-129.
18. Sullivan MF et al. 1986. Influence of oxidizing or reducing agents on gastrointestinal absorption of U, Pu, Am, Cm and Pm by rats. In : Uranium in drinking-water, http:/www.who.int/water sanitation health/. 2005. WHO/SDE/WHS0.30.4/118.
19. WHO. 2009. Calcium and Magnesium in drinking-water, Public health significance. WHO Library Cataloguing-in-Publication Data.
20. Wrenn, ME., Durbin, PW., Howard, B., Lipsztein, J., Rundo, J., Still, ET., Willis, DL. 1985. Metabolism of ingested U and Ra. Health Physics, 48:601-633.
21. www.slv.se. 2010-12-03
22. Yang, C-Y. 1998a. Calcium and magnesium in drinking water and risk of death from cerebrovascular disease. Stroke 29:411-414.
23. Yang, C-Y., Hung, C-F. 1998b. Colon cancer mortality and total hardness levels in Taiwan's drinking water. Arch. Environ. Contam. Toxicol. 35:148-151.
24. Yang, C.Y., Chiu, H.F., Cheng, M.F., Tsai, S.S., Hung, C.F., Tseng, Y.T. 1999a. Magnesium in drinking water and the risk of death from diabetes mellitus. Magnesium Research 12:131-137.
25. Yang, C-Y., Chiu, H-F., Tsai, S-S., Cheng, M-F., Lin, M-C., Sung, F-C. 2000a. Calcium and magnesium in drinking water and risk of death from prostate cancer. Journal of Toxicology and Environmental Health 60:17-26.

Evaluation of the monitoring activity performed for two Romanian companies which produce and supply drinking water

Irina Lucaciu[1], Liliana Cruceru[1], Cristiana Cosma[1], Margareta Nicolau[1], **Gabriela Vasile**[1], Jana Petre[1], Dumitru Staniloae[1], Lars John Hem[2], Goril Thorvaldsen[2] and Bjornar Eikebrokk[2]

[1] National Research and Development Institute for Industrial Ecology - INCD ECOIND, 90-92 Panduri Avenue, 050663 Bucharest-- 5, Romania
[2] STIFTELSEN SINTEF, Department of Water and Environment, Strindveien 4, 7034 Trondheim, Norway

Corresponding author e-mail: anandagabi@yahoo.com

In Romania, the water intended for human consumption is obtained from surface water resources (90-92%) and underground water resources. The treatment proceedings applied in Romania for potable water obtaining from natural sources are, generally, classic (pre-chlorination, coagulation/flocculation with aluminum sulphate, decantation/filtration on sand and disinfection with chlorine).

Under the EEA Financial Mechanism [1], a **project for monitoring of potable water quality** (from caching until production and distribution) supplied by two Romanian companies is developed in partnership, by INCD ECOIND (Romania) and STIFTELSEN SINTEF (Norway).

The overall goal of the project is to protect the population health from the adverse effects of any contamination of water intended for human consumption by ensuring that potable water produced and supplied by the Romanian Companies, fulfils the quality requirements imposed by the national and European regulations (EU Directive 98/83/EC), in force.

This paper presents the results of the complex monitoring program, developed during 12 months (October 2009 - September 2010) in 7 treatment plants from SC CUP Dunarea Braila and 5 treatment plants from SC ECOAQUA SA Calarasi. Sampling was performed every month, in the following locations: catchment (raw water), different phases of treatment process (treated water), distribution system and consumers (potable water). The total number of samples collected every month was 65 for Braila company and 52 for Calarasi company.

The control of water quality was performed in accordance with national and EU legislation and norms related to surface water and ground water used for potabilization **[2]** and also, for drinking water **[3]** and included all the imposed physical-chemical indicators (organic and inorganic) and also, microbiological indicators.

The overall assessment of the obtained results pointed out aspects which involve: improvement of the analytical control scheme, optimization /modernization of actual potabilization flow and remediation in the distribution system.

References:

1. G.O. no.24/2009 regarding financial administration of external founds related with Financial Mechanism SEE, Official Monitor of Romania, Part I, 601, 2009
2. G.O. no. 662/2005 - Norms for quality of surface water used for potabilization, Official Monitor of Romania, Part I, 616, 2005
3. Law 311/2004 (modified Law 458/2002) regarding drinking water quality, Official Monitor of Romania, Part I, 382, 2004

Drinking water quality monitoring systems in Poland

Jakub Bratkowski, Krzysztof Skotak, Janusz Swiatczak

National Institute of Public Health - National Institute of Hygiene - 00-791 Warsaw, 24 Chocimska str., Poland,

Corresponding author e-mail: jbratkowski@pzh.gov.pl

Quality of drinking water in Poland is under constant control, which covers resources of raw water (supervised by Ministry of Environment) used for public water supply, and water after treatment which is send to the consumers (supervised by Ministry of Health). Assessment of quality of drinking water is held by government and self-government institutions (including National Sanitary Inspection, National Inspection of Environment Protection) but also by drinking water suppliers. This presentation show, from legislation and administrative point of view, drinking water monitoring system in Poland held by Ministry of Health. This systems covers quality control in supply installations, and in case of any non-compliances found and possible threats, also in water intakes. Analysis are conducted in accredited laboratories of National Sanitary Inspection and also in approved by this institution private laboratories and these working for water suppliers.

In this presentation the authors present results from national monitoring of drinking water quality from recent years. The results includes also these from the scope of interests of COST Project and 637 Action - European Cooperation in the Field of Scientific and Technical Research (Al, As, Cd, Cr, Cu, Fe, Mn, Ni, Pb, Zn). Analysis were made according to the number of supplied people in administrative borders. The results are shown using charts and maps supported by GIS applications.

Section 4

Treatment Processes

Arsenic removal by traditional and innovative membrane technologies

A. Figoli[1], A. Criscuoli[1], J. Hoinkis[2], E. Drioli[1]

[1] *Institute of Membrane Technology, ITM-CNR, Via P. Bucci, 17/C 87030 Rende (Cs), Italy*
[2] *Karlsruhe University of Applied Sciences, Moltkestr. 30, 76133 Karlshruhe, Germany*

Corresponding author E-mail: a.figoli@itm.cnr.it

Abstract

Arsenic contamination of surface and groundwater is a worldwide problem in a large number of Countries (Bangladesh, Argentina, USA, Italy, New Zealand, etc.). In many contaminated areas a continuous investigation of the available arsenic removal technologies is essential to develop economical and effective methods for removing arsenic in order to comply with the new Maximum Concentration Level (MCL) standard (10 □g/l) recommended by the World Health Organization (WHO). Among the available technologies applicable for water treatment, membrane technology has been identified as a promising technology to remove arsenic from water. In this work, both traditional Pressure-driven processes, such as reverse osmosis (RO), nanofiltration (NF), ultrafiltration (UF), microfiltration (MF), and the innovative membrane process, membrane contactor, will be discussed in detail.

1. Introduction

Arsenic has become one of the major environmental problem in the world due to the exposition of millions of people to excessive arsenic through contaminated drinking water.

Arsenic contamination are from both natural and anthropogenic sources resulting in an increase of its concentration and distribution in the environment. Natural processes like erosion and weathering of crustal rocks lead to the breakdown and translocation of arsenic from the primary sulfide minerals, and the background concentrations of arsenic in soils are strongly related to the nature of parent rocks [1]. There is an extensive range of anthropogenic sources that may enhance concentration of arsenic in the environment. In fact, arsenic is used in industrial process, as in semiconductor, electronic, ceramic and glass industry, as well as in agriculture as wood preservative and component of insecticides and herbicides.

Among the two modes of arsenic input, the environment is mostly threatened by anthropogenic activities. Arsenic and its compounds are mobile in the environment.

Arsenic in drinking water affects human health and is considered one of the most significant environmental causes of cancer in the world [2]. When the toxic effects are considered, it is thus necessary to understand the level of arsenic in drinking water, and its chemical speciation, when establishing regulatory standards.

Consequently, in recent years, authorities have taken a more stringent attitude to arsenic in the environment and the new standard on the maximum contaminant level (MCL) of 10 □g L^{-1} arsenic in drinking water, recommended by the WHO [3], was accepted both within the European Union [4] and in the USA by the EPA (US Environmental Protection Agency) [5]. However, in some less development countries, such as Bangladesh, but also advanced ones as in Switzerland, the MCL of arsenic in drinking water is still 50 □g L^{-1}.

The purpose of this paper is to present an overview of the most established arsenic removal membrane technologies and reporting the new emerging membrane area of study (membrane contactor) which has been recently applied successfully to arsenic removal [6].

2. Species of Arsenic Present in Water

The chemistry of arsenic is a very extensive subject. Arsenic occurs in both inorganic and organic forms in natural water.

Inorganic arsenic is the result of dissolution from the respective mineral phase, such as arsenolite (As_2O_3), arsenic oxide (As_2O_5), orpiment (As_2S_3) or realgar (As_2S_4). In the natural environment; it may be present in two oxidation states, as

arsenate As(V) or arsenite As(III) depending on the governing pH and redox potential (Eh). In the predominant pH range in natural waters As(III) appears as neutral H_3AsO_3.

$$H_3AsO_3 \Leftrightarrow H^+ + H_2AsO_3^- \qquad pK_a = 9.22$$

$$H_2AsO_3^- \Leftrightarrow H^+ + H_2AsO_3^- \qquad pK_a = 12.3$$

where *Ka* is the dissociation constant. Pentavalent arsenic is thermodynamically stable and dominant in oxygenated waters, generally surface water, and exists as arsenic acid, which ionizes according to the following equations [7].

$$H_3AsO_4 \Leftrightarrow H^+ + H_2AsO_4^- \qquad pK_a = 2.2$$

$$H_2AsO_4^- \Leftrightarrow H^+ + HAsO_4^{2-} \qquad pK_a = 7.08$$

$$HAsO_4^{2-} \Leftrightarrow H^+ + AsO_4^{3-} \qquad pK_a = 11.5$$

3. Membrane Technology

In general the application of membrane techniques in the environmental protection involves a number of advantages, such as *a)* low energy required, *b)* easy to scale up, *c)* possibility of integrate membrane processes also with other unit process, *d)* separation carried out in mild conditions.

There are also some disadvantages, for example a decrease of capacity due to concentration polarisation and membrane fouling, which particularly concerns the processes of microfiltration and ultrafiltration. The limited lifetime of membranes and their low selectivity for a given separation problem may be regarded as disadvantageous. Membranes, in particular polymeric ones, are in many cases characterised by limited chemical or thermal resistance [8]. Furthermore, in physical membrane processes inorganic anions are not destroyed but normally concentrate and the concentrate disposal can be costly and difficult to be managed in many cases; therefore, post treatment of the concentrate stream or hybrid membrane-assisted technologies capable of converting anionic contaminants to harmless products are highly desirable [7].

3.1 Traditional Membrane Technology

The membrane technologies mainly employed in the arsenic removal are based on the use of different driving forces. Most commonly pressure driven processes such as reverse osmosis (RO), nanofiltration (NF), ultrafiltration (UF), microfiltration (MF) have been deeply investigated. The solution to be treated is usually passed across the filter membrane (cross-flow), where the pressure gradient forces the water (so called permeate) through the membrane, while basically being able to retain particulates down to solutes. In figure 1, it is reported the range of application of the different membrane driven processes [1].

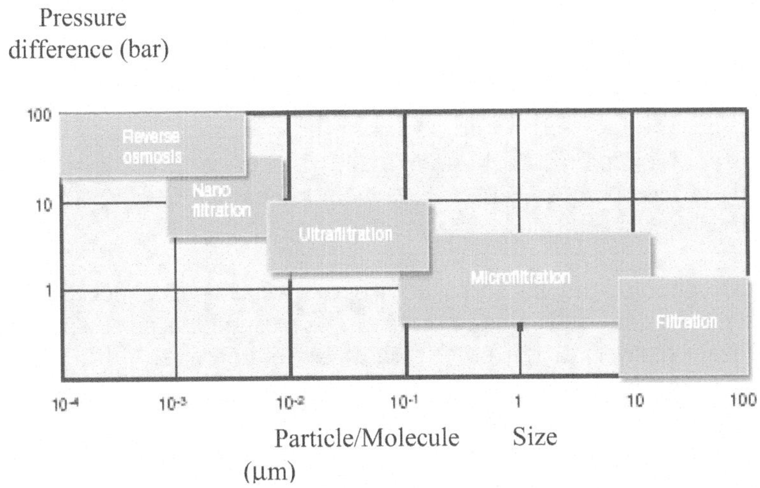

Figure 1. Overview of different pressure driven membrane processes.

In particular, reverse osmosis (RO) and nanofiltration (NF) are the most promising techniques since they remove dissolved arsenic along with other dissolved and particulate compounds. However, up to now, these membrane filtrations has needed bulky and sophisticated units with high use of energy. Only recently, a new generation of energy efficient techniques, so called low pressure RO, as well as new NF membranes, for brackish and tap water application, have been emerged on the market.

The latest results reported on the As removal by RO and NF are summarised in Table 1.

Table 1. Perfomance of RO and NF membranes for Arsenic removal.

Membrane and manufacturer	Water origin	Rejection (%) As(III)	As(V)	Flux (Kg $m^{-2}h^{-1}$)	References
		Reverse Osmosis			
LE, Dow Water Solutions	Arsenic spiked local	<80%		40***	[9]
XLE, Dow Water Solutions	tap water	>95%		60***	
XLE Dow Water Solutions	Arsenic spiked local	70-97%		28.3**	Geucke et al. (2009)
TW, Dow Water Solutions	tap water	96%		26.7***	[10]
SW, Dow Water Solutions		(>99%)		25.8****	
		Nanofiltration			
192-NF 300, Osmonics	Model water	-	93-99%	39.6***	[11]
	surface water	---	95%	37.1***	
NF-90 (Dow Chemical)	Tap water + As(III)	65%		51.6*	[12]
NF-200 (Dow Chemical)	and As(V)	98%		58.8*	
NF90 (Dow Chemical)	Synthetic water +	-		50*	[13]
N30F	As(V)		>91%	30-40*	

P=*7.5 bar , **4.5 bar, ***10 bar, ****15.2 bar.

3.2 Innovative Membrane Technology

In the last years, the membrane contactors (MC), considered as an emerging membrane technology, have been successfully applied to As removal. Up to now, few studies have been reported in this field but the increase of number of companies which use this technology and the specific application to As removal will strongly boost its development. This is also strictly related to the new standard on the maximum contaminant level (MCL) which should go even to lower value than 10 □m L^{-1} in the next years.

Membrane contactors are characterized by the fact that the membranes used are microporous hydrophobic or hydrophilic and do not take part in the separation process, simply acting as an inert barrier between phases, allowing their contact at each pore mouth while avoiding their mixing [14]. Using membrane contactors it is possible to carry out operations like liquid-liquid extractions, gas-liquid mass transfers, and membrane/osmotic distillation.

In particular, Direct Contact Membrane Distillation (DCMD) has been already used for treating water contaminated by arsenic. The microporous hydrophobic membranes are used for removing water vapor and volatile compounds from aqueous solutions.

Being the membrane hydrophobic, the liquid to be treated (warmer) cannot permeate through the pores and it is blocked at one side of the membrane in correspondence of the pores' mouths. By circulating the distillate stream (colder) at the other side of the membrane, a difference of temperature gradient is created across the membrane and both the water vapor and volatile species start to permeate through the membrane pores as shown in figure 2.

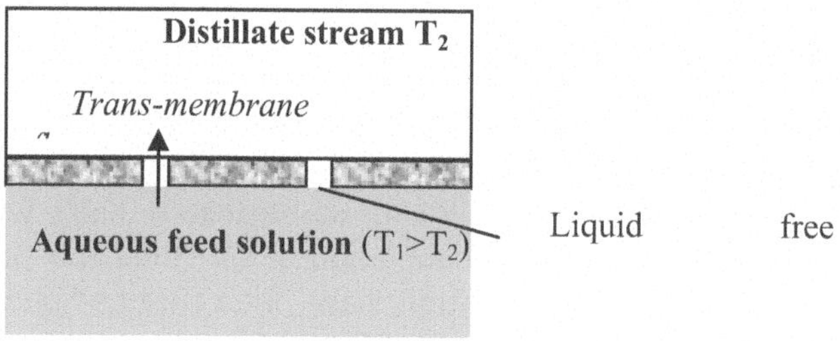

Figure 2. Permeation of water vapor and volatile species by DCMD.

The DCMD it is not limited by the osmotic pressure of the feed, while is strongly dependent on the temperature, because of its exponential relation with vapor pressure. Therefore, temperature polarization phenomena must be carefully controlled and minimized, for ensuring an efficient operation of the system. Temperature polarization consists into the creation of a temperature profile between the bulk of the phase and the membrane surface. The higher is the temperature polarization, the lower is the temperature at the membrane surface and, thus, the driving force available for the transport. The fact that the DCMD efficiency does not depend on the osmotic pressure of the feed is of big interest because there are not the limitations of the other membrane processes, such as reverse osmosis, which are not capable to treat high concentrated streams. Macedonio and Drioli [15] and Qu et al. [16], recently reported about the treatment of aqueous streams containing arsenic by DCMD and the most important result they achieved was that for all the investigated conditions, the rejection for both As(II) and As(V) was higher than 99.9%.

The possibility of obtaining a permeate practically free of arsenic even when As(III) was present in the feed solution is of extreme importance to avoid any pre-oxidation step to convert As(III) into As(V), with consequent benefits in terms of reduced environmental impact (no use of chemicals for the oxidation step) and the complexity of the overall system.

The interest in membrane contactor application has been also demonstrated by several companies which started to work on arsenic removal and recovery using this process. For example, in semiconductor field, where the redesign of the water purification and recovery step allows to produce ultra-pure water without presence of contaminants. Scarab, a Swedish company is actually working in this field [17]. In similar direction, membrane contactors have been applied by other Research Institutes as TNO, the Dutch Research Group, for seawater desalination and the Fraunhofer Institute Solar Energy-system which is mainly working on the water purification in remote area by membrane distillation using alternative energy sources such the solar energy.

Moreover, the interest of the use of this technology has been also pointed out by recent European projects (Innowa, Medina, Mediras) aimed at proving the potentialities of membrane contactors for water purification and recovery in many different fields.

4. Conclusions

Membrane technology have been applied more and more for drinking and water recovery in the last years, mainly due to the lowering of the maximum contaminant level of arsenic (from 0,05 mg L^{-1} to 0,01 mg L^{-1}) by the USEPA in 2006 and by the increase of water scarcity in many regions around the globe.

In Table 2, it is reported a summary of the results reported in different publications and "company" for arsenic removal by membrane technology.

Table 2. Comparison of the most used traditional membrane techniques with VMD for arsenic removal (revised from Figoli et al. [6]).

As type	UF	NF	RO	MC
As(III)	–*	–**	+/o**	++
As(V)	–*	+	++	++

++ very good, + good, o possibly effective, – not recommended
*viable option only with precipitation/coagulation as pre-step
**Pre-oxidation of As(III) to As(V) can achieve better performance
***Post-treatment neeed for mineral balance for drinking water

In particular, the removal efficiency for As(V) is reported to be remarkably higher than for As(III) by using traditional membrane technology such as RO and NF. Therefore, a pre-oxidation step is necessary for increasing the arsenic removal rate if the arsenic in the source water is primarily present as As(III).

It has been also reported that the use of membrane contactors has the main advantage to reject both As(III) and As(V) with the same efficiency (100%), Therefore, the development of MC units with higher trans-membrane fluxes and lower energy demand would also represent an interesting approach for arsenic removal and recovery from water.

References

[1] Figoli, A., Hoinkis, J., Bhattacharya, P., Drinking Water - Sources, Sanitation and Safeguarding, ISBN 978-91-540-6034-4, Sweden, 2009, Chapter 10, 68-91.

[2] National Research Council,Arsenic in Drinking Water, National Academy of Sciences, Washington, DC, USA, **2001**.

[3] World Health Organisation Guidelines for drinking-water quality, Addendum to Volume 1, Recommendations, Geneve, Switzerland, **1998**

[4] European Commissin Directive 98/83/EC, Brussels, related with drinking water quality intended for human consumption, Brussels, Belgium, **1998**.

[5] US Environmental Protection Agency. Panel 14: National Primary Drinking Water Regulations: Arsenic and Clarifications to Compliance and New Source Contaminants Monitoring, Washington DC, USA, **2001**, vol. 66, n. 194.

[6] Figoli, A., Criscuoli, A., Hoinkis, J., in Kabay, N., Bundschuh, J., Bhattacharya P., Bryjak M., Yoshizuka K., ISBN-13: 9780415575218, 2010, 131-145

[7] Kartinen, E. O., Martin, C. J., *Desal.*, 1995, 103, 79.

[8] Shih, M., *Desal.*, 2005, 172, 85.

[9] Deowan, A.S., Hoinkis, J., Pätzold, Ch., Low-energy reverse osmosis membranes for arsenic removal from groundwater. In: P. Battacharya, A.L. Ramanathan, J. Bundschuh, A.K. Keshari and D. Chandrasekharam (eds): *Groundwater for Sustainable Development -Problems, Perspectives and Challenges*. Balkema/Taylor & Francis, 2008, 275-386.

[10] Geucke, T., Deowan, S.A., Hoinkis, J., Pätzold, Ch., *Desal.*, 2009, 239, 198-206
[11] Saitúa, H., Campderrós, M., Cerutti, S., Padilla, A.P., *Desal.*, 2005, 172,173-180.
[12] Uddin, M.T., Mozumder, M.S.I., Islam, M.A., Deowan, S.A., Hoinkis, *J. Chem. Eng. Tech.*, 2007, 30, 1248-1254.
[13] Figoli, A., Cassano, A., Criscuoli, A., Mozumder, S.I., Uddin, T., Islam, A., Drioli, E., *Water Research*, 2010, 44, 97-104.
[14] Drioli E., Criscuoli, A., Curcio E., "Membrane contactors: fundamentals, applications and potentialities." ISBN: 0-444-52203-4, Elsevier, Amsterdam, 2006.
[15] Macedonio, F., Drioli, E., *Desal.*, 2008, 223, 396-409.
[16] Qu, D., Wang J., Hou, D., Luan, Z., Fan, B., Zhao, C., *J. Hazard. Mat.*, 2009, 163, 874-879.
[17] www.scarab.se/xzero/
[18] www.ise.fhg.de
[19] www.tno.nl

Treatment of arsenic containing drinking waters by electrochemical oxidation and reverse osmosis

Z. Lazarova[1], S. Sorlini[2] and D. Buchheit[1]

[1] Austrian Institute of Technology - AIT GmbH, Seibersdorf, Austria
[2] Department of Civil Engineering, Architecture, Land and Environment, University of Brescia, Italy

Corresponding author e-mail: zdravka.lazarova@ait.ac.at

Abstract

Removal of most distributed inorganic arsenic forms, As(III) and As(V), from drinking water was experimentally studied by using two methods - electrochemical oxidation and membrane separation. Transformation of As(III) into As(V) was performed by electrochemical oxidation. The effect of concentration parameters such as initial arsenic concentration (500 µg/L- 5000 µg/L), conductivity 700 µS/cm - 2000 µS/cm), and pH-value (5-10) on the transformation of the trivalent arsenic into pentavalent was investigated. Reverse osmosis by the new Dow Filmtec membrane BW30XFR was applied to separate the pentavalent arsenic from water. The hydrodynamic conditions (flow rate and pressure) were found at which arsenic concentration in the treated water lower than the limit value of 10 µg/L can be reached. The results showed that both methods can be successfully combined.

1. Introduction

The existence of arsenic in drinking waters represents one of the most serious acute problems of water pollution in many places around the world [1]. Different arsenic forms, organic and inorganic, have been found in the drinking water supplies. Many treatment technologies have been developed to remove the more toxic inorganic arsenic species dissolved in water [2-6]. Difficulties arise in their practical application, when the waters contain high levels of As(III). Generally, As(V) can be removed more efficiently than As(III) reaching the limit value of 10 µg/L (WHO). The reason is that at pH levels of 6-9 As(V) is more chemically active because it is a negatively charged ion. At the same conditions, As(III) is a fully protonated uncharged molecule. Therefore, for drinking water supplies containing significant concentrations of As(III), pre-oxidation of As(III) to As(V) is mandatory for high arsenic removal.

Our research study is aimed at development of innovative technology for removal of toxic inorganic arsenic compounds from drinking waters without any chemical additives. The main idea is to combine membrane separation with electrochemical oxidation for simultaneous transformation of As(III) to As(V), and its separation from the polluted drinking water by nanofiltration or reverse osmosis [3, 7].

2. Materials and Methods

2.1 Chemicals and Solutions

The experimental study was carried out using tap water with the following quality parameters: pH 7,20; conductivity 720 µS/cm; hardness 13.95 °dH. Its composition (cations and anions) is given in Table 1. As it can be seen, the tap water contained a lot of magnesium, calcium, potassium, as well as high concentrations of bicarbonate, sulphate, chloride, and nitrate. This water matrix was spiked with arsenic salts, pentoxide (As_2O_5) and sodium arsenite $Na_2HAsO_4.7H_2O$, to prepare aqueous solutions of trivalent and pentavalent arsenic, respectively.

2.2 Sampling and Analytical methods

All the water samples were acidified with phosphoric acid to pH<3 and stored in dark plastic flasks in refrigerator before analysis.

The pentavalent arsenic was analysed using an inductively coupled plasma mass spectrometry ICP-MS (Model ELAN 6100). For detection of the trivalent arsenic, a combination of HPLC system (LC 200, Perkin Elmer) with ICP-MS was applied. The ion chromatography column Hamilton PRP-X100 was used to separate both inorganic forms, As(III) and As(V). The limits of determination (LOD) and quantification (LOQ) were set at 10 µg As/L, and 20 µg As/L, respectively.

The conductivity of the water samples was measured using a conductometer (WTW-LF340), the water pH value - with a pH meter (VWR-pH100).

Table 1. Composition of the tap water used for preparing arsenic containing water solutions.

Cations	*Conc. mg/l*	Cations	*Conc. mg/l*	Anions	*Conc. mg/l*
Aluminium	*0,015*	Magnesium	*31,86*	Fluoride	*<0,1*
Calcium	*80,46*	Manganese	*<0,001*	Chloride	*35,6*
Cadmium	*<0,001*	Molybdenum	*0,0016*	Nitrate	*34,8*
Chromium	*0,0003*	Sodium	*14,21*	Sulfate	*101*
Copper	*0,0003*	Nickel	*0,0008*	$HCO3^-$	*280*
Iron	*0,0078*	Lead	*0,0010*		
Potassium	*1,276*	Zinc	*0,0251*		
Lithium	*0,0044*				

2.3 Experimental set-up

In Figure 1, the cross flow membrane separation system is shown schematically. It consists of a membrane cell for flat membranes (surface contact area 63,5 cm^2), high pressure pump, valves and pressure gauges, permeate meter, and vessel with a spiral cooling coil for the treated contaminated water. The flow rate range is between 100 L/h and 1000 L/h, the pressure at the outlet of the membrane cell - from 0 to 60 bar. Experiments were carried out in a closed loop in which permeate and retentate were continuously mixed in the feed vessel. The purpose was to keep the feed concentration practically constant and so as to simulate a continuous-time process.

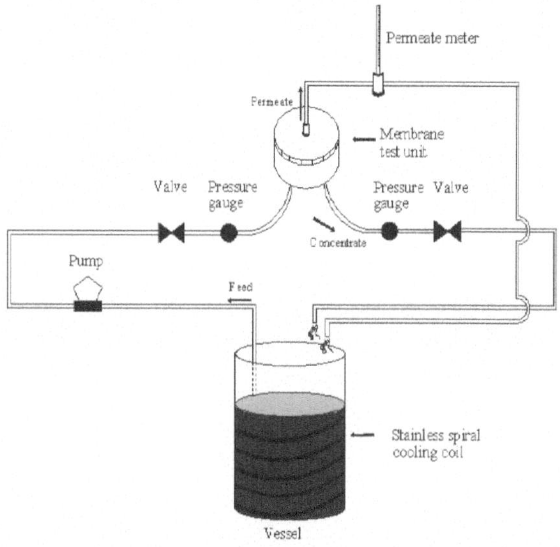

Figure 1. Experimental set-up for membrane separation

The membrane used in this study was a new composite Dow Filmtec membrane (BW30XFR) having a thin active film made of polyamide; the maximum allowable pressure recommended by the manufacturer was 41 bar.

A simplified diagram of the electrochemical plant used is presented in Figure 2. The electrochemical reactor consisted of 2 cells with electrodes made of mixed metal oxides based on silicium. The voltage applied was 10 V, the flow rate - 60 L/h (2x30 L/h). The volume of the treated water in the feed tank was 35 L. Kinetic experiments were performed in a close loop, and samples were periodically taken from the feed tank.

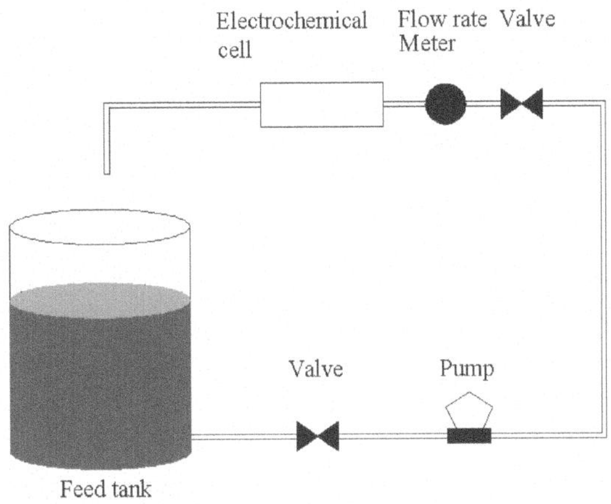

Figure 2. Experimental set-up for electrochemical oxidation

3. Results and Discussion

Two sets of experiments were performed as feasibility studies for future integration of both *reverse osmosis* and *electrochemical oxidation*. In the first study, the effect of As(III)-concentration, water conductivity, and pH level on the oxidation efficiency was investigated. The second study focused on the influence of transmembrane pressure difference and feed flow rate on the As(V)-retentation.

3.1 Electrochemical oxidation

The parameter study included the effect of

- initial arsenic concentration (500 µg/L - 5000 µg/L)
- conductivity (700 µS/cm - 2000 µS/cm)
- pH value (5-10)

In Table 1, results of kinetic studies at different trivalent arsenic initial concentrations, As°(III), are summarized. It can be seen that 35 L water volume containing up to 1000 As°(III) µg/L, with pH~8, and conductivity of 800 µg/L, is completely oxidized in less than 10 minutes. At higher As°(III) concentrations, longer oxidation time is needed to completely transform As(III) to As(V): at 2000 µg/L - more than 30 minutes, at 5000 µg/L - more than 2 hours.

Table 1. Effect of initial arsenic concentration (As°) concentration on the transformation of As(III) to As(V)

Experiment	As° (III) µg/L	Conductivity µS/cm	pH	Time min	As (III) µg/L	As (V) µg/L
				0	470	86
				2	300	160
1	500	800	8	5	180	300
				10	*DL	490
				30	DL	490
				60	DL	460
				0	1050	DL
				10	*DL	1000
2	1000	780	7,9	30	DL	1050
				60	DL	1050
				0	1900	34
				2	1600	150
3	2000	780	8,0	5	1500	320
				10	1300	490
				30	320	1500
				60	*DL	2000
				0	4700	170
				2	4600	210
				5	4000	320
4	5000	780	8,0	10	3700	610
				30	3200	1300
				60	1900	2200
				120	22	4200

*DL=lower than detection limit

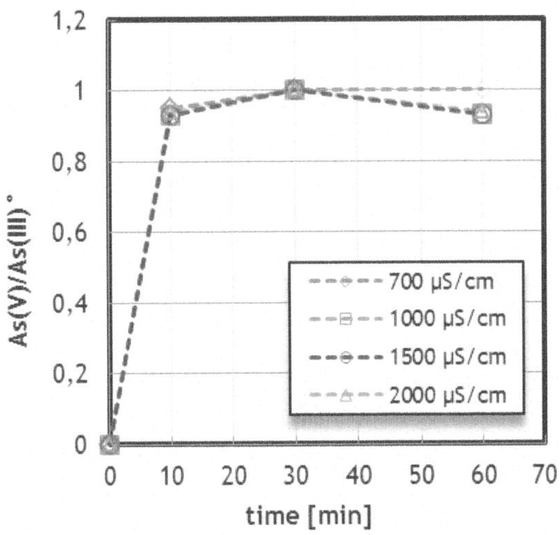

Figure 3. Effect of water conductivity on transformation of trivalent arsenic; As°(III) ~1000 µg/L; pH=7,9

In Figure 3, the effect of the water conductivity (from 700 µS/cm to 2000 µS/cm) on the time dependence of the dimensionless arsenic concentration, ratio of the pentavalent arsenic to the initial trivalent arsenic (As(V)/As°(III)), is shown. There is no visible influence of the water conductivity on the efficiency of the electrochemical oxidation process. This means that waters with low salt amounts (resp. low conductivities) could be successfully treated to transform As(III) to As(V).

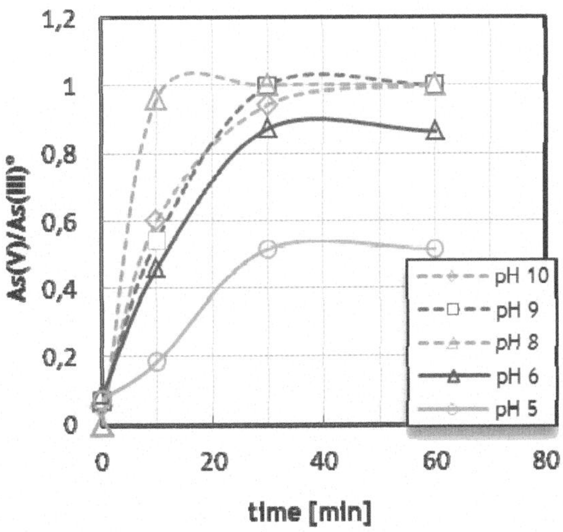

Figure 4. Effect of water pH on transformation of trivalent arsenic As°(III)~1000 µg/L; conductivity=800 µS/cm

Figure 4 illustrates how the pH value of the water solution containing As(III) influences the transformation kinetic and efficiency. It seems that the alkaline pH levels (8-10) facilitate the oxidation process.

3.2 Reverse osmosis

Reverse osmosis (RO) is a membrane separation process which needs the finest membranes and the highest pressures. In this study, water containing pentavalent or trivalent arsenic was treated using the membrane BW30XFR applying three pressures and four flow rates:

- pressure (at the outlet of the membrane cell): 20 bar, 30 bar, and 40 bar
- feed flow rate: 200 L/h to 500 L/h in 100 L/h steps (corresponds to linear velocities from 1,47 m/s to 3,71 m/s)

- temperature: 25°C
- initial arsenic concentration: 1000 µg/L

A linear dependence between permeate flux and pressure was found in all cases (Figure 5). The results showed that there is no influence of the flow rate on the permeate flux at fixed pressure.

The difference in the rejection of both As(III) and As(V) by the RO membrane can be seen in Table 2. At the lowest pressure of 20 bar, increasing the flow rate leads to higher rejection values. In all cases, there is lower rejection of As(III) in comparison to As(V) under the same conditions of pressure and flow rate.

In Figure. 6, the arsenic concentration in the permeates is shown as a function of time at three pressures: 20 bar, 30 bar, and 40 bar. It is proven that at the lowest pressure of 20 bar (see the first diagram) increasing the flow rate leads to higher arsenic rejection. However, it is not enough to reach the limit arsenic value of 10 µg/L (red horizontal line). At 30 bar (second diagram), only by applying the highest flow rate of 500 L/h is possible to decrease the arsenic concentration below the limit value. And at the highest possible pressure of 40 bar (last diagram), all the flow rates including the lowest one of 200 l/h can be used to purify successfully water polluted by the pentavalent arsenic (all tte permeate concentrations are lower than the limit value).

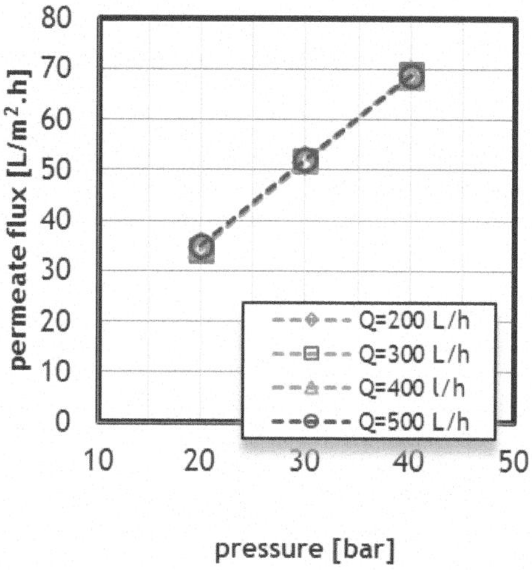

Figure 5. Effect of the pressure on the permeate flux at different flow rates

Table 2. Comparison of As(V) and As(III) rejection at different feed flow rates and pressures

Pressure bar	Feed flow rate L/h	Average As(III) Rejection %	Average As(V) Rejection %
	200	91,96	95,46
20	300	92,63	96,46
	400	93,11	97,40
	500	93,30	98,12
	200	93,14	98,73
30	300	90,00	98,38
	400	81,13	99,02
	200	94,32	99,36
40	300	94,90	99,27
	400	91,13	99,23
	500	-	99,17

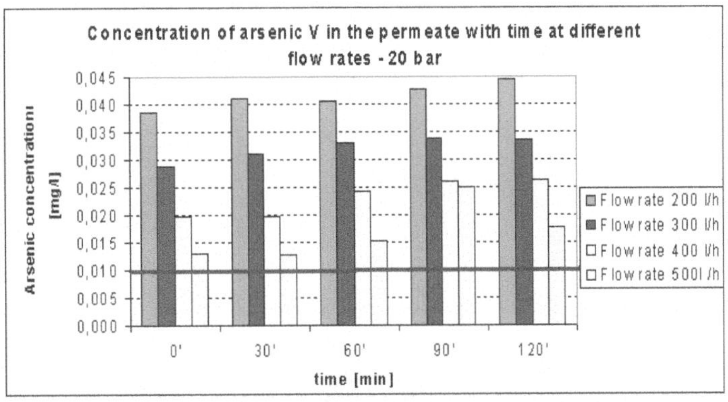

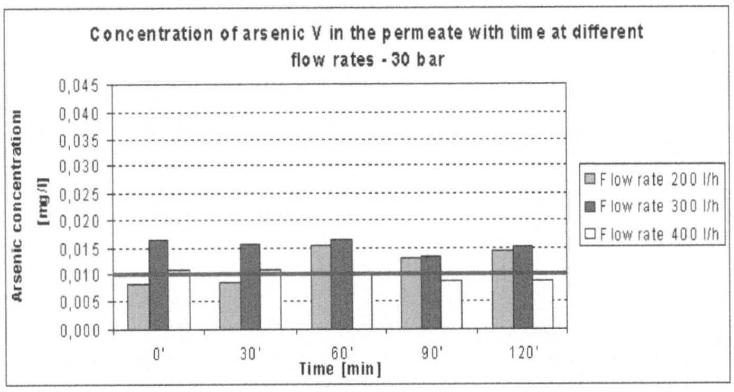

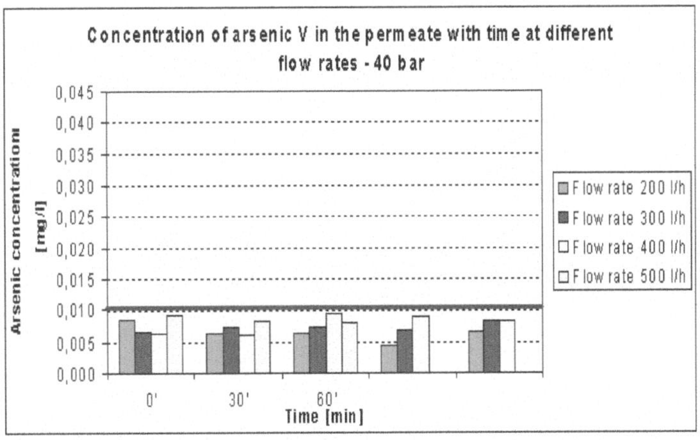

Figure 6. Arsenic (V) concentration in permeates after RO at different flow rates and pressures; As(V)°=1000 µg/L

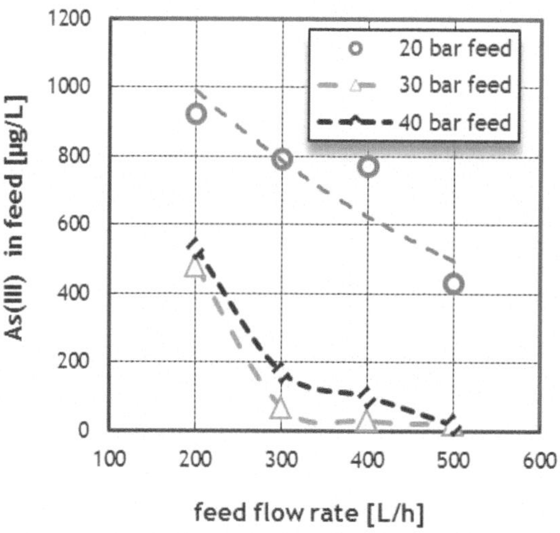

Figure 7. Transformation of As(III) into As(V) during the reverse osmosis:As(III)°=1000 µg/L; t= 1hour

In Figure 7, results are presented which prove that a large part of the trivalent arsenic in the feed solution was oxidized to pentavalent during the membrane separation process. The samples were taken after 1-hour recirculation of the feed water from the tank by the high pressure pump through the membrane cell back into the tank (feed water volume 5 L). The figure gives an answer of the question why do we receive relatively high rejection values also in the case of As(III). This is because a large part of As(III) in the feed tank was converted into As(V), especially at the high pressures and flow rates. The arsenic concentration in the corresponding permeate samples is shown in the Table 3.

Table 3. As(III)-concentration in the permeates after 1-hour recirculation (As°(III)=1000 µg/L)

Flow rate L/h	At 20 bar µg/L	At 40 bar µg/L
200	82	33
300	66	27
400	63	13
500	23	*DL

*DL=detection limit

4. Conclusions

Reverse osmosis

- Rejection of As(V) varies from 95.4% to 99,3% depending on the experimental conditions
- Rejection of As(III) varies from 81,1% to 94,9% depending on the experimental conditions
- Increasing the pressure leads to higher arsenic rejection: at 40 bar, the concentration of As(V) in all permeates was lower than the limit value of 10 µg/L (at As°=1000 µg/L)
- At high pressures (30-40 bar), oxidation of As(III) in the feed solution occurs during the membrane separation process
- At low pressure (20 bar), the flow rate influences the arsenic removal - higher flow rates are needed for higher retentation

Electrochemical oxidation for pre-oxidation of As(III)

- Very fast and efficient process: Complete oxidation of 35 L water containing up to 1000 µg/L As(III) is achieved in less than 10 minutes

- The increase of conductivity from 700 µS/cm to 2000 µS/cm doesn't influence the process; water with low conductivity can be successfully oxidized
- Alkaline pH values facilitate the oxidation process

References

[1] van Halem, D., Bakker, S., amy, G., and van Dijk, *Drink.Water Eng. Sci.*, **2009**, 2, 29-34.

[2] Lazarova Z., *Proceed. 1st intern. Conf. METEAU, Cost 637*, Antalya, 24-26, October, **2007.**

[3] Lazarova Z., *Proceed. 2nd Intern.Conf. METEAU, Cost637*, Lisbon, 29-31 October, **2008.**

[4] Selvin N., Upton J., Sims J., and Barnes, J., *Water Supply*, **2002**, *2(1)*, 11-16-

[5] Nquyen,T., Vugneswaran, S., Ngo, H., Pokhrel, D., and Viraraghavan, T., *Engineering in Life Science*, **2006**, 6(1), 86-90.

[6] Cakmakci, M., Baspinar, A., Balaban, U., Uyak, V., Kpyuncu, I., and Kinaci, C., *Desalination and Water Treatment*, **2009**, 9, 149-154.

[7] Uddin, M., Mozumder, M., M. A. Islam, M., Deowan, S., and Hoinkis, J., Chemical Engineering & *Technology*, **2007**, 30(9), 1248-1254.

The effect of fluidised bed softening on metal content in drinking water:11 years of experience from vombverket, sydvatten ab

B-M. Pott, S. Johnsson and K. M. Persson

Sydvatten AB, Skeppsgatan 19, SE 211 19 MALMÖ, Sweden

Corresponding author e-mail: britt-marie.pott@sydvatten.se

Abstract

Sydvatten is a municipally owned drinking water producer and wholesale drinking water supply company with a yearly production of approximately 70 M m^3 supplying 800 000 inhabitants in 14 municipalities in the south of Sweden with drinking water. Drinking water is produced at two waterworks with different raw water resources.

In 1997-1999, the waterworks Vombverket was supplemented by a softening plant through fluidized bed technology. Softening was regarded necessary in order to reduce copper corrosion in the houses, where the most common pipe material is copper, and to reduce the copper content in digestion sludge from the wastewater treatment plants of the municipalities. Softening has been in operation for several years with good results. The copper content in sewage sludge from wastewater treatment plants in Malmö has declined from 1300 mg/kg dry weight to less than 500 mg/kg dry weight. Low calcium content in tap water also decreases the use of shampoo, soap, detergent, dishwashing liquid, tea and coffee in the household with possible savings as a result. The introduction of central softening has decreased the content of magnesium, strontium, zinc and manganese in drinking water marginally. It has increased the sodium content with 25 mg/l due to addition of caustic soda for softening and pH-increase, and decreased the calcium content with 42 mg/l.

1. Introduction

Sydvatten is a municipally owned company that produces drinking water for 800 000 inhabitants of the Skåne region. The company was founded in 1966 and is at present one of Sweden´s largest producers of drinking water. The company supplies drinking water to the municipalities Höganäs, Helsingborg, Landskrona, Bjuv, Svalöv, Kävlinge, Eslöv, Lund, Lomma, Burlöv, Malmö, Staffanstorp, Vellinge and Svedala. Further on, the Skurup municipality has shares and an option to connect to the drinking water system if they want so in the future. The company produces drinking water from two raw water resources and has a third lake as reserve supply if anything happens with the two ordinary raw water supplies. The ordinary raw water sources are Lake Bolmen in Småland and Lake Vombsjön in Skåne. The reserve is Lake Ringsjön in Skåne. Sydvatten owns and operates the Bolmen Tunnel, a 80 km long tunnel leading raw water from the Lake Bolmen to Skåne. Sydvatten owns and operates two waterworks, Ringsjöverket and Vombverket, where the drinking water is produced and also owns the water mains for distribution of drinking water to the owner - municipalities. In total, the main network measures 300 km. Water from the waterworks is delivered to connection points in each municipality which distribute the water further on to the end consumer. Approximately 70 M m^3 per year of drinking water is produced, corresponding to about 2300 litres per second.

At the Vombverket waterworks, raw water is abstracted from the lake Vombsjön 30 km east of Lund and fed to a large glaciofluvial unconfined aquifer south of the lake to produce artificial groundwater. The catchment area of the lake is 450 km^2 and with a precipitation of 700-750 mm /a, the renewable water resource is at least 5 m^3/s. From the Swedish Geological Surveys description of the area, it is clear that the area south of Vombsjön mostly is composed of sand, pebbles and gravel. This area is rich in natural groundwater and very suited for artificial groundwater recharge. Since 1948, the area has been used for drinking water production for the cities of Malmö and later also Lund. Before the water from the lake is infiltrated in the aquifer it is sieved in four 500 µm micro-strainers to remove particles and reeds. A total number of 55 infiltration ponds cover a surface of 430 000 square meters. The water seeps slowly, with an average velocity of 0.4 m/d, through the alluvium of gravel and sand and recharges the aquifer forming groundwater. After a retention time of approximately three months in the aquifer, the groundwater is abstracted in 114 groundwater wells in the outskirts of the infiltration area. About 90% of the collected groundwater is recharged lake water, while 10% is naturally formed groundwater.

The water is pumped to two drip-aerators and two overflow aerators for aeration, carbon dioxide degassing and pre-oxidisation of iron and manganese present in the groundwater. Approximately 80% of the flow is led to in total eight pellets reactors for softening, each reactor with a volume of 78 m^3, while the remaining 20% is bypassed and mixed with softened water after the reactors. The pellet reactors consists of a cylindrical vessel partly filled with seeding material. The diameter of the seeding grain is small, mean grain size 0.56 mm, allowing rapid crystallization to take place since the surface is large. Water is pumped through the reactors in an upward direction at velocities between 60 and 100 m/h, maintaining the seeding material in a fluidised condition. At the bottom of each reactor, caustic soda is dosed. At the increased pH calcium carbonate becomes supersaturated and crystallizes on the seeding material, resulting in the formation of pellets. In the softening process step, total hardness of the mixed effluent, the supersaturation of the mixed effluent, the fluidised bed height, the discharged pellet diameter and the bed porosity must be monitored and managed. There are at least five operational parameters that can be changed, namely the water flow through the reactor, the water flow through the bypass, the caustic soda dosage the seeding sand dosage and the pellet discharge rate.

The softened water quality is directly controlled by the base dosage, but the state of the fluidised bed determines the performance of the reactor and therefore the necessary amount of base dosage. By controlling the fluidised bed height, discharged pellet diameter and bed porosity, these dosages can be minimised. The control of a softening reactor is therefore split into water quality control and fluidised bed control. A thorough theoretical overview of pellets reactor dynamics is presented by Rietveld [1] and Schagen et al. [2].

At regular intervals, pellets are removed through a set of valves in the bottom of each reactor. These pellets are re-used in industry and as limestone addition for white-washing lakes in south Sweden.

The water leaving the reactor is always supersaturated with respect to calcite formation. The softened water is mixed with the bypass water and led to a reactor where a small amount of ferric chloride is added for post coagulation of any remaining micro-crystals of limestone present in the water. The pH of the water is decreased by addition of sulphuric acid. The treated water is finally filtered in 26 rapid sand filters with a total filter surface of 720 m^2 and disinfected with addition of ammonium sulphate and sodium hypochlorite for secondary disinfection. In table 1, some data from the softening operation 2005-2009 are presented [3]

Table 1. Use of caustic soda and sand in the softening process

Parameter	2005	2006	2007	2008	2009
Produced drinking water (M m3)	24.4	28.7	27.6	28.9	29.5
Caustic soda (tonnes)	1260	1390	1287	1368	1386
Sand (tonnes)	229	266	248	275	280
Added caustic soda (mg/l)	51.6	48.4	46.6	47.3	47.0
Added sand (mg/l)	9.4	9.3	9.0	9.5	9.5

With eight parallel reactors installed, the reliability of the system and the flexibility in operation is guaranteed. Reactors can be switched on and off in case of flow changes, maintaining water production at an even level throughout the day and year. The waterworks Vombverket produces on average 900 l/s drinking water. Backwash water containing sludge from the 26 rapid sand filters is treated in a set of continuous Dyna-sand filters, thickened and then again filtered in two continuous Dyna-sand filters prior discharge of the water phase to a recipient. Limestone and iron sludge from the thickener is collected in a sedimentation pond that is emptied once a year. The sludge is used as a soil fertilizer and neutralizer.

The type of raw water determines the water quality and its variations. The variations in quality normally occur by seasonal variations in the source water (temperature of surface water, dilution during rain season, etc.). The temperature has a significant influence on the softening process. At low temperatures, the reaction rate is slow and crystallisation occurs higher in the reactor. Normal range of temperature at the Vomb waterworks is 7 to 14 degrees Centigrade and within this interval no practical effects of the temperature are noted with respect to the control of the softening process.

2. Water Quality Effects

Softening reduces the content of calcium in the drinking water. But the formation of pellets at a pH slightly above pH 9 also causes other metals to be co-precipitated with the limestone. In table 2, the mean value of different metals in drinking water from the waterworks Vombverket is presented for 1985-2009 [3].

The softening process reduces the hardness expressed as calcium content significantly (Table 2). Also the softening process reduces the content of copper, zinc, strontium and manganese in the drinking water, while of course the sodium content increases since caustic soda is added for the formation of carbonates, and caustic soda contains eqimolar concentrations of hydroxide and sodium. A simple mass balance shows that the increase in sodium corresponds to an addition of 1.15 mmol/l OH as caustic soda. From 2000 to 2009, the average caustic soda dose to the water has been slightly higher or 1.21 mmol/l. The difference is not significant, and the conclusion from change in content of sodium must be that all added sodium from the caustic soda is found dissolved in the drinking water. The reduction in magnesium is mainly due to a minor co-precipitaiton of magnesium carbonate with calcite in the softening process. The change from 6.2 mg/l to 5.7 mg/l magnesium ion is significant from an analytical chemistry point of view, but of course highly marginal.

The reduction of dissolved zinc, copper, strontium and manganese might be attributed to the low solubility product of these species in carbonate-rich water. In table 3, the solubility products of these elements as metal carbonates are presented [4].

Table 2. Metal content in produced water before and after introduction of the softening process

Species		1985-1998	1999-2009
Iron	µg/l	26	11
Manganese	µg/l	15	<10
Aluminium	µg/l	<2	<2
Arsenic	µg/l	<1	<1
Lead	µg/l	<0.5	<0.5
Cadmium	µg/l	<0.1	<0.1
Cobolt	µg/l	<1	<1
Copper	µg/l	2.6	1.3
Chromium total	µg/l	<0.5	<0.5
Mercury	µg/l	<0.1	<0.1
Nickel	µg/l	<1	<1
Silver	µg/l	<0.5	<0.5
Zinc	µg/l	1.8	<0.5
Selen	µg/l	<0,5	<0.5
Strontium	mg/l	0.22	0.15
Calcium	mg/l	76	34
Potassium	mg/l	3.2	3.0
Magnesium	mg/l	6.2	5.7
Sodium	mg/l	11	36

Table 3. Solubility products of some carbonates [4]

Solubility product, at 25°C.	Ksp $(mol/l)^2$
$Cu(II)CO_3$	1.4×10^{-10}
$MgCO_3$	6.82×10^{-6}
$MnCO_3$	2.24×10^{-11}
$SrCO_3$	5.6×10^{-10}
$ZnCO_3$	1.46×10^{-10}

The carbonate content in the pellets reactors is initially approximately 1,21 mmol/l and decreases with the formation of calcite. A carbonate content of 1.21 mmol/l gives a theoretical solubility of the metals according to table 4, indicating that none of the metals are present in the water in a supersaturated state.

Table 4. Concentration of metals at solubility equilibrium

Element	Theoretical equilibrium concentration (µg/l)
Cu	7.4
Mg	137000
Mn	10
Sr	41
Zn	7.9

Comparing table 4 and table 2, it can be seen that the metal ions are not present in a supersaturated state in the softened water, since the concentrations are at least one order of magnitude lower than the equilibrium concentrations. That a reduction still takes place could be explained through the fact that when caustic soda is dosed in the water, the local carbonate concentration will be several order of magnitudes higher prior full mixing in the reactor. Caustic soda is dosed as 35%, and gives on micro-level a pH higher than 11 in the reactor zone adjacent to the nozzles, forming a local carbonate concentration of at least 10 mmol/l. Some metal carbonate precipitation can occur on the pellets in this zone. Some metal carbonates may also occur statistically if calcite is formed rapidly, since other metals can be included in the crystals. This could be verified by controlling where in the pellets the other metals are present. In the reactors, the pellets are graded according to size, with the heaviest pellets closest to the bottom and the lightest fluidized at the top. A check of how the other metals are present in the pellets can prove which mechanism is dominant in the metal precipitation: if the other metals are enriched in the surface, the high local pH close to the heavy pellets at the bottom of the reactor is the main factor; if the other metals are evenly distributed in the pellets, the precipitation is mainly a co-precipiation with calcite. This study will however be conducted later and no data are present proofing the hypothesis of high local metal carbonate precipitation.

From a hygienic and health point of view, the reduction of dissolved metal ions in the drinking water will cause a minor reduction of metal content in the drinking water. In table 5, the total reduction of metal intake from drinking water due to softening is presented. The values corresponds to a consumer assuming a daily drinking water dose of 1.5 litre, and theoretically calculating the metal content to be 50% of reported less than concentration. This means that if the manganese content after softening is found to be <10 µg/l, then is assumed that the theoretical concentration is 5 µg/l. This is probably not true, since the content of for instance cadmium or mercury is less than half of the detection limit, but to have some figures to compare, this procedure is still done. As can be seen from table 5, the total change of metal intake is small, except for calcium and sodium. Recommended total daily intake of sodium in Sweden is 500 mg/d. The intake from drinking water is marginal and even after softening, the sodium content in drinking water from Sydvatten is low. The drinking water directive of EU (Directive 98/83/EC) states that sodium in drinking water should be below 200 mg/l. Swedish drinking water regulation states that the sodium content should be less than 100 mg/l (SLV FS 2001:30).

The softening process was built to produce at drinking water with less inherent copper corrosion effect on distribution pipes in the households. In figure 1, the copper content in digested sludge from Sjölunda wastewater treatment plant, the main WWTP in Malmö from the years 1976 to 2009 is presented [5, 6]. As can be seen in the figure, the content of copper in digested sludge decreased significantly after the introduction of the softening process.

In total, the softening process through pellet reactor softening changes the metal content marginally except for calcium and sodium. The softening has mainly led to a decrease in copper corrosion, reducing the daily copper exposure considerably. No negative changes in metal content can be observed.

Table 5. Daily exposure of metals from drinking water [3]

Species	Daily exposure (mg/d) Pre softening	Post softening	Change
Iron	0.039	0.0165	-0.0225
Manganese	0.0225	0.0075	-0.015
Aluminium	0.0015	0.0015	0
Arsenic	0.00075	0.00075	0
Lead	0.000375	0.000375	0
Cadmium	0.000075	0.000075	0
Cobolt	0.00075	0.00075	0
Copper	0.00390	0.00195	- 0.00195
Chromium total	0.000375	0.000375	0
Species	**Daily exposure (mg/d) Pre softening**	**Post softening**	**Change**
Mercury	0.000075	0.000075	0
Nickel	0.00075	0.00075	0
Silver	0.000375	0.000375	0
Zinc	0.0027	0.000375	- 0.00233
Selen	0.000375	0.000375	0
Strontium	0.33	0.225	-0.105
Calcium	114	51	-63
Potassium	4.8	4.5	-0.3
Magnesium	9.3	8.55	-0.75
Sodium	16.5	54	37.5

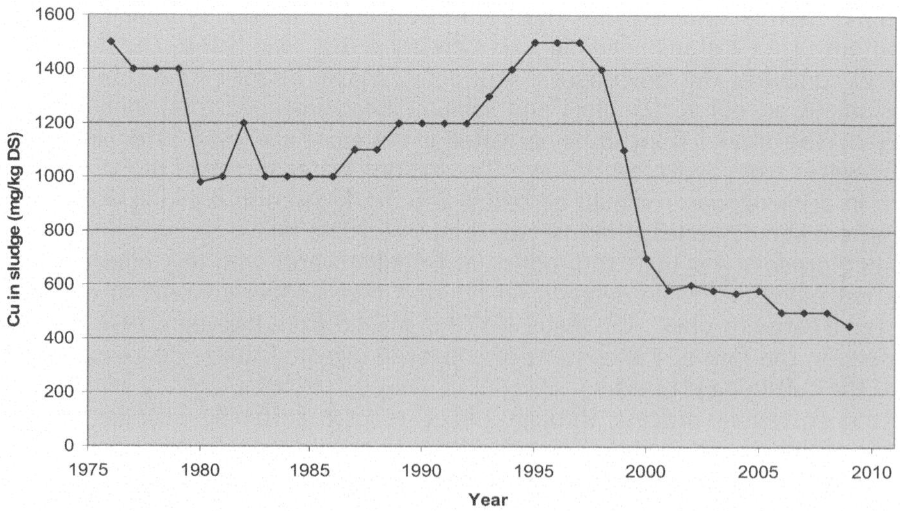

Figure 1. Copper content in digested sludge from Sjölunda WWTP in Malmö, 1975-2009. Data from environmental reports of the plant 2007 and 2009 [5, 6].

References

Rietveld, L., 2005. Improving operation of drinking water treatment through modeling. Ph.D. Thesis, Faculty of Civil Engineering and Geosciences, Delft University of Technology.
van Schagen, K.M., Rietveld, L.C., Babuska, R, Kramer O.J.I., 2008. Model-based operational constraints for fluidised bed crystallization. Water Research 42 (1-2) 327 - 337
Sydvatten: Production reports for Vombverket water treatment plant 1980 to 2009. Mainly unpublished data. Production report from 2008 available in Swedish at http://www.sydvatten.se/filearchive/3/3651/Produktionsrapport%202008.pdf Downloaded 2010-09-24
Handbook of Chemistry and Physics, 91st Ed. (Internet Version 2011), CRC Press/Taylor and Francis, Boca Raton, FL, USA. Table Solubility Products 8.127-8.129
VA-SYD. Environmental Reports for Sjölunda Wastewater Treatment Plant 2007 (in Swedish) http://www.vasyd.se/SiteCollectionDocuments/Vatten%20och%20avlopp/Avloppsvatten/Miljörapporter/Sjölunda_Miljörapport_2007.pdf Downloaded 2010-09-27
VA-SYD. Environmental Reports for Sjölunda Wastewater Treatment Plant 2009 (in Swedish) http://www.vasyd.se/SiteCollectionDocuments/Vatten%20och%20avlopp/Avloppsvatten/Miljörapporter/Sjölunda_Miljörapport_2009.pdf Downloaded 2010-09-27

Arsenic removal with chemical precipitation in drinking water treatment plants in Italy

S. Sorlini , F. Prandini and C. Collivignarellia
Department of Civil Engineering, Architecture, Land and Environment, University of Brescia, Brescia, Italy
Corresponding author e-mail: sabrina.sorlini@ing.unibs.it

Abstract
In Italy, arsenic is diffused in the groundwater in several areas with concentration also higher than 300 µg/L. A research performed in the University of Brescia investigated the main operations of upgrading employed by several drinking water treatment plants in Italy. The aim of this work was to elaborate an handbook about the arsenic diffusion in drinking water in Italy, the arsenic removal technologies applied in this country and the technical and operational aspects tied to these technologies. Sixteen drinking water treatment plants were analyzed in detail. Ten of them applied chemical precipitation process, one of the most common technology for arsenic removal in Italy. Treatment plants with arsenic removal by precipitation are generally composed of: aeration; biological filtration and/or pre-oxidation with chemicals; addition of chemicals for arsenic precipitation; sand filtration; possible adsorption with Granular Ferric Hydroxide (GFH) for arsenic removal and final disinfection. This paper reports information about the technical and operational aspects with focus on the process scheme, chemical dosage, residues management and costs.

1. Introduction

Arsenic is widely recognized as a dangerous contaminant and as a threat to some of the world's water resources.

Arsenic is an ubiquitous element found in the atmosphere, soils and rocks, natural waters and organisms. It is mobilized through a combination of natural processes such as weathering reactions, biological activity and volcanic emissions as well as through a range of anthropogenic activities. Most environmental arsenic problems are the result of mobilization under natural conditions. Among the various sources of As in the environment, drinking water probably poses the greatest threat to human health.

Well-known high-As groundwater areas have been found in many parts of the world, as Argentina, Bangladesh, Chile, China, Hungary, India (West Bengal), etc. [1].

Regulatory and recommended limits for arsenic in drinking water have been reduced in recent years following increased evidence of its toxic effects to humans. The World Health Organization (WHO) guideline value was reduced from 50 µg/L to 10 µg/L in 1993 although the recommendation is still provisional pending further scientific evidence [2].

Also in many parts of Italy groundwater contains arsenic concentrations higher than the national regulatory standard of 10 parts per billion (2001/31 Legislative Decree). Main arsenic affected groundwaters are located in several regions: Lombardia, Piemonte, Veneto, Trentino Alto Adige, Emilia Romagna, Toscana, Umbria, Lazio and Campania where concentration reaches values up to 500 µg/L.

A variety of treatment processes has been developed for arsenic removal from water, including precipitation, adsorption, ion exchange, membrane filtration, electrocoagulation, biological process. Precipitation/coprecipitation has been the most frequently used method to treat arsenic contaminated water [3].

Chemical precipitation process is traditionally realized by adding ferric or aluminum ions [4]. In this process, fine particles in water first aggregate into coagulates because added ferric or aluminum ions strongly reduce the absolute values of zeta potentials of the particles.

Then, arsenic ions (arsenate or arsenite) precipitate with the ferric or aluminum ions on the coagulates, and thus concentrate in the coagulates. After that, the coagulates are separated from water through filtration, eliminating arsenic from the water.

Alum and ferric salts dissolve upon addition to water, forming amorphous hydrous aluminum and ferric oxides (HAO and HFO, respectively), which are relatively insoluble in circumneutral pH ranges [5].

During coagulation and filtration, arsenic is removed through three main mechanisms [6]:

- precipitation: the formation of the insoluble compounds $Al(AsO_4)$ or $Fe(AsO_4)$;
- coprecipitation: the incorporation of soluble arsenic species into a growing metal hydroxide phase;
- adsorption: the electrostatic binding of soluble arsenic to the external surfaces of the insoluble metal hydroxide.

All these mechanisms can independently contribute towards contaminant removal. In the case of arsenic removal, direct precipitation has not been shown to play an important role, but high yields are due to coprecipitation and adsorption [5].

The coagulation is much more effective for the removal of As (V) than As (III). When only As (III) is present, oxidation of arsenite to arsenate is required. Arsenic removal from water achieved by coagulation process depends on initial arsenic concentration in water [7, 8]. The arsenic removal could reach 90% [5].

Generally, treatment plants with arsenic removal by precipitation are composed of several phases:

- aeration for water oxygenation and elimination of H_2S and CH_4;
- biological filtration and/or pre-oxidation with chemicals for the arsenic oxidation;
- addition of Fe or Al salts for arsenic precipitation;
- sand filtration for precipitates removal;
- possible addition of oxidants and iron salts for precipitation of As residual;
- possible adsorption for As residual removal;
- final disinfection.

Due to the importance of this problem in Italy, a working group called "Water for human consumption: arsenic removal" was activated in 2005 involving different subjects at national level: researchers, water technology companies, drinking water treatment plant managers and surveillance agencies.

The aim of this work was to elaborate a guideline for the choice of the best technology for arsenic removal and for the optimization of its operation for water treatment and residues management.

One activity of this working group was to perform an investigation about the main technologies produced or applied in Italy for arsenic removal and about the main features in the management of real treatment plants. The results of this investigation are presented in this paper with a specific focus on the water treatment plant with chemical precipitation of arsenic.

2. Investigation

The activity has regarded an investigation about the main technologies used in Italy in real scale treatment plants for arsenic removal with a study of technical and operational problems in drinking water treatment plants.

The specific aspects analysed in this survey are:

- general aspects: water characteristics, arsenic ionic form in raw water, adopted technology, arsenic removal yield, etc.;
- technical aspects: treatment plant description, chemicals dosage, hydraulic load, retention time, etc.;
- operational aspects: filter backwashing, media regeneration, etc.;
- residues treatment: residues characteristics, technical solution for their treatment and disposal, etc.;
- costs of technologies.

The investigation involved ten companies managing 43 drinking water treatment plants. The general information about these plants are shown in Table 1.

As shown in Table 1, different technologies are applied in Italian drinking water treatment plant in order to mitigate the problem of arsenic:

- chemical precipitation;
- adsorption;
- ione exchange;
- membrane filtration.

All these plants treat deepwater polluted by As, with concentration between 14 and 65 µg/L and a flow rate between 0.3 and 450 L/s.

For arsenic removal, the most common technology applied in Italy is the chemical precipitation: 34 drinking water treatment plants apply the chemical precipitation and two of them are completed with the adsorption with Granular Ferric Hydroxide (GFH) as tertiary treatment. The flow rate is variable from 2 to 450 L/s and arsenic concentration from 20 to 65 µg/L.

Table 1. Information about the drinking water treatment plant managers involved in the survey.

Company	Technology	Number of plants*	Flow rate (L/s)	As concentration (μg/L)
1	Chemical precipitation	2	450	24
2		1	13	20
3		3 (5)	60 9 13	38 50 22
4		1 (23)	28	47
5		1	20	20
6	Chemical precipitation + Adsorption	2	15 2	65 44
7	Adsorption	2 (5)	2 7	40 14
8		2	0.6 0.3	32 48
9	Ion exchange	1	2	30
10	Reverse osmosis	1	34	50

* The number in the brackets indicates the total plants managed by each company

3. Results

Among the plants with chemical precipitation (the total number is 34), ten plants were deeply analysed in this survey. Nine plants are located in the North and one in the Central part of Italy.

Table 2 shows the process schemes used in drinking water treatment plants involved in this investigation.

Table 2. Process schemes adopted in drinking water treatment plants analyzed in the survey.

Plant	Other contaminants	S	CP	A	POX	BF	SF	OX	CP	SF	ADS	DIS
1-2	NH_3, Fe, Mn, CH_4, H_2S	-	-	X	-	X	-	X	X	X	-	X
3	NH_3, Fe, Mn	-	-	-	X	-	X	X	X	-	GAC	X
4	NH_3, Fe, Mn, H_2S	-	X	X	-	X	-	-	X	X	-	X
5	NH_3, Fe, Mn, H_2S	X	X	X	-	X	-	-	X	X	-	X
6	NH_3, Fe, Mn, H_2S	-	-	X	-	-	-	-	X	X	-	X
7	NH_3, Fe, Mn, CH_4, H_2S	X	X	X	-	X	-	X	X	X	-	X
8	NH_3, Fe, Mn	-	-	X	-	X	-	-	-	-	-	X
9	NH_3, Mn	-	-	X	-	X	-	X	X	X	GFH	X
10	NH_3, Fe, Mn	-	-	X	-	X	-	-	X	X	GFH	X

S = Stripping; CP = Chemical Precipitation; A = Aeration; POX = Pre-Oxidation; BF = Biological Filtration; SF = Sand Filtration; OX = Oxidation; ADS = Adsorption; DIS = Disinfection; GAC = Granular Activated Carbon; GFH = Granular Ferric Hydroxide

It is worth to notice that aeration is applied in most of these plants (9 out of 10) for iron oxidation and/or water aeration before ammonia biological nitrification. In 8 out of 10 plants there is a biological filtration, generally applied for ammonia removal and, obviously, all these plants apply the chemical precipitation (with the exception of plant number 8, where high concentrations of the iron in raw water, about 5.0 mg/L, allow to avoid dosage of chemicals for arsenic precipitation) followed by sand filtration.

3.1 Process schemes

Process schemes adopted in drinking water treatment plants with arsenic removal can be divided into 4 typologies, as described below.

Type A.

This scheme is composed by a biological oxidation followed by a chemical precipitation and sand filtration (Figure 1). This solution is adopted in plants 4, 5, 6 and 8.

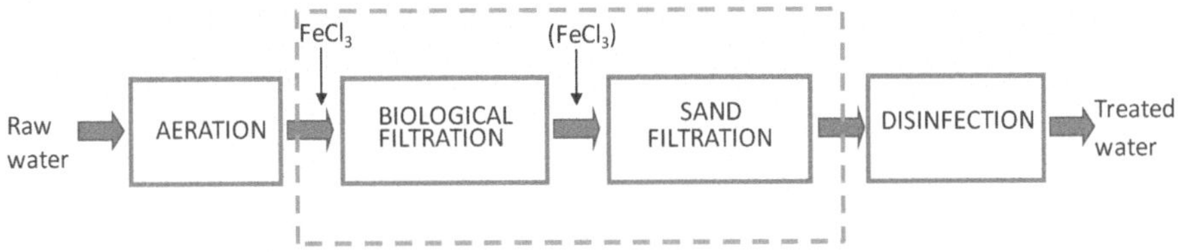

Figure 1. Process scheme A.

The plants are generally composed of several phases: first, there is an aeration for water oxygenation, iron oxidation and H_2S and CH_4 removal; then, there is the addition of $FeCl_3$ for arsenic precipitation; then, there is a biological filtration for arsenic oxidation (which precipitates on the same filter with ferric salts), NH_3 nitrification and Mn oxidation; then, a second step of chemical precipitation can be added, followed by sand filtration for insoluble compounds removal; at the end there is a final disinfection for microorganisms removal.

Type B.

This scheme is composed by biological oxidation followed chemical oxidation and chemical precipitation (Figure 2). This solution is adopted in plants 1, 2 and 7.

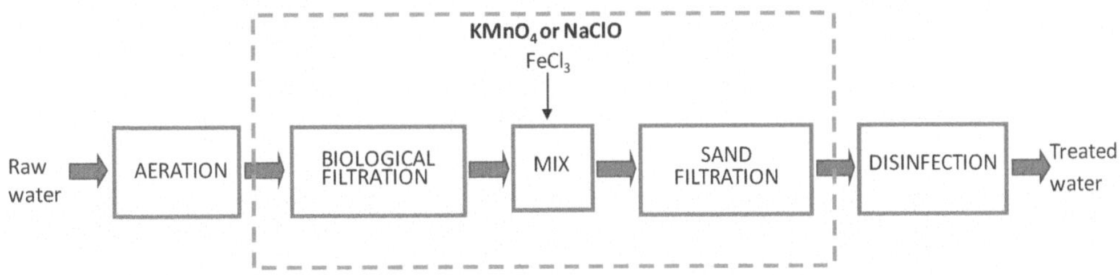

Figure 2. Process scheme B.

With respect to type A this new scheme applies a chemical oxidation after the biological filtration in order to guarantee a complete arsenic oxidation.

Type C.

This scheme is composed by chemical oxidation followed by chemical precipitation (plant 3 see Figure 3).

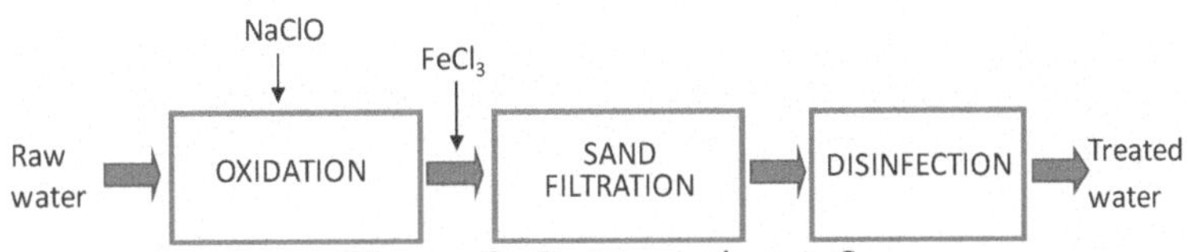

Figure 3. Process scheme C.

The only plant applying this kind of scheme (plant 3), treats water with about 20 µg/L of As, in addition to iron, manganese and ammonia. The treatment plant is composed of chemical pre-oxidation with NaClO, chemical precipitation with ferric chloride and sand filtration, and final disinfection. Unlike the previous plants, this scheme doesn't apply biological filtration but only chemical processes.

Type D.

The last type is similar to type A followed by adsorption on GFH material. The diagram is shown in Figure 4.

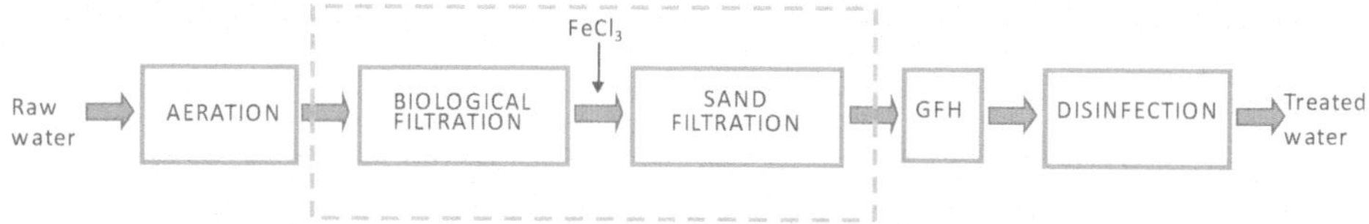

Figure 4. Process scheme D.

The treatment plants adopting this technology are plants 9 and 10 where arsenic concentration in raw water is higher than 40 µg/L: a refining step with an adsorption media is needed in order to comply with the Italian limit.

3.2. Reagent for arsenic chemical precipitation

Table 3 shows the main parameters involved in the chemical precipitation process.

Table 3. Main parameters in the precipitation process analyzed in the survey.

Plant	Q (L/s)	As_{IN} (µg/L)	As_{OUT} (µg/L)	Other contaminants	Dosage ($mgFeCl_3/L$)	As removal yield (%)	mgFe/mgAs removed
1-2	450	24	3	NH_3, Fe, Mn, CH_4, H_2S	6.7	82	109
3	13	20	8	NH_3, Fe, Mn	3.0	60	86
4	59.7	38.2	8.3	NH_3, Fe, Mn, H_2S	3.2-4.3	78	37-50
5	9.4	50.1	1.9	NH_3, Fe, Mn, H_2S	3.2-7.2	96	23-51
6	13.3	22.4	7.6	NH_3, Fe, Mn, H_2S	3.2-4.3	66	74-100
7	27.8	47	4-8	NH_3, Fe, Mn, CH_4, H_2S	4.3-6.6	83-91	38-58
8	20	20	3	NH_3, Fe, Mn	-	85	286
9	2	44	2.9 (6.2)^	NH_3, Mn,	3-3.8	93 (86)*	25-32
10	15	65	5 (15)^	NH_3, Fe, Mn	3-3.8	92 (77)*	17-22

^ concentration before GFH; * arsenic removal yield before GFH

The comparison among the plants shows that:

- ferric chloride is the only reagent used for arsenic precipitation in all the drinking water treatment plants analyzed;
- the adopted dosage is between 3 and 7.2 mg/L;
- the contact time adopting in chemical precipitation step is variable from 3 to 4.5 min;
- the iron specific dosage is 17-109 mgFe/mg $As_{removed}$ (average dose: 60 mgFe/mg $As_{removed}$);
- the chemical precipitation allows to reach As concentration in treated water of 3-15 µg/L and in case of tertiary treatment with GFH a final arsenic concentration of 2.9 and 5 µg/L;
- the arsenic removal yield is variable from 60 to 96%.

3.3 Residues management

Two general types of residues are potentially generated from the chemical precipitation: liquid and solid wastes.

The main residue is the water produced from filter backwashing whose volume (V_b) changes from 3 to 24% of the total volume of water treated in the plant (V_t). Generally, this operation is performed with a frequency of 3 times/week-3 times/day for biological and sand filters and from once a week to once a month for GFH.

In Table 4 the operational aspects of filters backwashing are reported.

Table 4. Operational aspects of filters backwashing.

<table>
<tr><th>Plant</th><th>Type of filter</th><th>Frequency</th><th>Phase of filter backwashing</th><th>Duration (min)</th><th>V_b/V_t (%)</th><th>V_b/V_t (%)</th></tr>
<tr><td rowspan="2">1-2</td><td>BF</td><td>once/48 h</td><td>Air+water</td><td>42</td><td rowspan="2">10</td><td rowspan="2">14</td></tr>
<tr><td>SF</td><td>once/20 h</td><td>Air+water</td><td>30</td></tr>
<tr><td rowspan="2">3</td><td>SF</td><td>twice/d</td><td>Water</td><td>10</td><td>5.6</td><td rowspan="2">11</td></tr>
<tr><td>GAC</td><td>twice/d</td><td>Water</td><td>10</td><td>5.6</td></tr>
<tr><td rowspan="2">4</td><td>BF</td><td>once/d</td><td>Air+water</td><td>44</td><td rowspan="2">5.3</td><td rowspan="2">5</td></tr>
<tr><td>SF</td><td>3 times/w</td><td>Air+water</td><td>52</td></tr>
<tr><td rowspan="2">5</td><td>BF</td><td>once/d</td><td>Air+water</td><td>25</td><td rowspan="2">16.2</td><td rowspan="2">16</td></tr>
<tr><td>SF</td><td>3 times/w</td><td>Air+water</td><td>25</td></tr>
<tr><td>6</td><td>BF</td><td>once/w</td><td>Air+water</td><td>40</td><td>3.0</td><td>3</td></tr>
<tr><td rowspan="2">7</td><td>BF</td><td>once/d</td><td>Air+water</td><td>20</td><td>10.0</td><td rowspan="2">21</td></tr>
<tr><td>SF</td><td>3 times/w</td><td>Water</td><td>20</td><td>11.0</td></tr>
<tr><td>8</td><td>BF</td><td>once/48 h</td><td>Water</td><td>20</td><td>7.8</td><td>8</td></tr>
<tr><td rowspan="3">9</td><td>BF</td><td>3 times/w</td><td>Air+water</td><td>25</td><td rowspan="3">24</td><td rowspan="3">24</td></tr>
<tr><td>SF</td><td>3 times/w</td><td>Air+water</td><td>25</td></tr>
<tr><td>GFH</td><td>once/month</td><td>Water</td><td>25</td></tr>
<tr><td rowspan="3">10</td><td>BF</td><td>3 times/d</td><td>Air+water</td><td>18</td><td>6.3</td><td rowspan="3">20</td></tr>
<tr><td>SF</td><td>3 times/d</td><td>Water</td><td>43</td><td>15.5</td></tr>
<tr><td>GFH</td><td>once/month</td><td>Water</td><td>25</td><td>0.065</td></tr>
</table>

BF= Biological Filter; SF= Sand Filter; GAC = Granular Activated Carbon; GFH = Granular Ferric Hydroxide

The water from filters backwashing can be directly discharged in to a public sewage (four plants) or treated in a dedicated line for residues treatment (six plants), as can be observed in Table 5.

This line can be constituted of a thickening tank (in one case; plant number 9) or can be more complex and arranged with a storage, flocculation, thickening and dewatering phases (in five case; plants number 1, 2, 4, 5 and 6). The treatment of filter backwashing in a dedicated line produced two separated flows: a liquid phase (called supernatant) that can be discharged into a water surface body and a solid phase (sludge) that can be disposed of in landfill for not hazardous wastes (see Figure 5).

Table 5. Residues management in the plants analyzed in the survey.

Plant		Residues	Backwashing treatment	Residues disposal
Precipitation	1-2	Water from filter backwashing	Floc.+Thick.+ Dewat.	Supernatant: surface water Sludge: landfill for no dangerous waste (NDW)
	3	Water from filter backwashing	-	Sewerage
		Exhausted GAC		Landfill NDW
	4	Water from filter backwashing	Floc.+ Sed.+Thick.	Supernatant: sewerage Sludge: landfill NDW
	5	Water from filter backwashing	Floc.+Sed.+ Thick.	Supernatant: sewerage Sludge: landfill NDW
	6	Water from filter backwashing	Floc.+Sed.+ Thick.	Supernatant: sewerage Sludge: landfill NDW
	7	Water from filter backwashing	-	Sewerage
	8	Water from filter backwashing	-	Sewerage
Precipitation + GFH	9	Water from filter backwashing	Sed.	Supernatant: surface water Sludge: landfill NDW
		Exhausted GFH		Landfill NDW
	10	Water from filter backwashing	- (equalization tank to be constructed)	Sewerage
		Exhausted GFH		Landfill NDW

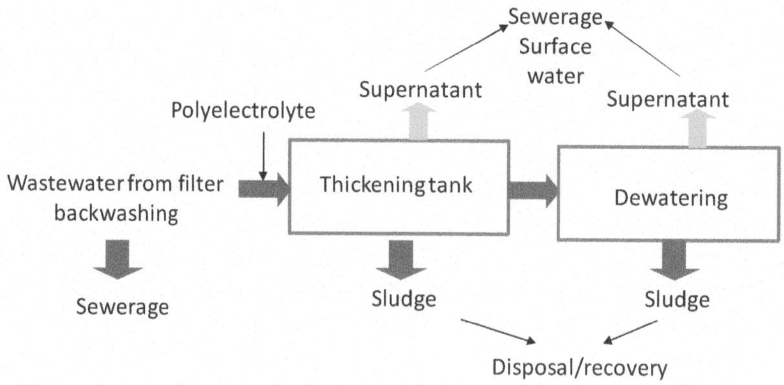

Figure 5. Scheme of the residues treatment.

3.4. Costs

A cost analysis has been done and the results are shown in Table 6.
Analysis cost results show that:

- in case of arsenic removal by chemical precipitation, costs ranged from 0.3 €cent/m^3 of treated water (only for ferric chloride dosage in plant n. 3) to 10.4 €cent/m^3 of treated water (plants n. 1 and 2 where energy, chemicals, residues disposal, labor, maintenance and equipment are included);

- for chemical precipitation followed by adsorption with GFH, costs are variable between 24.9 and 48.2 €cent/m^3, where reagents, energy, labor, equipment and residues disposal are enclosed.

Table 6. Costs of drinking water treatment plants analyzed in the survey.

	Plant	As_{IN} (μg/L)	As_{OUT} (μg/L)	Q (L/s)	Parameters	Cost (€cent/m^3)
Precipitation	1-2	24	3	450	Energy+chemical+residue disposal+labor+maintenance+ equipment	10.4
	3	20	8	13	$FeCl_3$	0.3
	4	38	8	60	Reagents+labor+sludge disposal	3.1
	5	50	2	9		
	6	2	8	13		
	7	47	6	28	Chemical+energy+labor	3.5
	8	20	3	20	Not available data	Not available
Prec. +	9	44	6	2	Chemical+energy+labor +equipment+GFH disposal	48.2
	10	65	15	15	Chemical+energy+labor +equipment+sludge and GFH disposal	24.9

4. Conclusion

- This survey concerned 10 companies of drinking water supply systems managing a total number of 43 drinking water treatment plants with arsenic removal.
- One of the main technology used in Italy is the chemical precipitation (34 applications analyzed in this study).
- Ten plants have been analyzed in detail: 8 plants with chemical precipitation and 2 supplemented with GFH filters. The flow rate is included from 2 to 450 L/s and arsenic concentration from 20 to 65 μg/L.
- Iron salts ($FeCl_3$) are always employed for arsenic precipitation with an average dosages of 60 mgFe/$mgAs_{removed}$.
- In four plants arsenic is removed by biological oxidation combined with a chemical precipitation; this solution can offer a simultaneous removal of ammonia and iron, manganese and arsenic oxidation. In some cases (3 plants) the chemical oxidation (with NaClO or $KMnO_4$) and chemical precipitation are applied after the biological process.

- Moreover, when the initial arsenic concentration is higher than 35-40 µg/L the biological oxidation combined with the chemical precipitation is followed by a tertiary step with GFH (two plants).
- The precipitation process is simple to operate, it needs simple equipments for the upgrading of existing plants and it is cheap. Its efficiency can reach 90-95% of arsenic removal.
- The operation of this process needs the chemicals dosages and filter backwashing.
- The main residues generated are the water of filter backwashing that has a volume from 3 to 24% of the total volume of the treated water. This residue can be discharged into public sewage or treated in a proper line for residues treatment. In this case a flocculation, thickening and dewatering phases are generally employed.

Acknowledgments

- The authors gratefully acknowledge all the subjects who collaborated to the activity of the working group "Water for human consumption: arsenic removal".
- A special acknowledgment is addressed to companies managing the ten drinking water treatment plants with chemical precipitation, deeply investigated in this research: A2A S.p.A. (Brescia), AIMAG S.p.A. (Mirandola, MO), AEM Gestioni S.r.l. (Cremona), Padania Acque Gestione S.p.A. (Cremona), SISAM S.p.A. (Castelgoffredo, MN), Acque S.p.A. (Ospedaletto, PI).

References

[1] Smedley, P.L., Kinniburg, D.G., A review of the source, behavior and distribution of arsenic in natural waters, *Appl. Geochem.*, **2002**, 17, 517-568.

[2] WHO, *Guideline for drinking water quality, first addendum to third edition - Volume 1, Recommendations,* **2006**, Geneva.

[3] Song, S., Lopez-Valdivieso, A., Hernandez-Campos, D.J., Peng, C., Monroy-Fernandez, M.G., Razo-Soto, I., Arsenic removal from high-arsenic water by enhanced coagulation with ferric ions and coarse calcite, *Water research,* **2006**, 40, 364 - 372.

[4] Hering, J.G., Chen, P.Y., Wilkie, J.A., Elimelech, M., Liang, S., Arsenic removal by ferric chloride, *J. AWWA*, **1996**, 88 (4), 155-167.

[5] WHO, UN synthesis report on arsenic in drinking-water, **2001**.

[6] Edwards, M., Chemistry of arsenic removal during coagulation and Fe-Mn oxidation, *Journal of American Water Works Association*, **1994**, 86(9), 64-78.

[7] Thirunavukkarasu, O.S., Viraraghavan, T., Subramanian, K.S., Chaalal, O., Islam, M.R., Arsenic removal in drinking water—impacts and novel removal technologies, *Energy Source*, **2005**, 27, 209-219.

[8] Jiang, J.Q., Removing arsenic from groundwater for the developing world—a review, *Water Sci. Technol.*, **2001**, 44 (6), 89-98.

Assessment of trace metal concentrations in the different processes at water treatment plants of EPAL

André Miranda, João Miguel Paiva and Maria João Benoliel, EPAL - Empresa Portuguesa das Águas Livres, S.A., Rua do Alviela, 12, 1170-012, Lisboa, Portugal; telephone: +351218552750

Corresponding Author E-Mail: Andremir@Epal.Pt

Abstract

EPAL - Empresa Portuguesa Das Águas Livres, S.A., is the largest water supplier in Portugal, Responsible For the production and distribution of drinking water to approximately 25% of the Portuguese population (about 2.8 million people). Following the Bonn Charter Goal For "Good Safe Drinking Water That Has The Trust Of Consumers" (Iwa, 2004) and the World Health Organization (Who) Guidelines, Epal Implemented The Water Safety Plans (Wsp), From The Catchment To The Consumers' Taps, which were integrated in the risk-based management of the company.[1], [2] And [3] This paper reports the strategy followed for the evaluation of the water quality risks (health and aesthetic) due to the presence of metals in raw water and/or inefficient treatment and for the definition of control measures. Total and dissolved metal concentrations were studied after each process installed in each water treatment plant (Wtp) of Epal. metal concentrations were analysed by Icp-Oes And Icp-Ms, except for mercury samples which were analysed by Cold Vapour Atomic Absorption Spectroscopy (CVAAS).

Variations and profiles concentrations of selected metal are presented for each process of the water treatment. concentrations of Ag, Be, Cd, Co, Cr, Hg, Sb, Se, Sn and Tl were lower than the quantification limits in all collected samples (including in the raw water). Ba, B, Li, Mo And V concentration profiles showed that theses metals are relatively unaffected by the treatment processes. As, Cu, Fe, Mn, Ni And U Present A similar concentration profile, in which the different steps of treatment are determinant to the content of these metals in the drinking water produced.

1. Introduction

EPAL - Empresa Portuguesa Das Águas Livres, S.A., is the largest water supplier in Portugal, and has been operating since 1868. EPAL is responsible for the direct water supply to about 500,000 inhabitants in the city of Lisbon and for the bulk water suppy to 32 municipalities north of the Tagus River, corresponding to approximately 25% of the Portuguese population (2.8 million people). The production system includes two water treatment plants (WTP - Asseiceira And Vale Da Pedra) which catch and treat surface water from Castelo De Bode Dam And Tagus River, respectively. It also includes 19 groundwater sources. the transport and distribution network has more than 2100 km of pipes and includes 45 reservoirs (75 compartments). the average water volume supplied daily is 600 million litres.

Asseiceira WTP is located 120km north of Lisbon near the city of Tomar. It is responsible for 67,4% of the overall production of drinking water. The water is abstracted in the Castelo De Bode Dam, in the River Zêzere, which has a total storage capacity of 1100hm3. The WTP has a production capacity of 625.000m3/day and comprises the following process operations: pre-chlorination, correction of aggressiveness and remineralisation with calcium hydroxide and carbon dioxide, coagulation / flocculation with aluminium sulphate and polyelectrolyte, dissolved air flotation, ozonation, filtration in double layer filters with sand and anthracite, ph adjustment with calcium hydroxide and pos-chlorination. Figure 1 Presents a schematic diagram Of The Water Treatment Plant.

Vale Da Pedra Wtp is located 50km north of Lisbon near the village of Valada Do Ribatejo. It is responsible for 22,7% of the overall production of drinking water. The catchment is in the River Tagus, The Largest River In Portugal. It has a production capacity of 225000 m3/day and comprises the following process operations: pre-chlorination, ph adjustment with carbon dioxide or sulphuric acid, coagulation / flocculation with aluminium sulphate and polyelectrolyte, decantation, filtration in monolayer filters with sand, ph adjustment with calcium hydroxide and pos-chlorination. Figure 2 Presents a schematic diagram Of The Water Treatment Plant.

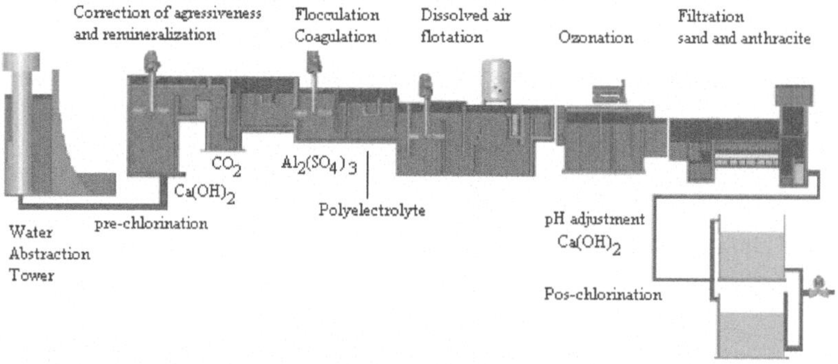

Figure 1 - Water Treatment Process Of Asseiceira.

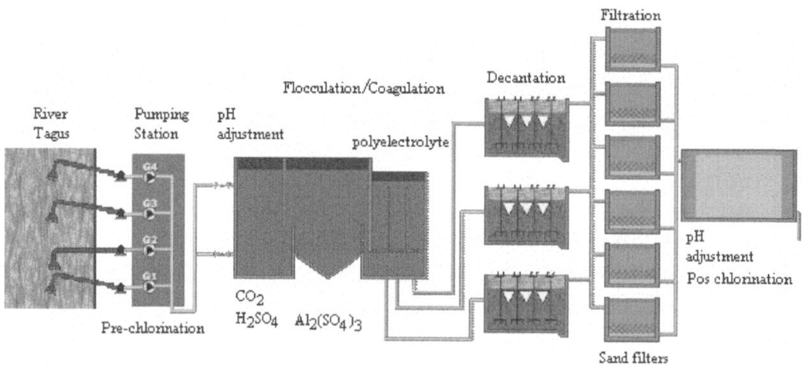

Figure 2 - Water Treatment Process Of Vale Da Pedra.

Inorganic chemicals in natural water usually occur as dissolved salts such as carbonates and chlorides attached to suspended material or as complexes with naturally occurring organic matter. Conventional water treatment plants are designed to remove suspended solids and bulk organics by the use of coagulation, flotation and filtration processes. [4] And [5]

This Study Was Developed To Support The Risk Assessment Of The Production Processes And Had The Goals Of Determining The Behavior Of Metals In The Treatment Steps And Identifying The Critical Ones To Be Monitored.

2. Materials and Methods

Two sampling campaigns were realized; one in the beginning of October Of 2009 (end of summer) and the other was in January Of 2010 (winter). The samples were collected at the entry of each WTP, after each process and in the reservoirs installed at the end of the treatment plants. At each sampling point samples were taken for ph, alkalinity and conductivity, for mercury analysis (preserved in the field with K2cr2o7 In Hno3), for total metal analysis (preserved in the field with 2% of conc. hno3) and for dissolved metal concentrations (filtered in the field just after collection with a teflon filter of 0.45 Mm and acidified with 2% of conc. Hno3).

The analyses were carried out in the central laboratory of EPAL. For mercury, samples were digested with Kmno4 (5% in ultrapure water), K2s2o8 (10% in ultrapure water) and concentrated H2so4 in an ultrasound bath for 30 minutes, at 50°c. The digested samples were analysed by cold vapour atomic absorption spectroscopy in a perkin elmer fims 400. Samples for total and dissolved metals were microwave digested with 10% of concentrated Hno3. Metal concentrations were determined by Icp-Ms in a thermo Xseriesii Or By Icp-Oes In A Thermo Iris Intrepid.

Table 1 presents, the method used, quantification limits, accuracy and method uncertainty for each metal.

Table 1 - Analytical methods used

Element	Units	Method	LoQ	Accuracy	Method uncertainty *
Ca	mg/L	ICP-OES	5,00	0,53%	12%
Mg	mg/L	ICP-OES	1,00	2,0%	15%
Na	mg/L	ICP-OES	5,00	2,8%	10%
K	mg/L	ICP-OES	1,00	5,4%	10%
Al	µg/L	ICP-OES ICP-MS	40,0 5,00	1,3% 0,97%	8,0% 6,0%
Ba	µg/L	ICP-OES	5,00	2,4%	8,0%
Cu	µg/L	ICP-MS	1,00	1,0%	8,0%
Fe	µg/L	ICP-OES	20,0	0,61%	8,0%
Mn	µg/L	ICP-MS	0,50	0,28%	6,0%
Zn	µg/L	ICP-MS	4,00	3,3%	6,0%
B	µg/L	ICP-OES	20,0	0,25%	12%
Li	µg/L	ICP-MS	1,00	0,13%	6,0%
Be	µg/L	ICP-MS	0,50	0,24%	6,0%
V	µg/L	ICP-MS	0,50	0,46%	6,0%
Cr	µg/L	ICP-MS	1,00	1,6%	6,0%
Co	µg/L	ICP-MS	0,50	0,35%	6,0%
Ni	µg/L	ICP-MS	1,00	4,6%	6,0%
As	µg/L	ICP-MS	0,50	2,0%	6,0%
Se	µg/L	ICP-MS	2,00	2,4%	12%
Mo	µg/L	ICP-MS	0,50	1,9%	6,0%
Ag	µg/L	ICP-MS	0,50	3,6%	6,0%
Cd	µg/L	ICP-MS	0,50	0,16%	6,0%
Sn	µg/L	ICP-MS	0,50	2,0%	6,0%
Sb	µg/L	ICP-MS	0,50	1,5%	6,0%
Tl	µg/L	ICP-MS	0,50	3,1%	6,0%
Pb	µg/L	ICP-MS	0,50	1,3%	6,0%
U	µg/L	ICP-MS	0,50	2,3%	6,0%
Hg	µg/L	CV-AAS	0,20	2,1%	10%

3. Results and Discussion

3.1 Metals in Asseiceira Water Treatment Plant

Raw water from Castelo De Bode Dam is a low mineralization water with Ph varying between 6,94 and 7,90, low alkalinity (12,3 To 16,4 Mg/L Caco3) and conductivity (between 58 and 85 ¤S/Cm). The water characteristics are relatively constant during the year. Ph, alkalinity, conductivity and calcium levels in drinking water are mainly controlled by reagents added in the treatment process, showing similar profiles along production steps. Figure 3 Represents The variation of calcium. We can observe that the major variations correspond to the steps of remineralisation and final Ph adjustment as it was expected.

Results for Na, Mg And K are constant during the Water Treatment Process, as it can be seen in Figure 3, for sodium.

Ba, Li And Zn present a similar profile. They are unaffected by the treatment steps and the concentration of total and dissolved metals shows no differences.

Results for total and dissolved Fe, B, Be, V, Cr, Co, Ni, Se, Mo, Ag, Cd, Sn, Sb, Pb, Tl, U and Hg are lower than the Limits Of Quantification (LOQ) in raw water and in all steps of the Water Treatment Process.

In raw water from Castelo De Bode Dam, As and Cu are mainly in the dissolved fraction and the average concentrations are 1,09□0,05 □G/L For As And 2,3□0,2 µg/L for Cu. Profiles show a decrease between concentrations in raw water and in water after flotation step (concentrations become lower than LOQ). In Figure 4 we can see the variation of the concentrations of arsenic inside the WTP.

The concentration of aluminium in the WTP is controlled by the coagulation reagent dose as it was expected. However, it is important to mention that al in raw water is lower than 15µg/L and in the treated water the residual concentration of dissolved Al is around 20µg/L.

In raw water, Mn is mainly in the suspended fraction as we can observe in Figure 5; total Mn is near 4µg/L whilst dissolved concentrations of this metal are lower than 0,5 µg/L. There is a significant reduction of total Mn concentration after flotation. Then the values are lower or near LOQ.

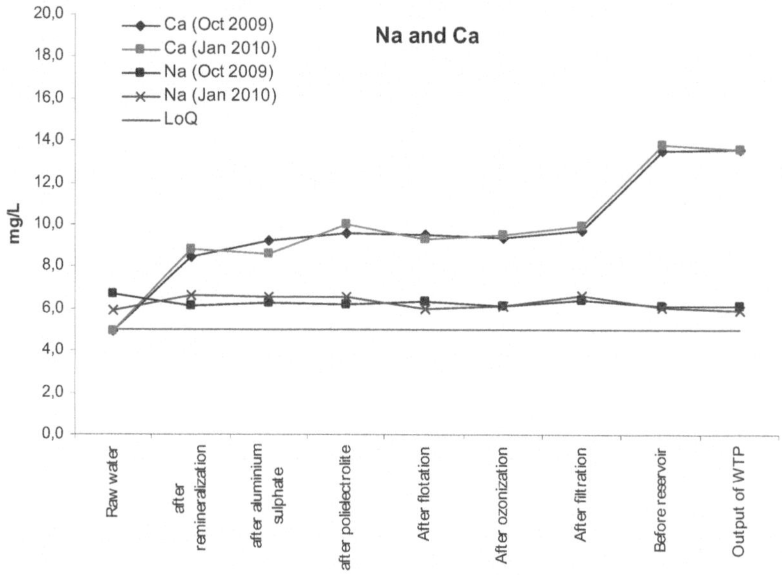

Figure 3 - Variation Profile For Calcium And Sodium In Asseiceira WTP

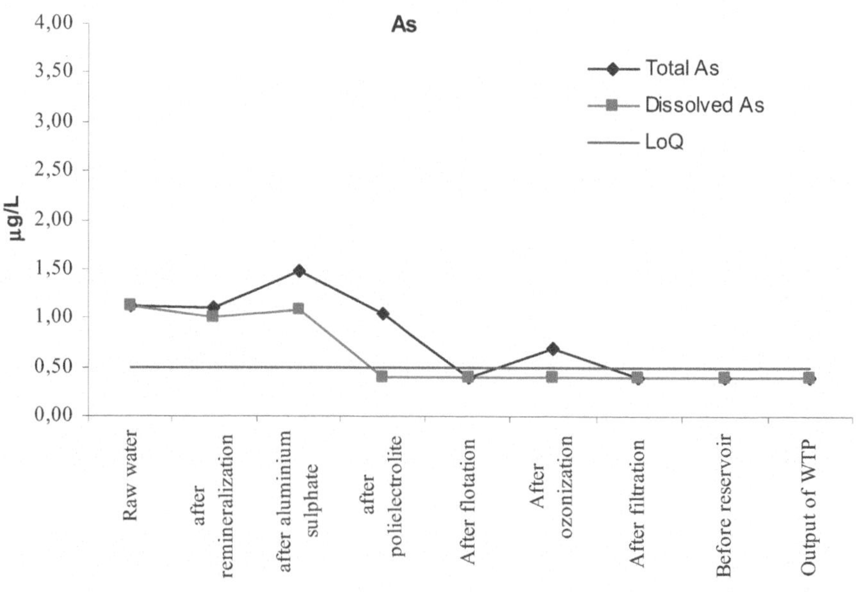

Figure 4 - Variation Profile For Arsenic In Asseiceira WTP.

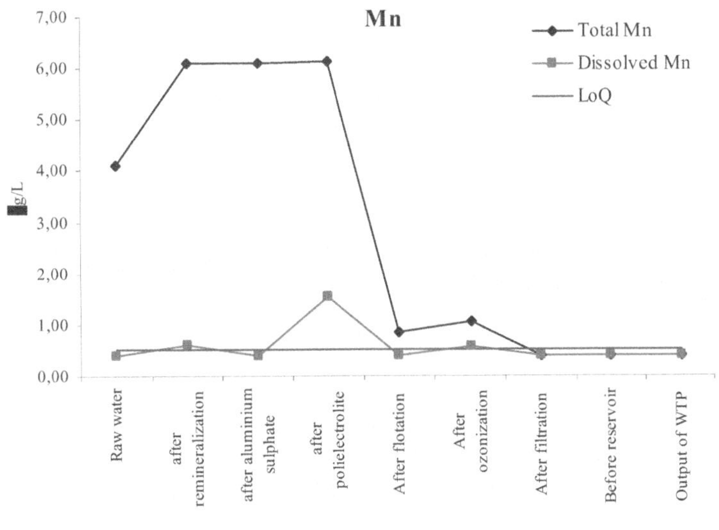

Figure 5 - Variation Profile For Manganese In Asseiceira WTP.

3.2 Metals In Vale Da Pedra Water Treatment Plant

In raw water from River Tagus, Ph can range between 7,6 and 8,3, conductivity varying between 137 and 618 µs/Cm and alkalinity is in the range of 40 To 200 Mg/L Caco3. One can observe a seasonal effect caused by the low level of water in the Tagus River in summer and strong precipitation events in winter.

Results of Ph, alkalinity, conductivity, sodium, potassium, magnesium and calcium are relatively constant during the treatment, in the two sampling campaigns.

Results for total and dissolved Be, Cr, Co, Se, Ag, Cd, Sn, Sb, Tl and Hg are lower than the limits of quantification in raw water and in all steps of the treatment plant as it was in results from Asseiceira WTP.

Despite the fact that Mo and B have low levels in raw water, it seems that they are unaffected by the treatment process. Results from total and dissolved concentrations are similar, which means that these metals are only present in the dissolved fraction.

Dissolved Cu, V, As, Ni, Pb and U concentrations are near the limits of quantification, but the variation of total metal content shows a decrease between raw water and water after decantation, which indicates that the suspended fraction is removed by the treatment. Figure 6 Represents the variation of uranium concentration profile.

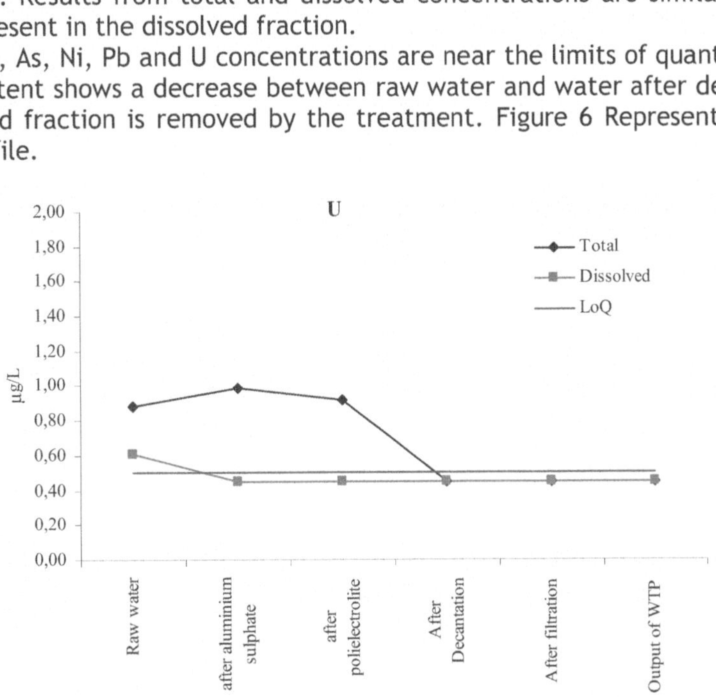

Figure 6 - Variation Profile For Uranium In Vale Da Pedra WTP.

In river tagus water, the concentration of aluminium is very variable and can be up to 730µg/l, whilst dissolved concentrations are low (10 to 20µg/l). The profile of aluminium in the treatment process shows a significant decrease in the water after decantation, as it was expected. The coagulant reagent added is efficiently removed by the treatment process but there is still a residual concentration of aluminium (around 32µg/l) in the drinking water produced.

In general, total fe and mn concentrations can be up to 600µg/l for fe and 30µg/l for mn. However, dissolved fe and mn concentrations are very low in River Tagus water (lower than 20,0µg/l for fe and lower than 1,00µg/l for mn). The variation of these metals inside the WTP shows a significant decrease until the end of decantation step as a result of physical process separations of non dissolved fractions.

4. Conclusions

Metal concentrations in drinking water produced in the two water treatment plants of EPAL are below the parametric values established in the ec directive 98/83/ec, 3rd November, and most of them are lower than the quantification limits.

In general, the dissolved fraction of the studied metals is not retained in the water treatment plants, as it was expected. mo, li, ba and b are unaffected by the Water Treatment Processes because they are mainly in the dissolved fraction in raw water. They should be monitored in raw waters at least twice a year (summer and winter), both total and dissolved concentrations. Alert concentrations should be established according to their toxicity.

be, cr, co, se, ag, cd, sn, sb and tl were not found in raw waters. However a risk assessment for Castelo de Bode Dam and River Tagus should be frequently revised and monitoring programs defined for the catchment areas. Sediment analysis in the catchment areas is advised in order to identify trends.

The water treatment plants are very efficient removing mn, fe, al and trace concentrations of as, ni, v, u and cu. however, a residual concentration of al is present in treated water, coming from the coagulation reagents used in the treatment process.

Dissolved metal concentrations in water after decantation in Vale Da Pedra Water Treatment Plant and water after flotation in Asseiceira Water Treatment plant should be monitored because these processes are responsible for the major variations of metal content in the water.

The chemical products used in the treatment plants and are in contact with water, must also be controled in order to guarantee that the levels of toxic substances are according the defined in the european standards for these treatment products.

6. REFERENCES

1. Iwa, (2004) - The Bonn Charter For Safe Drinking Water, September
2. Who/Iwa (2009) - Water Safety Plan Manual. Step-By-Step Risk Management For Drinking-Water Suppliers.
3. Who (2006), Guidelines For Drinking-Water Quality.
4. E.G.Wagner, R.G.Pinheiro, Upgrading Water Treatment Plants, Spon Press, London, 2001.
5. Royal Society of Chemistry (2007), Sustainable Water: Chemical Science Priorities, Summary Report.
6. Nhmrc (2004) Australian Drinking Water Guidelines.
7. N. Liu, T. Ni, J. Xia, M. Dai, C. He, G. Lu, Environmental Monitoring And Assessment, Epub 2010 Aug 13.
8. Maleki, H. Izanloo, M. Zazoli, B. Roshani, Asian Journal Of Water, Environment And Pollution, Vol. 3, 2 (2006), 107-110.
9. M. Lasheen, G. El-Kholy, C. Sharaby, I. Elsherif, S. El-Wakeel, Management Of Environmental Quality: An International Journal, Vol. 19, 3 (2008), 367-376.

Arsenic removal by energy-efficient small-scale reverse osmosis units

J.Hoinkis and S.A. Deowan

Karlsruhe University of Applied Sciences, Moltkestr.30, 76133 Karlsruhe, Germany

Corresponding author e-mail: jan.hoinkis@hs-karlsruhe.de

A variety of membrane techniques among them nanofiltration (NF) and reverse osmosis (RO), may be used for arsenic removal (for overview see [1]). NF and RO have the advantage using very "dense" membranes in such a way that other dissolved contaminants can be retained along with arsenic, resulting in a very high water quality. Some time ago several companies brought small-scale marine RO units (known as watermakers) to the market. They are applied to produce drinking water from seawater on boats and it is a well-proven technology, which works reliably at remote locations under difficult conditions (e.g. high salt concentration). Some of them can be powered by sustainable energy sources, such as PV or wind wheels, or can by operated manually.

This work is reporting on laboratory tests using watermakers for arsenic removal. The study was conducted using different commercially available RO membranes and arsenic-spiked local tap water [2]. Initial findings of pilot studies currently running in rural Bihar, India will be also presented. The experiments shall provide a basis for developing a simple, low-cost RO desalinator for rural areas in developing countries, which can be operated decentrally by sustainable energy sources.

The findings indicate that the arsenic rejection is significantly higher for As(V) than for As(III) for all of the tested RO membranes. This is in agreement with preliminary laboratory-scale screening tests and other published results [1,3]. As for trivalent arsenic the arsenic values in permeate can be kept below the MCL of 10 µg/L only up to a feed concentration of approximately 200-300 µg/L. However, the As(III) rejection can be significantly improved by using a double pass unit (at feed concentration around 500 µg/L As in permeate can be kept below MCL). As(V) can be rejected efficiently by a single pass system up to a feed concentration of 2000 µg/L, without crossing the MCL level in permeate.

References

[1] A. Figoli,A. Criscuoli, J. Hoinkis, Review of membrane processes for arsenic removal from drinking water, In N. Kabay et Al. (Eds): The Global Arsenic Problem: Challenges for Safe Water Production, Arsenic Series, CRCpress,Vol3.

[2] T.Geucke, S.A. Deowan, J.Hoinkis, Ch.Pätzold, Performance of a small-scale RO desalinator for arsenic removal, Desalination, Vol. 239, Issues 1-3, 2009, 198-206

[3] A.S. Deowan, J. Hoinkis, Ch. Pätzold, Low-energy reverse osmosis membranes for arsenic removal from groundwater, in: Groundwater for Sustainable Development-Problems, Perspectives and Challenges" eds. P. Bhattacharya, A.L. Ramanathan, J. Bundschuh, A. K. Keshari and D. Chandrasekharam, Balkema/Taylor&Francis, 2008, 375–386

Arsenic oxidation treatment by H_2O_2 and UV radiation

S. Sorlini and F. Gialdini

Department of Civil Engineering, Architecture, Land and Environment, University of Brescia, Via Branze 43, 25123 Brescia, Italy

Corresponding author e-mail: sabrina.sorlini@ing.unibs.it

Abstract

Arsenic is a widespread contaminant in the environment. The intake of water containing high concentration of arsenic gives serious effects on human health, such us skin and lung cancer. In the European Union, and in Italy, the arsenic limit in drinking water is 10 µg/L. Several treatments are available for arsenic removal. For some processes the removal yields can be improved after the oxidation treatment. Most full scale applications are based on conventional oxidation processes but, if water contains arsenic and organic refractory contaminants, the Advanced Oxidation Processes could be considered. The aim of this work was to investigate the effectiveness of the arsenic oxidation using hydrogen peroxide, UV radiation and their combination in distilled and in real water. Some tests were performed on real groundwater polluted with arsenic and Terbuthylazine (TBA). Good arsenic and TBA oxidation yields can be reached in presence of H_2O_2 combined with a high UV radiation dose.

1. Introduction

Arsenic is a widespread contaminant in the environment. Occurrence of arsenic in nature can be related both to natural and anthropogenic causes. The most abundant species of arsenic in groundwater are arsenite [As(III)] and arsenate [As(V)]. The intake of water containing high concentration of arsenic produces serious effects on human health, like cancer of skin and lungs. The World Health Organization (WHO) revised the guideline for arsenic from 50 to 10 µg/L in 1993. In the European Union, and in Italy, the arsenic standard level is now set to 10 µg/L.

Several treatments are available for arsenic removal, such as coagulation with ferric salts, adsorption on ferric hydroxide or activated alumina, reverse osmosis and anion exchange. For some processes the removal yields can be improved after arsenic oxidation, especially for coagulation, adsorption on activated alumina and anion exchange. For this reason it is often necessary to proceed with a pre-oxidation of As(III) to As(V) before its removal from water.

Most full scale applications are based on conventional oxidation (potassium permanganate, chlorine, ozone, etc.), however the advanced oxidation processes (AOPs) could be applied successfully to the remediation of water contaminated with arsenic and/or organic refractory contaminants.

Arsenic oxidation with AOPs was investigated in previous research studies. Pettine et al. studied the influence of pH on arsenite oxidation by H_2O_2 in aqueous solutions [1]. Zaw and Emett investigated the iron/UV and sulphite/UV based oxidation processes for As (III) removal [2]. Yang et al. studied the photocatalytic reactions for oxidation of As(III) to As(V) using a 400 W medium pressure mercury lamp, at an initial As(III) concentration of 40 mg/L and several H_2O_2:As mole ratios. Only an excess of H_2O_2 (H_2O_2/As mole ratio = 4:1) can complete the arsenic oxidation in 10 minutes, in the dark, under specific experimental conditions. In the presence of the UV light and H_2O_2, the oxidation was completed in less than 10 minutes, even at a low hydrogen peroxide concentration (H_2O_2:As(III) molar ratio = 1:4) [3]. Bissen and Frimmel showed that 90% of As(III) was oxidized within 90 seconds in a water sample containing 40 µg/L As(III) when irradiated with a high pressure mercury UV lamp [4]. Ghurye and Clifford showed that UV radiation alone is not effective at arsenic oxidation and only high UV doses (up to 46,080 mJ/cm^2) can reach yield up to 73% of oxidation to As(V). The authors employed a low pressure mercury lamp with an incident irradiance of 32,000 µW/cm^2 at 254 nm to treat water contaminated with 50 µg/L As(III) [5].

Some groundwater sources could be simultaneously contaminated by arsenic and other organic micropollutants generated by human activities. In agricultural areas many groundwater reach significant concentration of pesticides, herbicides, etc.

Terbuthylazine (TBA) is a herbicide that belongs to the chloro-triazine family and it is selective for several crops (maize, sorghum, potatoes, etc.). Degradation of TBA in natural water depends on the presence of sediments and biological activity. In studies investigating the photolysis of TBA in aqueous solution, a half-life of >1 month was estimated using a sun-like light source, in agreement with a reported half-life of about 3 months in natural sunlight. However, for very high sunlight intensities (full midday sunlight), the half-life was 39 hours [6].

TBA is widespread in superficial and deep water in Italy [7] because its massive employment in agriculture and its high mobility and persistence in water. In Italy, the main occurrence of TBA is in water surface, in particular in the River Po where concentration between 0.03 and 0.25 μg/L were observed (year 2005). In groundwater, in 1821 point investigated, only 2.7% of samples exceeds the Italian standard regulation (0.1 μg/L for anti parasitic).

There is no evidence that TBA is carcinogenic or mutagenic. A TDI approach was therefore used in the derivation of a guideline value for TBA in drinking-water of 7 μg/L [8].

Several treatment processes for TBA removal were investigated. Low removal percentages, below 30%, were obtained for TBA employing oxidation chlorine. Preoxidation by ozone can reach 45% of degradation. The activated carbon can remove up to 60% of TBA [9]. Pesticide degradation is possible through different photochemical processes that require an artificial light source (generally a high pressure mercury or a xenon arc lamp) or natural sunlight. Most of these methods require long treatment periods with high energy photons and rarely achieve a complete degradation of the pollutant. The most common reactions observed when a contaminant is irradiated with UV light are dechlorination, substitution of chlorine atoms by hydroxyl groups, and formation of radical species [10]. A previous work showed that in presence of H_2O_2 and UV light the degradation of TBA quickly leads to the formation of ammeline [11]. This study shows that a total TBA degradation (initial TBA concentration = $2.18 * 10^{-5}$ mol/L) is observed after 5 minutes of irradiation employing a 125 W high pressure mercury lamp and $2.18 * 10^{-5}$ mol/L of H_2O_2.

This study shows the arsenic oxidation using hydrogen peroxide, UV radiation and their combination in distilled and in real groundwater samples spiked with arsenic to an initial concentration of 0.1 mg/L. Nevertheless, the advanced oxidation process for arsenic oxidation only is not sustainable because some conventional oxidation processes more suitable and cheaper can be employed. Otherwise, the AOPs could be usefully applied in water contaminated with arsenic and other organic refractory contaminants. For this reason, some oxidation tests were performed also in real groundwater samples polluted with arsenic and TBA.

2. Materials and Methods

The experimental study was performed with a collimated beam apparatus equipped with a low pressure mercury lamp that delivered an incident irradiance of 200 μW/cm^2 at 254 nm (Figure 1).

In order to quantify the contribution of any potential oxidation of As(III) to As(V) due to either H_2O_2 or UV radiation to the overall UV/H_2O_2 process, individual preliminary tests were performed.

The effect of UV radiation alone on As(III) was examined in distilled water and groundwater (Table 1) at four fluence levels (300, 600, 1,200, and 2,000 mJ/cm^2). A similar test was performed with H_2O_2 only (5 mg/L). The UV/H_2O_2 process was applied by combining the same above indicated conditions for UV and H_2O_2 alone. The initial As(III) concentration was set to 0.1 mg/L.

The effect of UV/H_2O_2 process on TBA oxidation was performed in real groundwater spiked with 10 μg/L of TBA at three fluence levels (300, 1,200, and 2,000 mJ/cm^2) and two H_2O_2 doses (5 and 10 mg/L).

After each exposure time, the residual hydrogen peroxide was quenched with a bovine catalase solution to a final concentration of 0.2 mg/L, in order to prevent any potential thermal oxidation of As(III) to As(V).

Total arsenic concentration was determined by Hydride Generation Atomic Absorption Spectrometry (HG-AAS). As(III) was analyzed in the water sample filtered through an As(V)-selective resin. Therefore, As(V) was calculated as difference between total As and As(III). TBA was analyzed with SPME-GC-MS.

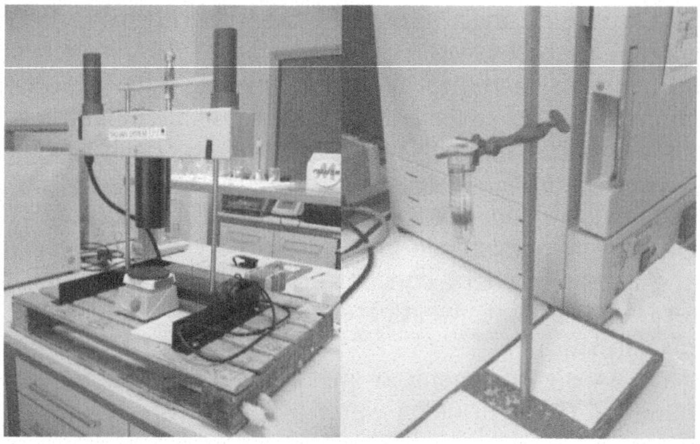

Figure 1. Collimated beam apparatus (on the left) and filtration sistem on Serdolite (on the right).

Table 1. Groundwater characteristics.

Parameter	Value	Parameter	Value
Redox potential	178 mV	Total As	14.68 µg/L
pH	8.12	As(III)	14.55 µg/L
Conductivity	368 µS/cm	Total Fe	0.13 mg/L
Turbidity	<0.4 NTU	Total Mn	128 µg/L
Total carbon	70.86 mg/L	NH4+	1.42 mg/L
Inorganic carbon	66.75 mg/L	UV254 Absorption	0.067 1/cm
TOC	4.10 mg/L	DUV254 Absorption	0.060 1/cm

3. Results and Discussion

In distilled water, the experimental results indicated that As(III) oxidation with hydrogen peroxide and UV radiation applied separately is very low. The maximum oxidation yield is obtained when an UV dose of 2,000 mJ/cm^2 is employed. In the advanced oxidation process (UV/H_2O_2), As(III) removal is relatively constant (~50%) over the UV dose range of 300-1,200 mJ/cm^2. Only with an UV dose of 2,000 mJ/cm^2 the oxidation yield is significantly increased up 70% (Figure 2).

In groundwater, the experimental results indicate that oxidation with only hydrogen peroxide is a very slow process, as observed in distilled water. Oxidation with UV radiation alone is a slow process too, except for the high doses (2,000 mJ/cm^2). The combination of H_2O_2 with different UV doses can efficiently oxidize As(III). A promising oxidation yield (62%) is obtained at 600 mJ/cm^2 in the presence of 5 mg/L H_2O_2. The application of higher UV doses does not appear to improve this result (Figure 3).

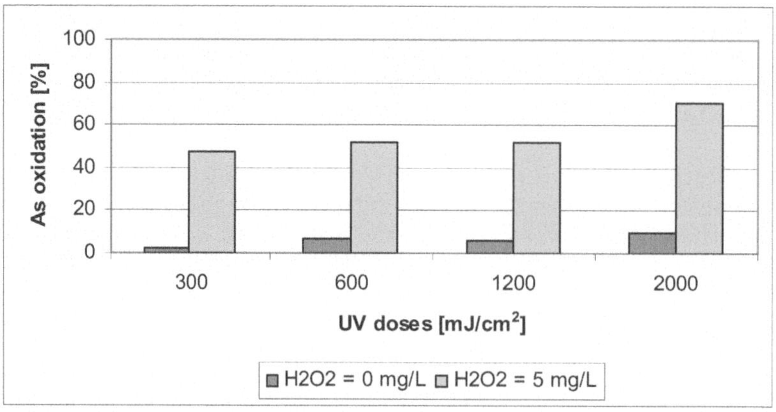

Figure 2. As(III) oxidation in distilled water.

As concerns TBA oxidation, in presence of a fixed hydrogen peroxide concentration, an increase in the UV dose from 1,200 mJ/cm^2 to 2,000 mJ/cm^2 can increase the TBA oxidation from 70-80% to 90-100%. In particular, doubling the H_2O_2 concentration (from 5 to 10 mg/L) the TBA removal does not increase too much. A noticeable TBA oxidation is possible in presence of UV dose (2,000 mJ/cm^2) without hydrogen peroxide. Moreover, the increase of the H_2O_2 concentration in presence of UV radiation (both 1,200 and 2,000 mJ/cm^2), can increase the TBA removal. In presence of hydrogen peroxide in concentration variable from 5 to 15 mg/L, the increase of the UV doses from 1,200 to 2,000 mJ/cm^2 can increase the TBA oxidation yields (Figure 4 and Figure 5).

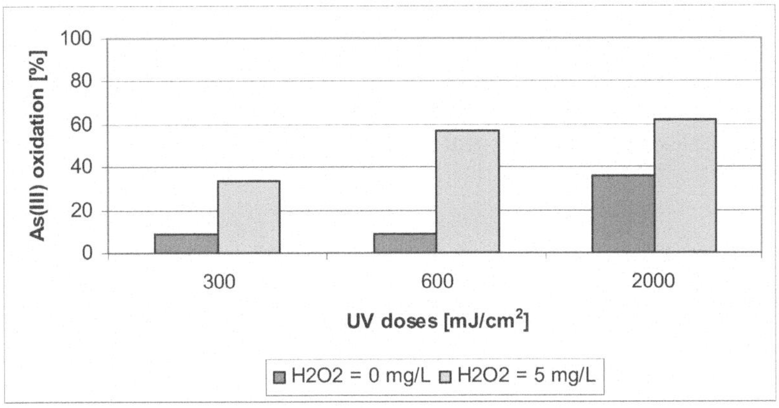

Figure 3. As(III) oxidation in real water.

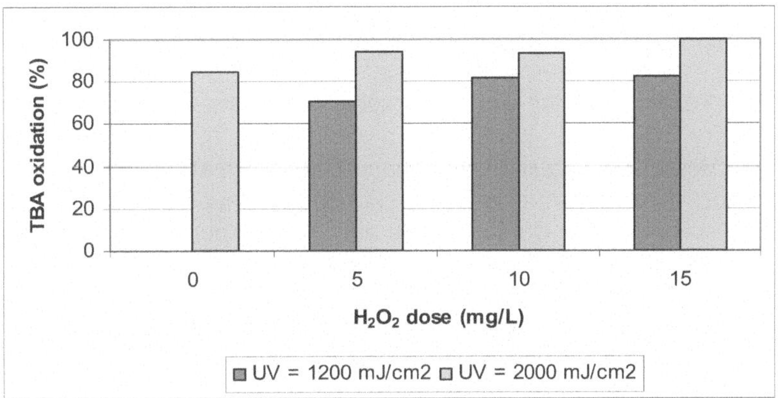

Figure 4. TBA oxidation in real water.

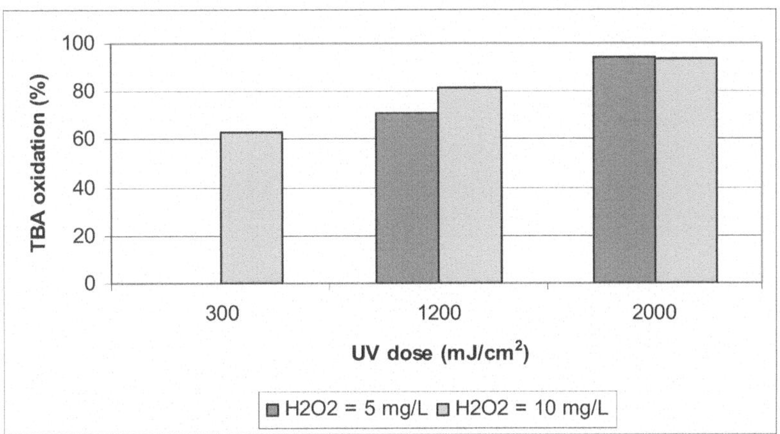

Figure 5. TBA oxidation in real water.

4. Conclusions

Hydrogen peroxide and UV radiation alone are not effective at the arsenic oxidation. Good arsenic oxidation yields can be reached in presence of hydrogen peroxide combined with a high UV radiation dose (2,000 mJ/cm^2).

More promising results are obtained by combining the two oxidants, even if only the application of high UV doses (2,000 mJ/cm^2) can simultaneously guarantee good arsenic and Terbuthylazine oxidation yields.

Acknowledgments

The authors thank the staff of Trojan UV for the collaboration in the research, in particular dr. Mihaela Stefan and dr. Domenico Santoro to have provided useful information about basic operational aspects of the collimated beam and theory of UV dose calculation. Thanks to engineers Ottavio Franceschini and Valerio Zani too, for supporting this experimental work.

References

[1] Pettine, M., Campanella, L., Millero, F.J., Arsenite oxidation by H_2O_2 in aqueous solutions, *Geochim. Cosmochim. Acta*, **1999**, 63, 2727-2735.

[2] Zaw, M., Emett, M., Arsenic removal from water using advanced oxidation processes, *Toxicol. Lett.*, **2002**, 133, 113-118.

[3] Yang, H., Lin, W. Y., Rajeshwar, K., Homogeneous and heterogeneous photocatalytic reactions involving As(III) and As(V) species in aqueous media, *J. Photochem. Photobiol. A Chem.*, **1999**, 123, 137-143.

[4] Bissen, M., Frimmel, F. H., Arsenic-a review. Part II: oxidation of arsenic and its removal in water treatment, *Acta Hydrochim. Hydrobiol.*, **2003**, 31(2), 97-107.

[5] Ghurye, G., Clifford, D., As(III) oxidation using chemical and solid-phase oxidant, *J. AWWA*, **2004**, 96(1), 84-96.

[6] WHO, Terbuthylazine (TBA) in Drinking-water. Background document for development of Guidelines for Drinking-water Quality. Originally published in Guidelines for drinking-water quality, 2nd ed. Addendum to Vol. 2. Health criteria and other supporting information, **2003**.

[7] APAT, *Residui di prodotti fitosanitari nelle acque. Rapporto annuale*, **2005**.

[8] WHO, Guidelines for Drinking Water Quality. Third edition incorporating the first and second addenda. Volume 1. Recommendations, **2008**.

[9] Ormad, M.P., Miguel, N., Claver, A., Matesanz, J.M., Ovelleiro, J.L., Pesticides removal in the process of drinking water production, *Chemosphere*, **2008**, 71, 97-106.

[10] Chiron, S., Fernandez-Alba, A., Rodriguez, A., Garcia-Calvowat, E., Pesticide chemical oxidation: state-of-the-art, *Wat. Res.*, **2000**, 34(2), 366-377.

[11] Sanlaville, Y., Guittonneau, S., Mansour, M., Feicht, E.A., Meallier, P., Kettrup, A., Photosensitized degradation of Terbuthylazine in water, *Chemosphere*, **1996**, 33(2), 353-362.

Brown lakes - causes, effects and remedial measures

Heléne Annadotter[1], Ingegerd Rosborg[2] and Johan Forssblad[1]
[1]Regito AB, SE-28 022, Vittsjö, SWEDEN
[2] Royal Institute of Technology, KTH, Stockholm, Sweden.

Corresponding author e-mail: rosborg@spray.se

The color of surface water has increased in several parts of Europe during the past decades. This has created serious problems for a large number of drinking water producers since brown water is difficult or impossible to purify to drinking water. Brown color of the surface water may be due to a content of humic substances and/or various iron compounds. Apart from the difficulties with drinking water treatment, the animals in the brown lakes are suffering because of the low light transparency. One example is water fowl such as Black-throated loon (*Gavia arctica*) and Common merganser (*Mergus merganser*). These species have disappeared from a large number of lakes in Sweden due to the increased water color. These birds feed on fish that they catch by diving in the water. In dark water, the birds are unable to detect the fish.

The EU water framework directive requires that lakes in EU be restored to a good ecological status. This includes lakes that have been transformed from clear water lakes to brown water lakes during the past century.

It is imperative to find measures to reduce the color of the surface water. We have studied the levels of color and iron in several Swedish lakes, rivers, brooks and manmade drainage ditches. The levels of color and iron were increased in waters that were affected by drained peat bogs, drained coniferous forests and clear-cut areas.

In the presentation, we will highlight the mechanisms that affect the levels of iron and other coloring compounds in surface waters. The green house effect is one out of a number. Restoration of brown lakes and remedial measures to decrease the content of iron and color will be presented.

Applied technologies and possibilities of modernisation of groundwater treatment plants in Poland

Joanna Jeż-Walkowiak, **Alina Pruss** and Marek M. Sozański

Poznan University of Technology, Institute of Environmental Engineering, Poznan, POLAND

Corresponding author e-mail: alina.pruss@put.poznan.pl

Abstract

In the paper the quality of Polish groundwater is presented with respect to iron and manganese content and the diversity of its concentration in raw water. The applied in Poland technologies for groundwater treatment are presented. The analysis of technological parameters of applied devices is done. The effectiveness of applied technologies is established with respect to reduction of iron and manganese concentration as well as the achievement of chemical stability of the treated water. The presented data allowed to show the possibilities and methods of intensification of water treatment processes.

1. Introduction

Groundwater is the most valuable source for drinking water production because of mostly stable water quality parameters and low contamination. In Poland 56% of total produced drinking water come from groundwater sources. Quaternary water-bearing layer dominates over Cretaceous, Jurassic and Tertiary water. Groundwater supplies 87% of polish drinking water treatment plants. [1] In uncontaminated groundwater, iron and manganese cause the greatest difficulties for use of these waters for municipal and industrial purposes. Present in water iron and manganese ions impart a metallic taste and odor, stain laundry and household fixtures. Iron and manganese may discolor industrial products such as textiles and paper. Precipitates can clog pipes and support the growth of iron and manganese bacteria, which can cause taste and odor problems [2,3].

WHO set the standard for iron and manganese content on the level of 0,3 mgFe/L and 0,1 mgMn/L [4]. The European Community set more restricted standards for iron and manganese concentration in drinking water. According to Directive 98/83/EC the maximum level for iron and manganese in drinking water is equal to 0,2mgFe/L and 0,05 mgMn/L [5]. New Polish Standards [6] lower the maximum iron and manganese concentration in drinking water from 0,5 to 0,2 mgFe/L and from 0,1 to 0,05 mgMn/L. Therefore, we can expect that in the near future, many drinking water plants in Poland, that use groundwater as a source, will have to improve the removal of iron and manganese. The paper presets results of national questionnaire of drinking water supply systems in Poland [1].

2. Groundwater quality

Quality of groundwater depends on hydro-geological properties of water-bearing layer as well as physical, chemical and biological processes in the ground.

According to review of groundwater quality in Poland it appeared that [1]:

- total iron content in groundwater ranges from traces to 30 mg/L,
- manganese content ranges from traces to 2,3 mg/L,
- alkalinity ranges from 3,0 to 6,0 mval/L,
- 98% of water treatment plants treat water with pH of 6,5-9,5,
- 94% of water treatment plants treat water with total hardness of 60-500 mg $CaCO_3$/L,
- 62,4% of water treatment plants treat water with color up to 15 mgPt/L,
- 34,6% of water treatment plants treat water with turbidity up to 1,0 mgPt/L,
- COD-$KMnO_4$ ranges from 0,3 to 15 mgO_2/L,
- Ammonium concentration ranges from traces to 9,5 mg/L,
- Most of the groundwater has corrosive properties due to high CO_2 content.

Relationship between percentage of volume of treated water and iron content is present on figure 1. Almost 90% of volume of treated water has the iron content up to 4 mgFe/L.

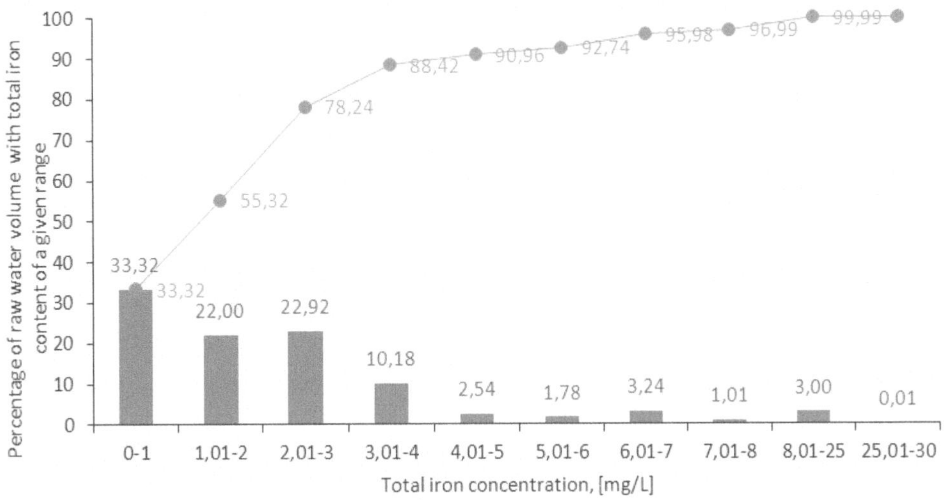

Figure 1. Total iron concentrations in groundwater in Poland.

Relationship between percentage of volume of treated water and manganese content is presented on figure 2. Almost 88% of volume of treated water has the manganese content up to 0,6 mgMn/L.

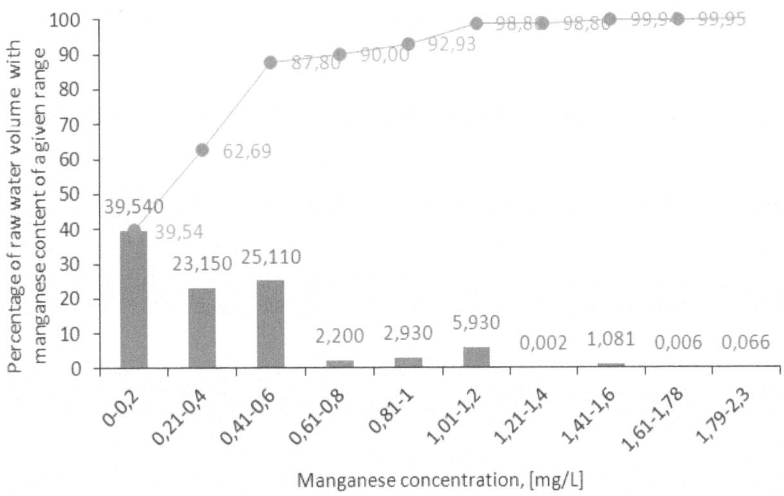

Figure 2. Manganese concentrations in groundwater in Poland.

Only a small amount of raw groundwater in Poland does not need treatment before its consumer use. In uncontaminated groundwater, iron and manganese cause the greatest difficulties for use of these waters for municipal and industrial purposes. Iron and manganese content and corrosive properties determined the treatment technology of the most water treatment plants.

3. Technology of groundwater treatment

The technological train of groundwater treatment in Poland usually consist of aeration or chemical oxidation and rapid filtration.The goal of aeration process is to introduce oxygen to water and remove carbon dioxide - a chemical compound responsible for corrosive properties of treated water.

At small water treatment plants (WTP) in Poland usually pressure aerators are applied. The pressure aerators should be applied for waters with alkalinity equal and higher than 5 mval/L. In practice pressure aeration is used too often due to simplicity of design and operation. Presser aerators applied for low alkalinity water have insufficient effectiveness of CO_2 removal leading to corrosive properties of treated water. More than 40% of volume of raw water are aerated with pressure aerators (fig. 3).

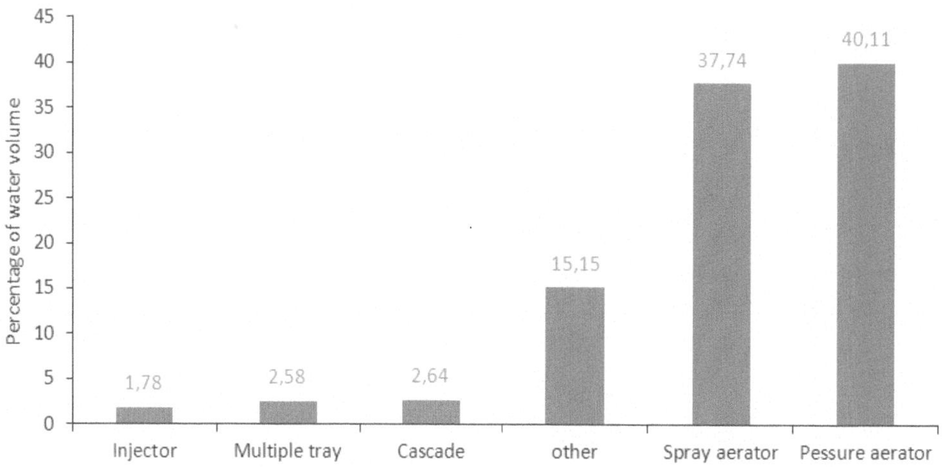

Figure 3. Types of aerators used for aeration of groundwater in Poland.

Filtration is the most important process in groundwater treatment. Filtration material and proper operational parameters are essential for obtaining high efficiency of water treatment technology. Filter bed should be characterized by high efficiency of iron and manganese removal and causing no difficulties during operation.

Manganese can be removed from groundwater without chemical oxidation in filtration process trough oxidative media. The term „oxidative media" describes filtration materials in which catalytic and heterogenic oxidation reactions of iron and manganese take place. Manganese ore is an example of a natural oxidative filter material, which consists mostly of MnO_2 - the strong oxidant. Most of the oxidative beds used are obtained by covering the grains with active coatings, mainly Fe_2O_3 and MnO_2 [2,7,8,9].The results of the research show the influence of internal structure parameters (specific surface and pore volume) and MnO_2 percentage content in oxidative filter media on the effectiveness of manganese removal [7]. Up to a certain point, an increase in the above parameters results in improved effectiveness of manganese removal.

Mass capacity is a parameter allowing evaluation of filter performance. The mass of iron and manganese oxides kept in filter may be calculated according to the formula:

$$PM = t \cdot (c_{in} - c_{out}) \cdot v_f$$

where: t - time of filtration [h], c_{in}, c_{out}, - inlet and outlet concentration of iron and manganese [mg/L], v_f - filtration rate [m/h].

Mass capacity equal at least 2250 g/m^2 characterizes good performing filter.

On figure 4 the percentage of water volume filtrated trough one or two stage filter in gravity and pressure systems are presented. Most of the water (over 68%) a treated in one stage filtration in gravity or pressure filters. Over 88% of water is filtrated trough naturally activated silica sand.

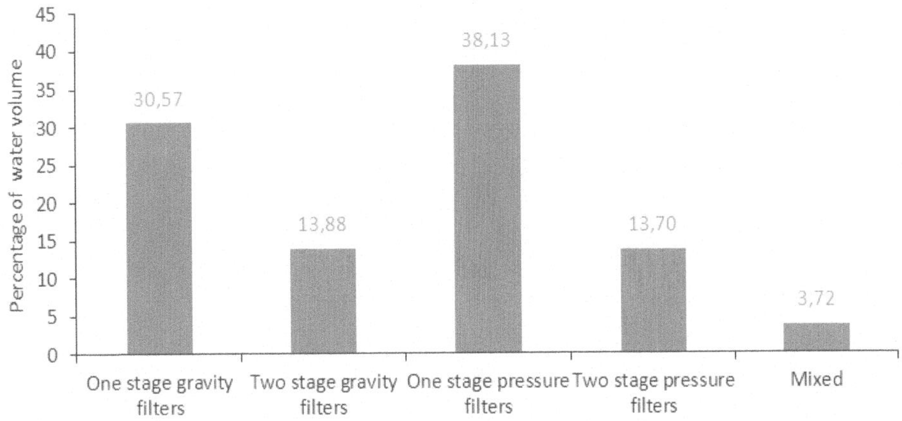

Figure 4. Filter types applied in WTP in Poland.

4. Quality of treated groundwater

Analysis of drinking water produced by groundwater treatment plans in Poland [1] show that:

- 94% of treated groundwater volume has total iron concentration lower than the limit 0,2 mgFe/L,
- only 72% of treated groundwater volume has manganese concentration lower than the limit 0,05 mgFe/L,
- major amount of treated water are characterized by corrosive properties,
- technological problems occur at WTP supplied with raw water with elevated values of iron, manganese, ammonium and color. In such waters complexes of metals and organics may be expected leading to difficulties in classic treatment trains of aeration and filtration.

5. Possibilities of modernization

Manganese removal cause problems at 34% of groundwater treatment plants. An economically viable solution for improving treatment in these case can be the replacement of the media in old filters by new media that are capable of removing iron and manganese more effectively. Oxidative media may be a good choice in this case. Economic issues such as price and operational cost (related for instance to duration of filtration cycles, water and air use for backwashing and persistence of material) are also important.

To enhance chemical stability of treated water the analysis of aeration systems should be done. Collected data show that pressure aeration is applied too often. Over 55% of pressure aerated water volume has alkalinity lower than 5 mval/L (fig.5), so in that case the spay or tray aerators should be applied instead of pressure aerators.

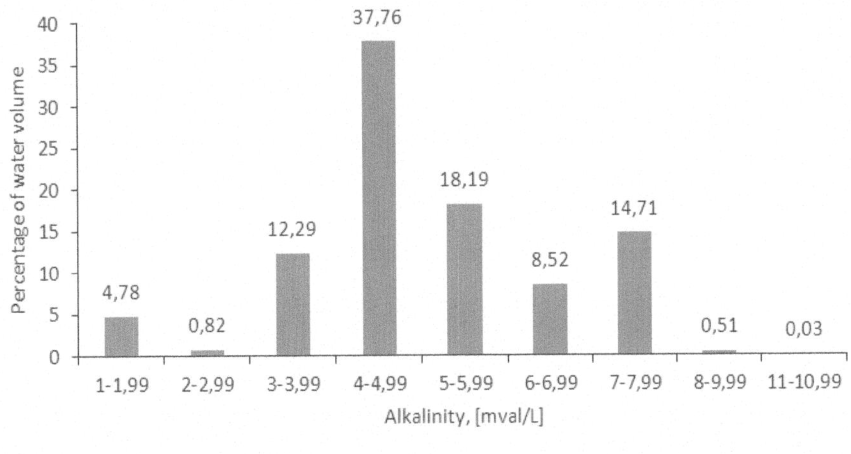

Figure 5. Frequency of applying pressure aerators for water of a given alkalinity.

Metal-organic complexes are not removed from groundwater by classic aeration - rapid filtration system. To enhance the effectiveness of treatment the coagulation and chemical oxidation processes should be introduced to technological train.

Mass capacity is a parameter characterizing performance of filters. On figure 6 the mass capacity of one stage filtration is presented.

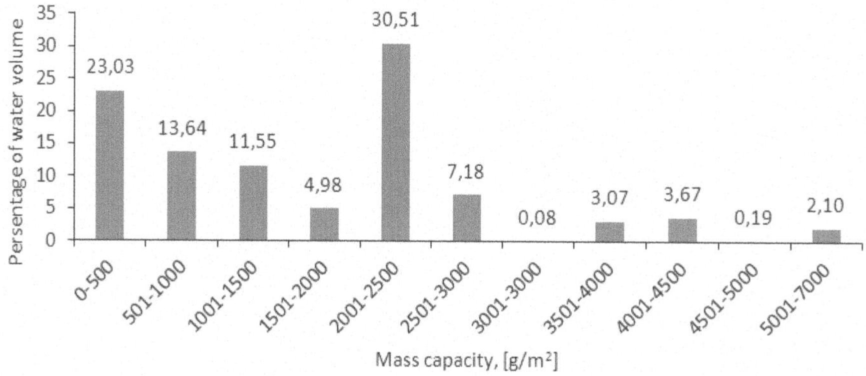

Figure 6. Mass capacity of one stage filtration beds.

Presented data show that over 30% of one stage filter have a proper mass capacity indicating good filter performance. Many filtration systems have very low technological effectiveness achieving mass capacity up to 500 g/m^2. Low mass capacity may be caused by wrong granulometric parameters of filtration material. The second reason of low mass capacity of filter bed may be wrong backwash parameters. To enhance filter performance the sieve analysis of filtration materials has to be done as well as backwash intensity and time should be established properly.

6. References

[1] *"Waterworks in Poland- tradition and present time"*, Marek M. Sozański et al., Polish Foundation of Water Resources Protection, Poznań-Bydgoszcz, Poland, 2002r.

[2] MWH, 2005. *Water Treatment Principles and Design* (Second Edition, Revised by J.C. Crittenden, R.R. Trussell, D.W. Hand, K.J. Howe and G. Tchobanoglous). John Wiley & Sons, Inc., Hoboken, NJ.

[3] Sly, L. I., Hodgkinson, M. C., Arunpairojana, V., (1990), "Deposition of Manganese in a Drinking Water Distribution Systems", Appl. Environ. Microbiol., 56, 3, p. 628-639.

[4] *Guidelines for drinking-water quality. Third edition.* Volume 1. Recommendations. World Health Organization, Geneva 2004.

[5] DWD, 98, Council Directive 98/83/EC on the quality of water intended for human consumption. Official Journal L 330, 05/12/1998 p. 0032-0054

[6] Regulation of Polish Ministry of Health referred to requirements for drinking water, Dz.U.07.61.417. Poland, 29 March 2007.

[7] Jeż-Walkowiak J., (2000) Characteristic of oxidative filter materials for manganese removal from groundwater, Proceedings of 4th International Conference on Water Supply and Water Quality, Kraków, Poland, page 263-272.

[8] *Water Treatment Plant Design*, American Water Works Association, McGraw-Hill Publishing Company, New York 1990r.

[9] Sommerfeld E.O., *"Iron and manganese removal handbook"*, American Water Works Association, USA, 1999.

Heavy metals (Pb, Cr) removal from aqueous solution by modified clinoptilolite

M. Zabochnicka-Świątek and E. Okoniewska

Czestochowa University of Technology, Faculty of Environmental Protection and Engineering, Brzeznicka 60a, 42-200 Częstochowa, Poland; tel. +48 34 3250917
Correspondig author e-mail: mzabochnicka@is.pcz.czest.pl

Abstract

One of the most required characteristics of natural zeolites is the ability to selectively sort molecules. Zeolites preferentially adsorb certain molecules while excluded others. The ion exchange functions occurs when cations present in the solution are exchanged for cations in the structure (sodium, potassium, magnesium and calcium). Each zeolite has distinction exchange selectivity and adsorption capacity that can be more effective after modification. Physical and chemical regeneration can recover adsorption capacity of used zeolite. The scope of this study was to modify the natural zeolite clinoptilolite for its lead and chromium (III) retention capacity. Clinoptilolite was pretreated by sodium chloride solution and sodium hydroxide solution under thermal and microwave irradiation. The removal of heavy metals were investigated by conducting as series of batch experiments. The lead and chromium (III) ions retention capacity of thus obtained modified clinoptilolites were found to be good material for these heavy metals removal from aqueous solutions. For lead, the zeolite treated under microwave irradiation and NaCl had the highest heavy metal adsorption capacity value, followed by the zeolite obtained by thermal process and NaCl and by thermal process and NaOH. For chromium (III), the clinoptilolite treated under thermal process and NaOH had the highest heavy metal adsorption capacity value, followed by the zeolite obtained by thermal process and NaCl and microwave irradiation and NaCl. Based on the obtained results, it was concluded that the thermal and/or microwave treated natural zeolite was a promising adsorbent for heavy metals such as lead and chromium (III) removal from water. Due to their low cost and high adsorption capacity, the modified clinoptilolite has the potential to be utilized for cost-effective removal of lead and chromium (III) from aqueous solution.

1. Introduction.

Chromium may occur in the environment due to weathering of naturally occurring minerals present in ultrabasic rocks. Chromium is a major contaminant in groundwater of several countries, as a results from anthropogenic source such as tanning, steel works, plating, corrosion control [1].

Anthropogenic sources of lead include the mining and smelting of ore, manufacture of lead-containing products, combustion of coal and oil, and waste incineration. Lead has been found to be acute toxic to human beings when present in high amounts in drinking water). Lead is known to damage the kidney, liver and reproductive system, basic cellular processes and brain functions. The toxic symptoms are anemia, insomnia, headache, dizziness, irritability, weakness of muscles, hallucination and renal damages [2].

Zeolites are crystalline hydrated aluminosilicates with a three dimensional framework structure. This structure is formed by AlO_4 and SiO_4 tetrahedra joined by a common oxygen atom. One of the most required characteristics of natural zeolites is the ability to selectively sort molecules. Zeolites preferentially adsorb certain molecules while excluded others. The ion exchange functions occurs when cations present in the solution are exchanged for cations in the structure (sodium, potassium, magnesium and calcium) [2].

Clinoptilolite has received the extensive attention due to its attractive sorption capacity for several cations and especially for heavy metals [3]. Each zeolite has distinction exchange selectivity and adsorption capacity that can be more effective after chemical or physical modification [4].
After the process of heavy metals adsorption on zeolite the regeneration can recover adsorption capacity of used zeolite [5].

This paper presents heavy metals: Pb (II), Cr (III) removal from aqueous solution by modified clinoptilolite. The scope of this study was to modify the natural zeolite clinoptilolite for its lead and chromium (III) retention capacity. Clinoptilolite was pretreated by sodium chloride solution and sodium hydroxide solution under thermal and microwave irradiation. The removal of heavy metals were investigated by conducting as series of batch experiments.

2. Materials and methods

The clinoptilolite (type: zeoflocc) was of Hungarian origin and the granulation of clinoptilolite was of 0,0-0,125 mm and the mineralogical composition was: 55% of clinoptilolite, 6% of quartz, 13 % of montmorillonit, 26% of ash and volcanic glass.

2.1. Modification of clinoptilolite

The modified natural clinoptilolite was used as a adsorbent. Thermal-treated zeolite was prepared as follows - 25.0 g of clinoptilolite and:

- 100mL NaCl solution of 2.5 mol/L were placed in a 500mL conical flask - T+NaCl,
- 100mL NaOH solution of 2.5 mol/L were placed in a 500mL conical flask - T+NaOH.

The suspension was stirred under 65^{0}C for 24 h, and then the solid was separated by centrifugation at 4000 rpm for 10 min, washed with distilled water until the pH value was fixed. Then the resulting solid were dried at 105^{0}C for 12 h in the air, sieved and stored for further use in the adsorption experiments.

Microwave-treated zeolite was was prepared as follows: 25.0 g of clinoptilolite and 100mL NaCl solution of 2.5 mol/L were placed in a 500mL conical flask - M+NaCl. The flask was placed into the microwave oven at 240W for 15 min. Then the solid was separated, washed, dried and sieved following the same procedure to thermal-treated zeolite.

Inorganic chemicals used in the study, such as NaOH, NaCl, $Pb(NO_3)_2$, $Cr(NO_3)_3$ were all analytical grade reagents.

2.2. Adsorption studies

To study the adsorption capacity of clinoptilolite for heavy metals (Pb, Cr) removal from water 2g of clinoptilolite and 100 mL portions of heavy metal solution (single metal solutions) were placed in a suitable container. The corresponding concentration of each metal (Pb, Cr) was: 0.1 mg/L, 0.5 mg/L, 1.0 mg/L, 5.0 mg/L.

The experiments were performed in triplicates at temperature of 20°C (±5°C). The samples were agitated continuously for 2 hours and the mixtures were equilibrated for 22 hours. Next, final pH was recorded and concentrations of lead and chromium (in the liquid phase) were analyzed using ICP-AES. The solutions were separated by solid phase centrifugation at 4000 rpm for 10 min, before chemical analysis.

Blank solutions without clinoptilolite were prepared in order to examine the possible precipitation of heavy metals.

All the experiments were performed three times and the average experimental relative error was found to be 3.5% for metal concentration measurements.

The sorption capacity of clinoptilolite for Pb and Cr was calculated according to the following formula:

$$A = \frac{(C_0 - C_k)}{m} V \qquad (1)$$

where:
A - the adsorption capacity (mg/g);
C_0 and C_k - the initial and final metal concentration (mg/L);
V - the sample volume (L);
m- the clinoptilolite weight (g).
Metal removal from the solution was determined as follows:

$$Uptake(\%) = \frac{C_0 - C_k}{C_0}100 \qquad (2)$$

Where C_0 and C_k are the initial and final metal concentrations (mg/L).

3. Results and discussion

The initial and final pH values of tested samples for adsorption of Pb by modified clinoptilolite was presented in Table 1.

Table 1. The initial and final pH values of tested samples for sorption of Pb.

C_0Pb (II) [mg/l]	T + NaOH		T + NaCl		M + NaCl	
	Initial pH	Final pH	Initial pH	Final pH	Initial pH	Final pH
0.1	10.3	9.5	8.7	8.5	8.4	7.7
0.5	10.2	9.4	8.5	8.0	8.1	7.6
1.0	10.0	8.9	8.3	8.0	7.8	7.4
5.0	6.2	8.6	3.1	4.3	3.1	3.6

For the initial concentration of lead from 0.1 mg/L to 1.0 mg/L the final pH values were lower than the initial pH. The final pH values were within the range of 7.6 to 9.5 for the initial concentrations of lead within the range of 0.1 mg/L and 1.0 mg/L and from 3.6 to 8.6 for the concentrations of lead: 5.0 mg/L. The selectivity series for final pH values were following: T+NaOH > T+NaCl > M+ NaCl.

The initial and final pH values of tested samples for adsorption of Cr(III) by modified clinoptilolite was presented in Table 2. For the initial concentration of Cr(III) from 0.1 mg/L to 1.0 mg/L the final pH values were lower than the initial pH. The final pH values were within the range of 7.4 to 9.6 for the initial concentrations of Cr(III) within the range of 0.1 mg/L and 1.0 mg/L and from 3.8 to 8.5 for the initial concentrations of Cr(III): 5.0 mg/L. The selectivity series for final pH values were following: T+NaOH > T+NaCl > M+ NaCl.

Table 2. The initial and final pH values of tested samples for sorption of Cr(III).

C_0 Cr (III) [mg/L]	T + NaOH		T + NaCl		M + NaCl	
	Initial pH	Final pH	Initial pH	Final pH	Initial pH	Final pH
0.1	10.1	9.6	8.4	8.2	8.3	7.7
0.5	10.0	9.6	7.8	7.7	8.2	7.4
1.0	9.6	9.4	7.7	7.5	7.8	7.5
5.0	5.6	8.5	3.2	3.9	3.0	3.8

The adsorption capacity of lead by modified clinoptilolite is presented in figure 1.

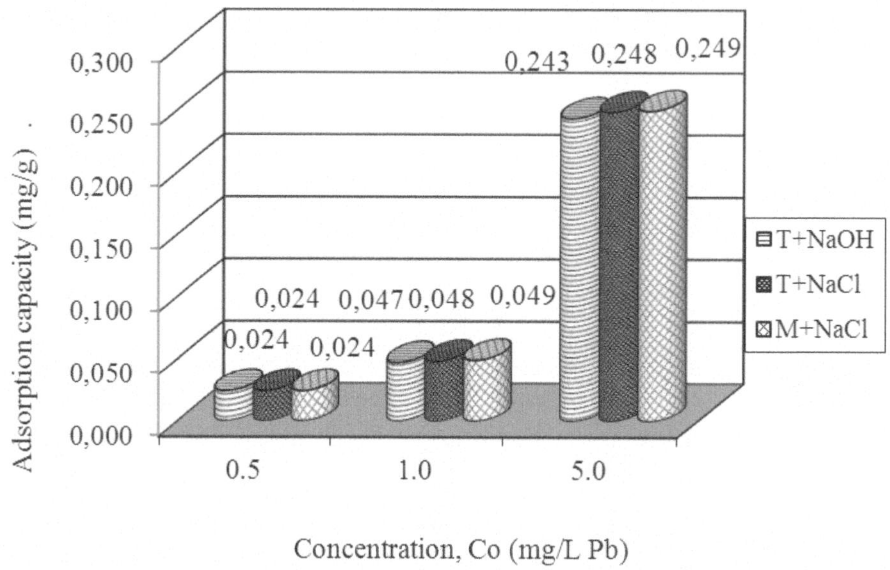

Figure 1. The adsorption capacity of clinoptilolite for lead.

The adsorption capacity of modified clinoptilolite was slightly influenced by the method used for modification. According to the initial concentration, the differences between the amount of adsorbed metal were observed. Modified clinoptilolite shows the highest adsorption capacity towards lead after modification by M+NaCl and the lowest after modification by T+NaOH.

The highest adsorption capacity of clinoptilolite towards lead of 0.249 mg/g was found for the initial concentration of 5.0 mg/L.

The adsorption capacity of chromium by modified clinoptilolite is presented in figure 2.

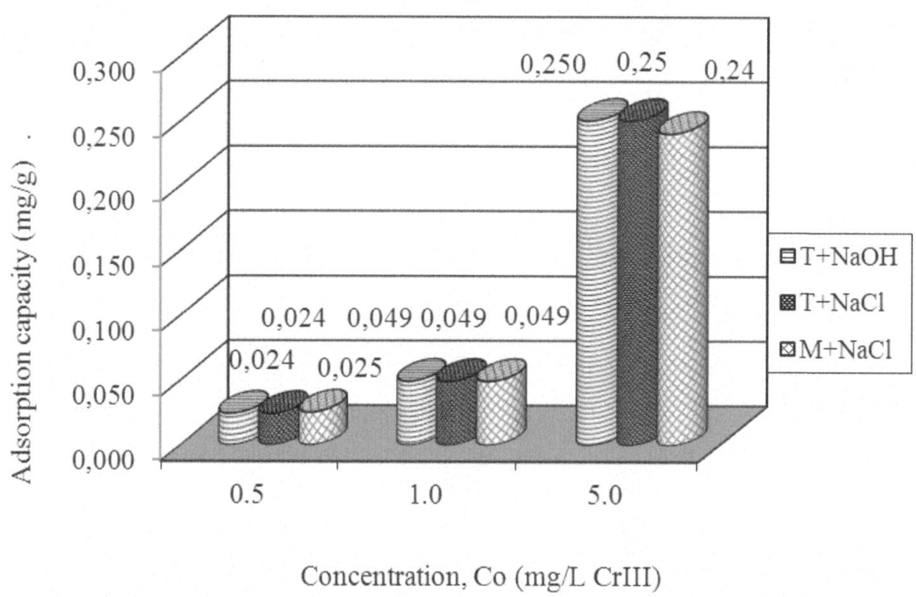

Figure 2. The adsorption capacity of clinoptilolite for chromium.

According to the data presented in figure 2 the highest adsorption capacity of clinoptilolite towards chromium of 0.25 mg/g was found for the initial concentration of 5.0 mg/L. Modified clinoptilolite shows the highest adsorption capacity towards chromium after modification by T+NaOH and the lowest after modification by M+NaCl. In figure 3 the uptake (%) of lead is presented.

Based on the conducted investigations, lead was adsorbed with higher efficiency from the aqueous solutions. The uptake of lead reached by modified clinoptilolite was in the range of 81% - 99.7% in all concentration. The highest removal level of lead reached by modified clinoptilolite was 99.8% in the initial metal concentration of lead: 5.0 mg/L for T+NaOH and the lowest of 81% in the metal concentration of 0.1 mg/L for T+NaCl. There was observed the decrease in pH value in T+NaCl and M+NaCl modification methods resulted in decrease in Pb uptake.

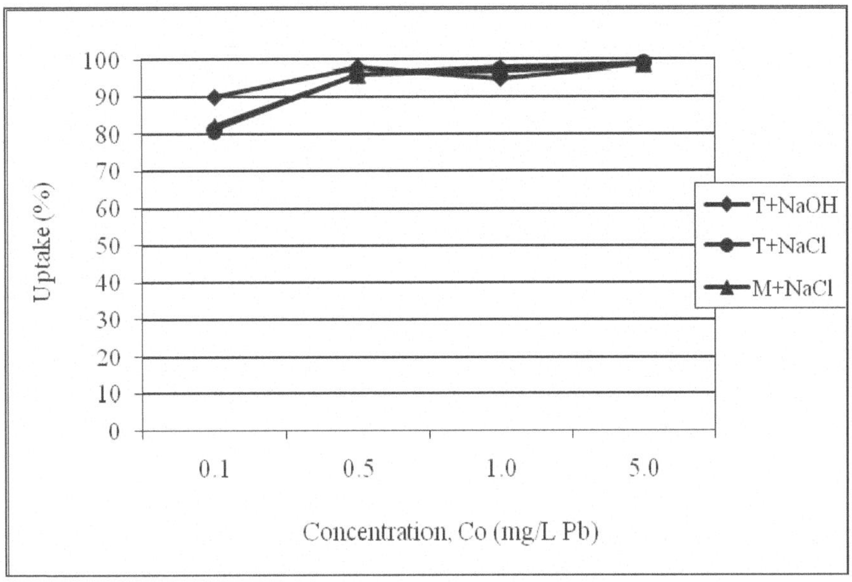

Figure 3. Percentage removal of lead by modified clinoptilolite.

In figure 4 the uptake (%) of chromium is presented.

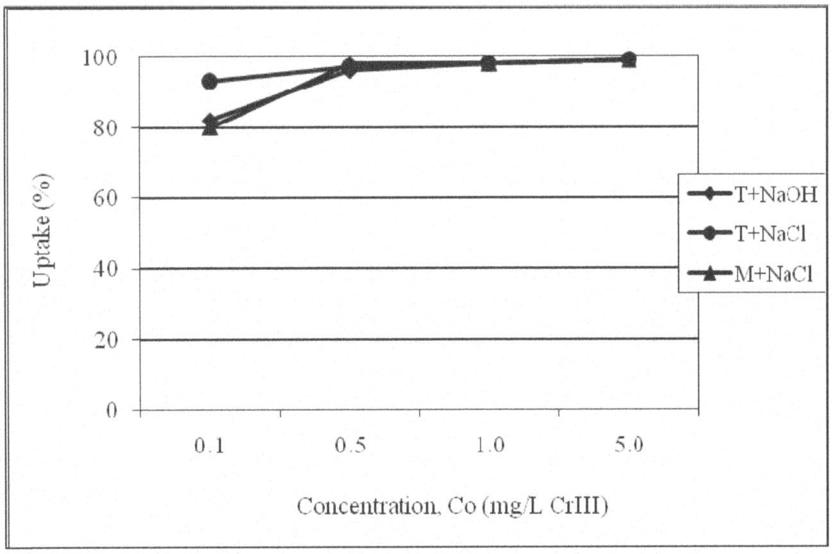

Figure 4. Percentage removal of chromium by modified clinoptilolite.

Chromium was adsorbed with higher efficiency from the aqueous solutions in all concentration. The uptake of chromium reached by modified clinoptilolite was in the range of 80% - 99.2%. The highest removal level of chromium reached by modified clinoptilolite was 99.2% in the initial metal concentration of chromium: 5.0 mg/L for T+NaOH and the lowest of 80% in the metal concentration of 0.1 mg/L for M+NaCl. There was observed the decrease in pH value in T+NaCl and M+NaCl modification methods resulted in decrease in Cr(III) uptake.

The obtained results indicate that choosing the best modification method are important factor for adsorption of lead and chromium (III) by clinoptilolite which is in line with the literature. The results are in agreement with the research of Wanga and Peng [6] and Schmidt and Anielak [7] in which modification of clinoptilolite was found important in affecting sorption of heavy metals.

Modified natural clinoptilolite was found to be a good sorbent material for lead and chromium (III) [8] removal from water solutions.

4. Conclusion

In the present study, the adsorption capacity of natural zeolite clinoptilolite modified for its lead and chromium (III) retention capacity has been investigated. Clinoptilolite was pretreated by sodium chloride solution and sodium hydroxide solution under thermal and microwave irradiation. According to the obtained results the following conclusions can be drawn:

1. Modification of natural zeolites can be done in several methods making the modified zeolites achieving higher adsorption capacity for heavy metals removal.
2. The value of the adsorption capacity of modified clinoptilolite for heavy metal is connected to the method used for modification.
3. The lead and chromium (III) ions retention capacity of thus obtained modified clinoptilolite was found to be good material for these heavy metals removal from aqueous solutions.
4. For lead, the zeolite treated under microwave irradiation and NaCl had the highest heavy metal adsorption capacity value, followed by the zeolite obtained by thermal process and NaCl and by thermal process and NaOH.
5. For chromium (III), the clinoptilolite treated under thermal process and NaOH had the highest heavy metal adsorption capacity value, followed by the zeolite obtained by thermal process and NaCl and microwave irradiation and NaCl.

To sum up, it was concluded that the thermal and/or microwave treated natural zeolite was a promising adsorbent for heavy metals such as lead and chromium (III) removal from water. Due to their low cost and high adsorption capacity, the modified clinoptilolite has the potential to be utilized for cost-effective removal of lead and chromium (III) from aqueous solution.

Acknowledgments

This study was carried out within the frame of the BW 401/206/08 State Committee for Scientific Research (KBN) in Poland.

References

1. Kumar A.R., Riyazuddin P., Comparative study of analytical methods for the determination of chromium in groundwater samples containing iron. Microchemical Journal 93 (2009) 236-241.
2. Zabochnicka-Świątek M., Stępniak L., The potential applications of aluminosilicates for metals removal from water, 2nd International Conference Metals and related substances in drinking water, Cost Action 637, Lisbon, Portugal 29-31 October 2008, Proceedings 168-179.
3. Zabochnicka-Świątek M., Okoniewska E., Adsorption capacity of clinoptilolite for heavy metals (Cd, Cr, Zn) removal from water, 3rd International Conference Metals and related substances in drinking water, Cost action 637, Ioannina, Greece 21-23 October 2009, Proceedings pp.130-136.
4. Lei L., Li X., Zhang X., Ammonium removal from aqueous solutions using microwave-treated natural Chinese zeolite, Separation and Purification Technology 58 (2008) 359-366.
5. Zabochnicka-Świątek M., Czynniki wpływające na pojemność adsorpcyjną i selektywność jonowymienną klinoptylolitu wobec kationów metali ciężkich, Inżynieria i Ochrona Środowiska, Częstochowa 2007, 10, 1, 27-43.
6. Wanga,S., Peng Y., Natural zeolites as effective adsorbents in water and wastewater treatment, Chemical engineering Journal 156 1 (2010) 11-24.
7. Schmidt R., Anielak A.M. Adsorpcja Cr (III) na klinoptylolicie naturalnym i modyfikowanym manganem, VIII Ogólnopolska Konferencja Naukowa, 503-513, 2007.
8. Maryuk O., Gładysz- Płaska A., Rudaś M. Majdan M., Adsorpcja jonów chromu (VI) na powierzchniowo modyfikowanym klinoptylolicie, Przemysł Chemiczny 84/5 (2005) 360-363.

Water cleaning from toxic elements using phytofiltration with *Elodea canadensis*.

Maria Greger[1], Arifin Sandhi[1], Daniel Nordstrand[1], Claes Bergqvist[1] and Hanna Nyquist-Rennerfelt[1,2]

[1]Department of Botany, Stockholm University, 106 91 Stockholm, Sweden

[2]Sweco Environment AB, Box 340 44, 100 26 Stockholm, Sweden

Corresponding author e-mail: maria.greger@botan.su.se

Abstract
Using plants to remove toxic elements from various polluted waters has successfully been shown by so called phytofiltration. Best results are found when using submerged plants, and among the submerged species tested *Elodea canadensis* is suitable. The aim was to investigate the removal efficiency of a number of toxic elements and essential basic cations by *E. canadensis*. The importance was to find out if removal of toxic elements could be performed without a significant alteration of the element concentration of the essential basic cations. In five different studies we investigated removal of Cd, Cu, Zn, Pb, Sb, As, Mg, Ca, Na, K as well as changes in alkalinity (HCO_3^-) and pH by *E. canadensis*. The data showed that plants removed 51% Zn, 33% Cu, 41% Cd, 32% Pb, 75% As and 8% Sb. There was no removal of Mg, Ca, Na and K, and no or only a small change in HCO_3^- and pH. Thus, it seems possible to remove toxic elements from water by submerged plants and simultaneously keep the essential elements in the water for a good water quality.

1. Introduction

Natural waters as well as waste waters contain toxic elements to different degrees, sometimes in too elevated concentrations to be regarded as drinking water for humans and animals. Examples on toxic elements, which can be elevated are arsenic (As), antimony (Sb), cadmium (Cd), zinc (Zn) and copper (Cu), and they shall not exceed certain limit values in drinking waters [1]. Other elements, like calcium (Ca), magnesium (Mg), potassium (K) are essential and water is an important source of intake of these elements.

To get a healthy drinking-water cleaning is often necessary. Passive treatments of polluted waters with elevated levels of metals and metalloides using wetlands have been used since the early 80s [2]. In the wetlands, plants contribute to the metal removal process by decreasing the retention time of the water and by that increase the sedimentation of the metals [3]. They also provide a carbon source that enhances the sulphate reduction in the sediment [4], which may increase the binding of metals. Plants add oxygen and provide physical sites for microbial activity [5]. The oxygen originates to a high extent from the photosynthesis activity, which additionally increases the pH by the CO_2 uptake and change of the carbon equilibrium towards bicarbonate (HCO_3^-) [6]. The increased pH influences the precipitation of metals and the plant metal uptake [7]. Plants take up both essential and non-essential metals [8, 9], and in constructed wetlands, aquatic plants are able to remove both metals [10, 11, 12] and nutrients [13, 14]. The importance of plant metal uptake in a wetland in relation to other removal processes is, however, debated [4, 15, 16, 17].

There are several types of plants in a wetland; emergent, floating-leaved, submerged and free-floating plants. Among those, highest accumulation of heavy metals is found for submerged and free-floating plants [18]. One of the investigated metal accumulating submerged species is water weed, *Elodea canadensis* [12, 18]. It is commonly found in polluted waters and is a known metal accumulator [19, 18]. This plant is a weed that easily grows in temperate areas and is considered suitable to use in water cleaning, phytofiltration, in cold climate.

The aim was to find out if *E. canadensis* is suitable to use for removal of toxic elements from waters without decreasing essential element concentrations. A removal of toxic elements is desirable but not that of such elements as Ca and Mg. The influence on pH and the bicarbonate is also important to follow in phytofiltration.

2. Material and Methods

Plant Material and Growth Conditions. A greenhouse cultivation of Canadian waterweed, *Elodea canadensis*, from the Department of Botany at Stockholm University was used in these studies. The plants were cultivated in a greenhouse that maintained a light intensity of 110 ± 50 µmol m^{-2} s^{-1}, a light/dark cycle of 16/8 h and a water temperature of 22 ± 2°C.

The first and second experiment was performed in a climate chamber, which maintained light intensity at 80 ± 10 µmol m^{-2} s^{-1} in a 16/8 h light/dark regime. They were grown in various containers in different waters with a water temperature of 22 ± 2°C. The third, forth and fifth experiments were run in the greenhouse in the same conditions as above.

Removal by Plants of Cd, Cu, Zn and Pb from Artificial Storm Water. Containers (50L), 40 cm wide, 60 cm long, and 30 cm deep, were provided with staggered walls placed cross-wise to create a serpentine flow of 240 cm in length through the system and an inlet and outlet for the polluted water. Four kg of soil were used as a bottom layer in each box, on top of which a 3-cm-thick layer of sand was placed. Fifty-four plants (giving a biomass of 5 g per litre of water, were placed in the boxes. During 28 days, artificial storm water, containing 1.3 Zn, 0.1 Cu, 0.1 Cd and 0.3 Pb mg L^{-1} and a pH of 7.0 was running through the system with a flow rate of 3.5 ml min^{-1}. Water samples (2ml) were collected daily from the inlet and outlet for further analyses on Cd, Cu, Zn and Pb.

Removal of Sb in Sb Contaminated Water. Each of one-litre containers was provided with one plant (1 g biomass/L) as well as water containing 0.1 and 1 mg Sb L^{-1} added as antimony(III)chloride. pH of the solution was 5.9 and 4.7, respectively. The uptake of Sb was run for 24 hrs and thereafter analysed.

Normal plant density in a natural storm water treatment pond was determined by measuring the biomass in an area of 50cm x 50cm and a depth of 80cm. This biomass was calculated to be 8 g per litre of water. The biomass data and the uptake data on Sb from above mentioned experiment was then used to recalculate the removal (%) of Sb from a fictive storm water containing 0.1 and 1 mg Sb L^{-1}.

Removal of As from Polluted Water. Prior to the experiment water was polluted with soil (120 mg As/kg) from a glass-works, Flygfors, Sweden (56•48'N, 15•45'E) for 2 days giving water containing 30 µg As/L. Each of one-litre containers was provided with the polluted water (without soil added) and 3 g biomass of plants per L, which corresponding to 3 plants per litre. Experiment was run for 4 days with aeration. Samples, 2 ml each, was collected before and after and analysed on total As. pH of the solution was 5.9.

Removal of Essential Elements from Storm Water. Storm water was collected from Flemingsberg, Stockholm, Sweden (59•13´N, 17•59'E). Plants with a biomass of 6 g per litre were placed in 14 L containers provided with in and outflow system. The water was running 10 days through the system with a flow rate of 2 ml min^{-1}. The pH of the water was 7.4. Water samples (2ml) were collected from the inlet and outlet for further analyses on Ca, K, Mg, Na, pH and alkalinity (HCO_3^-).

Influence on pH and Bicarbonate. Tap water from the Department of Botany was used in this study. Plants with a biomass of 0.3 g per litre were placed in 14 L containers provided with in and outflow system. The water was running 3 days through the system with a flow rate of 2 ml min^{-1}. Water from the inlet and outlet was analysed on pH and alkalinity (HCO_3^-).

Analysis of Elements, alkalinity and pH. Plant samples for Sb analyses were wet digested according to Krachler et al. [20] and for Zn, Cu, Cd and Pb in HNO_3:$HClO_4$, 7:3. Water samples and the digested plant samples were analysed with atomic absorption spectrophotometer on Mg, K, Ca, Na, Zn by flame, on Cd, Cu, Pb by graphite furnace and on Sb with a HG-AAS (Varian SpectrAA-100). Standards were added to the samples to eliminate interaction with the sample matrix. Bicarbonate concentration was determined by titration method. pH was analysed using a pH-meter.

Calculations and Statistical Treatments. Removal of elements was calculated as

$$\text{Removal (\%)} = 100-(([M]_{outlet}/[M]_{inlet}) \times 100) \quad (1)$$

or

$$\text{Removal (\%)} = 100-(([M]_{after}/[M]_{before}) \times 100) \quad (2)$$

where [M] is the element concentration in outlet and inlet water, respectively, or after and before plant treatment, respectively.

Removal of heavy metals by plants during the whole time period (28 days) in the soil-containing wetland was calculated as,

$$\text{Removal by plants (\%)} = (M_{plant} / M_{added}) \times 100 \qquad (3)$$

where M_{plant} is the total content of the metal in the whole plant biomass and M_{added} is the total amount of the metal supplemented to the wetland.
Mean values are based on 3-4 replicates. Statistical differences between data was calculated using Student's t-test and Turkey-Kramer HSD test with the computer program JMP (SAS Institute INC.)

3. Results and Discussion

Wetlands did efficiently remove heavy metals from the water; 74% Zn, 79% Cu, 50% Cd and 92% Pb was removed from the artificial storm water (Tab. 1). When only plant uptake was taken into consideration the removal of metals from the water was lower; 50% Zn, 35% Cu, 49% Cd and 30% Pb was removed by plant uptake (Tab. 1). The rest was removed by sedimentation, which also increased by the presence of the plants, since plants retard the water velocity and promotes precipitation and sedimentation [3]. Of the total removal, plants did the majority of it in the case of Cd and Zn, 82 and 69%, respectively, while less in the case of Cu and Pb, 42 and 35%, respectively (Tab. 1). It is well known that Pb and Cu more easily binds to organic matter than the other two heavy metals and have a shorter retention time in water than Zn and Cd [21]. The latter may increase the metal uptake time by the shoot.

Table 1. Concentration of Cd, Cu, Zn and Pb in water before and after growing plants for 28 days in a flow system with artificial storm water. Maximum acceptance levels [22] in drinking water, MAL, and reference fresh water [23], RFW, are also given. SE < 10%

Metal	Concentration before, mg L^{-1}	Concentration after, mg L^{-1}	Total removal, %	Removal by plants, %	MAL, mg L^{-1}	RFW, mg L^{-1}
Cd	0.10	0.05	50	41	0.005 - 0.01	0.0002
Cu	0.09	0.02	79	33	1	0.003
Zn	1.19	0.30	74	51	5	0.005
Pb	0.28	0.02	92	32	0.05	0.003

In the case of metalloids, 75% of As but only up to 8% of Sb was removed through uptake by the plants (Tab. 2). The big difference could be due to the removal time; the removal of As was performed during 96 hrs while that of Sb only 24 hrs. In the As setup, matrix effects relating to the exchange of solutes between soil and water could also influence a higher uptake of As into plants, compared to the Sb setup where Sb was added to the water as a salt. The reason for the higher removal of Sb at high Sb concentrations was likely due to the big difference in pH, making Sb anion easier to take up at lower pH, in opposite to that of heavy metal cations [8].

Table 2. Removal of As and Sb by *Elodea canadensis* growing for 4 days and 24 hrs, respectively, in closed systems. Maximum acceptance levels [22] in drinking water, MAL and reference fresh water [23], RFW, are also given. SE < 10%

Metal	Concentration before, mg L^{-1}	Concentration after, mg L^{-1}	Total removal, %	MAL, mg L^{-1}	RFW, mg L^{-1}
As	0.03	0.02	75	0.05-0.2	0.5
Sb	0.10	—	1	0.01	0.0002
Sb	1.00	—	8	0.01	0.0002

Of the analysed essential basic cations only Ca tended to be removed by the plants, but not significantly (Tab. 3). This means that also Ca/Mg ratio tended to decrease, although not significant. The Na/K ratio was not affected.

Table 3. Concentration of Ca, Mg, K and Na in water before and after growing plants for 10 days in a flow system containing storm water. Maximum acceptance levels [22] in drinking water, MAL and reference fresh water [23], RFW, are also given.

Metal	Concentration before, mg L^{-1}	Concentration after, mg L^{-1}	Changes, %	MAL, mg L^{-1}	RFW, mg L^{-1}
Ca	22 ± 3	16 ± 3	16 ± 5	100	0.1 - 215
Mg	22 ± 1	24 ± 1	0	30	0.1 - 225
Ca/Mg	0.99 ± 0.16	0.65 ± 0.09	24 ± 2		
K	40 ± 5	38 ± 1	0	No limit	0.04 - 38
Na	71 ± 1	74 ± 1	0	100	0.1 - 845
Na/K	1.82 ±0.23	1.84 ± 0.13	0		

It could be mentioned that in this experiment, where the essential cations were analysed the pH did not change and the bicarbonate increased with about 3 % (not shown). In the next experiment, pH of the water decreased during the treatment (Tab. 4), likely due to a less active photosynthesis. The differences between the two studies was mainly due to the biomass; 6 g per litre and 10 days in the previous treatment (not shown) and 0.3 g per litre and 3 days in the experiment shown in tab. 4. It is well known that photosynthesis increase pH of the water and that this is due to a change in equilibrium in the carbonic acid system [6]. The plant density can therefore influence the pH and bicarbonate content in the water.

Table 4. pH and HCO_3^- concentration in tap water before and after growing plants for 3 days in a flow system. Levels [22] in drinking water, AL and reference fresh water [23], RFW, are also given. *significant change.

Metal	Concentration before, mg L^{-1}	Concentration after, mg L^{-1}	Change, %	AL, mg L^{-1}	RFW, mg L^{-1}
HCO_3^-	358 ± 3.77	357 ± 357	0	No limit	0.002 - 326
pH	7.74 ± 0.02	7.44 ± 0.04	3.8 ± 0.5*	6.5 - 9.5	2.9 - 8.6

4. Conclusion

The conclusion we can draw from this work is that phytofiltration using *E. canadensis* to remove toxic metals and metalloids without any effect on the concentration of essential base cations is possible. For an efficient removal it is though necessary to have an optimal plant biomass density in the water and optimal growth conditions. This may lead to preventing a decrease in pH and in turn a decreased loss of CO_2 and thereby also bicarbonate. Pre-treating drinking water before the inlet into the water purification plant using the phytofiltration technique could prove useful in order to remove metals and metalloids from the water.

Acknowledgments

This work was financed by Georange AB, New Boliden AB, C.F. Lundström and Kurt and Alice Wallenberg foundations. The great help during the laboratory work by Tommy Landberg is very much appreciated.

References

(1) I. Pais, and J.B. Jones, The handbook of trace elements, St. Lucie, 1997.

(2) M. Demchik and K. Garbutt, Wetlands and aquatic processes: J. Environ. Qual., 28 (1999) 243-249.

(3) J. Skousen, A. Sexstone, K. Garbutt and J. Scencindiver, Acid mine drainage treatment with wetlands and anoxic limestone drains: Appl. Wetlands Sci. Techn. (1992) 263-281.

(4) W.J. Mitch and K.M. Wise, Water quality, fate of metals, and predictive model validation of a constructed wetland treating acid mine drainage: Water res. 32 (1998) 1888-1900.

(5) L. St-Cyr, P.G.C. Campbell and K. Guertin, Evaluation of the role of submerged plant beds in the metal budget of a fluvial lake: Hydrobiologia 291 (1994) 141-156.
(6) J.K. Cronk and M.S. Fennessy, Wetland plants, biology and ecology, CRC press, Florida, 2001.
(7) M.T. Javed and M. Greger, Cadmium triggers *Elodea Canadensis* to change the surrounding water pH and thereby Cd uptake: Int. J. Phytorem. *(In press).*
(8) H. Marschner, Mineral nutrition of higher plants, Academic press, London, 1995.
(9) S. Clemens, Toxic metal accumulation, responses to exposure and mechanisms of tolerance in plants: Biochimie 88 (2006) 1707-1719.
(10) J.S. Dunbabin and K.H. Bowmer, Potential use of constructed wetlands for treatment of industrial wastewaters containing metals: Sci. Tot. Environ. 111 (1992) 151-168.
(11) M.O. Doyle and M.L. Otte, Organism-induced accumulation of iron, zinc and arsenic in wetland soils: Environ. Pollut. 96 (1997) 1-11.
(12) A. Samecka-Cymerman and A.J. Kempers, Biomonitoring of water pollution with *Elodea Canadensis*. A case study of three small Polish rivers with different levels of pollution: Water Air Soil pollut. 145 (2003) 139-153.
(13) A.S. Juwarkar, B. Oke, A. Juwarkar and S.M. Patnik, Domestic wastewater treatment through constructed wetlands in India: Water Sci. Technol. 32 (1995) 291-294.
(14) L. Yang, H-T. Chang and M.L. Huang, Nutrient removal in gravel- and soil-based wetland microcosms with and without vegetation: Ecol. Eng. 18 (2001) 91-105.
(15) A. Sobolewski, A review of processes responsible for metal removal in wetlands treating contaminated mine drainage: Int. J. Phytorem., 1 (1999) 19-51.
(16) M. Kamal, A. Ghaly, N. Mahmoud and R. Côté, Phytoaccumulation of heavy metals by aquatic plants: Environ. Inter. 29 (2004) 1029-1039.
(17) L.G. Vardanyan and B.S. Ingole, Studies on heavy metal accumulation in aquatic macrophytes from Sevan (Armenia) and Carambolim (India) lake systems: Environ. Int. 32 (2006) 208-218.
(18) Å. Fritioff and M. Greger, Aquatic and terrestrial plant species with potential to remove heavy metals from stormwater: Int. J. Phytorem., 5 (2003) 211-224.
(19) M.A. Kähkönen, M. Pantsar-Kallio and P.K.G Manninen, Analysing heavy metals concentrations in different parts of Eloda Canadensis and surface sediment with PCA in two different boreal lakes in southern Finland: Chemosphere, 35 (1997) 1381-1390.
(20) M. Krachler, M. Burow and H. Emons, Development and evaluation of an analytical procedure for the determination of antimony in plant materials by hydride generation atomic absorption spectrometry: Analyst 124 (1999) 777-782.
(21) U. Förster and G.T.W. Wittman, Metal pollution in the aquatic environment, Springer-Verlag, Berlin, 1979.
(22) Livsmedelsverkets föreskrifter om dricksvatten, SLVFS 2001:30.
(23) http://www.slu.se/vatten-miljo, http://info1.ma.slu.se/RI2005/

Selectively facilitated transport of Zn(II) through a novel polymer inclusion membrane containing Cyanex 272 as a carrier reagent

Abdurrahman Yilmaz[1], Gulsin Arslan[1,], Ali Tor[2], Ilker Akin[1], Yunus Cengeloglu[1] and **Mustafa Ersoz**[1]

[1]Department of Chemistry, Selcuk University, Konya 42075, Turkey
[2]Department of Environmental Engineering, Selcuk University, Konya 42075, Turkey

Corresponding author e-mail: 71arslan@gmail.com

This paper describes the selectively facilitated transport of Zn(II) through a novel polymer inclusion membrane (PIM) containing Cyanex 272 as a carrier reagent. The prepared PIM was characterized by using FTIR and atomic force microscopy (AFM) techniques and contact angle measurements. The effects of Zn(II) (in feed phase), HCl (in stripping phase) and Cyanex 272 (in membrane) concentrations on the transport were investigated. When the feed phase contained $1x10^{-4}$ M Zn (II) at pH 3.4, 99% of Zn(II) was transported through the PIM (prepared with 1.0% carrier solution) by using 0.5 M of HCl as a stripping phase. Furthermore, Zn(II) was preferably transported in the presence of various metal ions (i.e., Ni(II), Cu(II), Pb(II) and Co(II), etc.). The results also showed that transport efficiency of the PIM was reproducible and it could be efficiently used in the long-term separation processes.

Peculiarities of Fe(III) sorption from drinking water onto chitosan

Ona Gylienė

Institute of Chemistry of the Center for Physical Sciences and Technology,

A.Goštauto 9, Vilnius 01108, Lithuania

Corresponding author e-mail: gyliene@ktl.mii.lt

Abstract:
The high sorption capacity, high sorption rates of iron ions onto chitosan nanofibers and the easily regeneration enables to use this sorbent in flowing mode. The sorption mechanism of iron ions remarkably differ from that of heavy metal ions. FT-IR investigations showed that the sorption proceeds in the first order due to interactions of iron ions with chitosan hydroxyl groups. Despite the presence in solutions Fe(II) or Fe(III) on the surface of chitosan only the Fe(III) ions are sorbed. It could be accepted that the chitosan nanofibers act as catalyst for the conversion Fe(II) to Fe(III).

1. Introduction

Treatment methods used for preparation of drinking water from natural aquifers do not ensure the complete removal of contaminants because of pollution of the environment with products of anthropogenic nature. Among these pollutants the heavy metals are considered to be most dangerous. In order to improve the drinking water quality domestically different means are proposed. For this purpose sorbents, such as activated carbon and synthetic ion exchangers are widely used. In recent years biosorbents have been intensively investigated. The main advantages of biosorbents are their biocompatibility, biodegradability and renewal of raw material sources.

The biosorbent chitosan is distinguished for its high selectivity in the sorption of pollutants of anthropogenic nature, especially heavy metal ions; meanwhile the innocuous metals are not sorbed. The regularities of heavy metal sorption by chitosan are widely studied. The chitosan sorption ability depends on its physical and chemical properties. The deacetylation degree (the number of amino groups in polymer molecule) has the decisive influence. The amino groups interact with heavy metal ions forming the complexes on the sorbent surface. Metal ions may be bound with several nitrogen atoms from the same or from different chains of chitosan ("bridge model") or with the single amino group ("pendant model"). The residual sites could be occupied by -OH group of chitosan or by H_2O or hydroxyl group [1, 2]. However, the sorption of other pollutants, such as iron ions, which present, as a rule, in drinking water in large amounts and organic compounds, is much less investigated [3, 4]. Besides, iron and aluminum salts are widely used as a coaguliants for preparation drinking water in centralized treatment facilities. The hazardous effects of iron ions on human health are evidently shown in works [5, 6]. Our investigations were performed with the purpose to evaluate possibility to use the chitosan for removal of iron ions from drinking water.

2. Experimental

For investigations chitosan in form of fibers and flakes was used. Batch and fixed bed column studies were carried out to investigate the sorption of iron ions under equilibrium and dynamic conditions. The load in batch reactor was 10 g of dry sorbent per liter of solution in sorption and desorption experiments. Adsorption was investigated at room temperature. The pH was adjusted with NaOH or H_2SO_4 solutions. After sorption, chitosan was filtered and rinsed with cold deionized water and dried at 70 °C.

For dynamic sorption experiments column of 20 cm length, 2 cm of diameter filled with 2 g of chitosan under superficial velocity 0.05 cm/s was used.

The model solutions, containing Fe(II) dissolved in distilled water and tap water containing different concentrations of Fe(III) have been used.. The sorbed quantities of iron ions were determined from the changes in the concentration of the solutions. Fe(II) in solutions was determined photometrically using indicator o-phenantroline. The total amount of Fe(II) and Fe(III) ions was determined after reduction of last to Fe(II) with hydroxylamine.

Column regeneration and possibility of reusing of regenerated chitosan were also investigated. Quality of chitosan was evaluated by determination its sorption ability, sorption rate and FT-IR investigations

3. Results and Discussion

Experiments performed under equilibrium conditions with model solutions containing Fe(III) ions indicated that pH values have the decisive influence on the uptake of Fe(III) ions by chitosan. With increase in pH the sorption rate dramatically increases; meanwhile the influence of pH on Fe(II) sorption is moderate (Figure. 1). Sorbed amounts of iron ions onto chitosan depend strongly on the form of chitosan. The sorption ability of chitosan fibers is much higher (up to 10 times) than that of chitosan flakes. The reason probably is the considerably larger surface of nanoscale fibers in comparison with the surface of flakes.

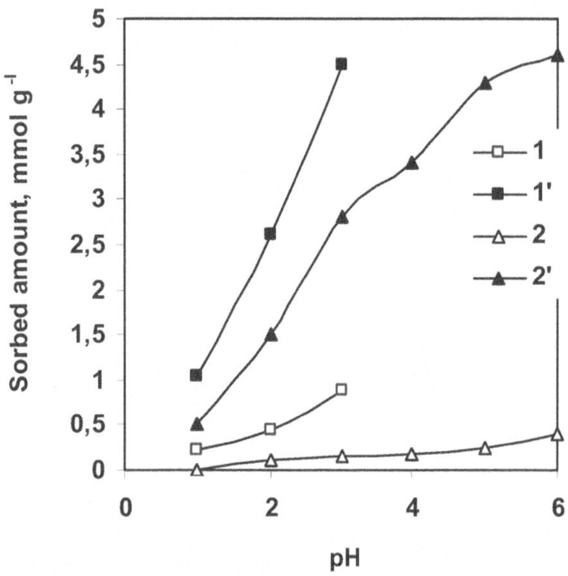

Fig. 1. pH influence on Fe(III) (1and 1') and Fe(II) (2 and 2') sorption on chitosan flakes (1 and 2) and chitosan fibers (1' and 2')

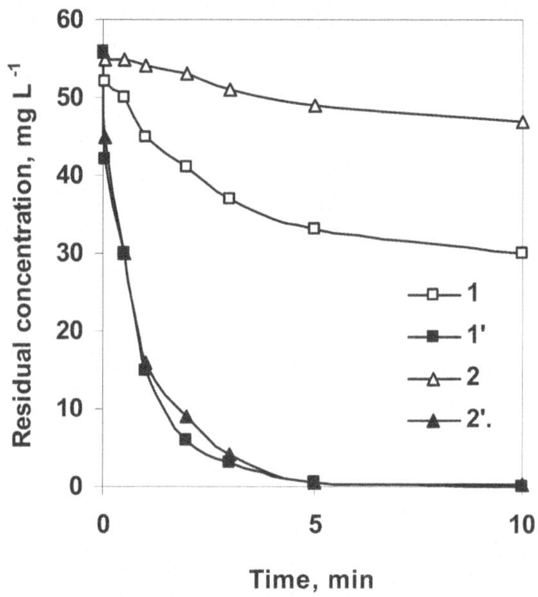

Fig. 2. Fe(III) (1 and 1') and Fe(II) (2 and 2') uptake by chitosan flakes (1 and 2) and chitosan fibers (1' and 2') at pH 4

Chitosan fibers distinguish by extremely high sorption rates, which exceed the sorption rate onto flakes about hundred times (Figure. 2). High sorption rates are not characteristic for biosorbents. It is worth to note that the sorption rate for Fe(II) is approximately the same as for Fe(III) on both sorbents - flakes and fibers. The reason is that on the surface of sorbent the Fe(III) only is absorbed. It is unexpected especially for chitosan fibers, where the sorption rates are very high. It could be accepted that chitosan acts as the catalyst in conversion Fe(II) into Fe(III).

The sorption ability of chitosan depends strongly on the concentration of metal ion in solution (Figure. 3). The decreased value of pH during sorption is an indication of the sorption of iron cations only, meanwhile the anions of the dissolved iron salt are not sorbed by chitosan. These results are in contradiction to the results of sorption of heavy metal ions, when the pH value during the sorption increased, thus partly removing the anions together with cations from solutions.

Unexpected results were obtained when the ionic strength of solutions was increased. The addition of sodium sulphate or sodium chloride to solutions containing Fe(II) ions caused the increase in both the sorption rate and sorption ability of chitosan. Usually an increase in ionic strength causes a decrease in sorption ability for many other pollutants. Thus, this effect of background electrolyte indicates, that the reason of sorption is not the electrostatic interactions. Obviously, the mechanism of Fe(III) sorption onto chitosan basically differs from that of other heavy metal sorption. The reason for the unusual course of Fe(III) sorption could be the affinity of Fe(III) to the OH groups of chitosan. In this case the microprecipitation of the insoluble Fe(III) compounds onto the surface of chitosan is possible. The different way of Fe ions sorption onto chitosan in comparison with heavy metals was confirmed by FT-IR investigations, which showed the interactions of hydroxyl groups of chitosan with iron ions after sorption; meanwhile the amine and amide peaks remain unchanged.

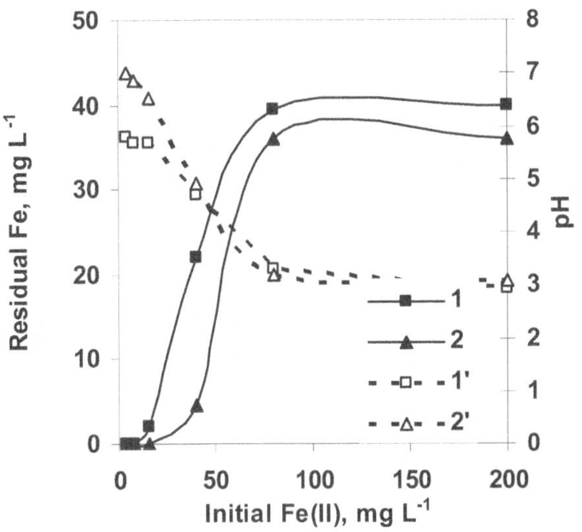

Fig. 3. Influence of background electrolyte on Fe(II) sorption onto chitosan flakes under equilibrium conditions: 1 and 1' in distilled water solutions; 2 and 2' in solutions containing 20 g/L Na_2SO_4

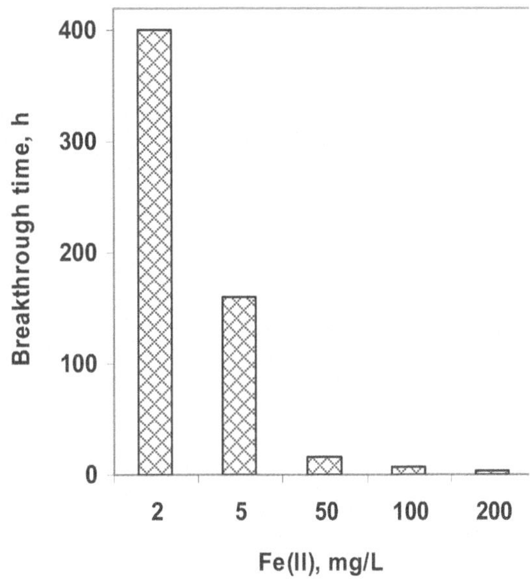

Fig. 4. Dependence of sorption breakthrough time on initial iron ion concentrations

The extremely high sorption rates onto chitosan nanofibers and high sorption capacity suggest the possibility to use the purification of the solutions from iron ions in dynamic conditions. The effectiveness of purification column depends on the concentration of iron ions in water. With increase in iron concentration the working time of column decreases. The results are presented in Figure 4.

Chitosan after iron ion sorption could be easily regenerated by treatment in dilute (1:100) H_2SO_4 solutions. After 10 cycles of regeneration the sorption capacity and sorption rate remain unchanged though some changes in constitution of chitosan nanofibers is visible.

4. Conclusions

Thus, the extremely high sorption rate of Fe(III) onto chitosan, its increase with increase in ionic strength and high sorption capability enable to use sorbent chitosan in flowing mode and easily to adapt it for practical use.

References

1. R. Rhazi, J. Desbrieres, A. Tolaimate, M. Rinaudo, P. Vottero and A. Alagui, Contribution to the study of the complexation of chitosan and oligomers, Polymer **43** (2000) 1267-1276.
2. E. Guibal, Interaction of metal ions with chitosan based sorbents: a review. Separation and Purification Technology **38** (2004), 43-74.
3. Ke-long Huang, Ping Ding, Su-qin Liu, Gui-yin Li and Yan-fei Liu, Preparation And Characterization Of Novel Chitosan Derivatives: Adsorption Equilibrium of Iron(III) Ion, Chinese Journal of Polymer Science**26** (2008), 1–11.
4. O. Gyliene and S. Visniakova, Heavy metal removal from solutions using natural and synthetic sorbents, Environmental Research, Engineering and Management, 2008 (1), 28-34.
5. I Rosborg,., B. Nihlgård, and L. Gerhardsson, Inorganic chemistry of well water in one acid and one alkaline area of south Sweden. WASP **142** (2003) 261-277.
6. I Rosborg,., B. Nihlgård, and L. Gerhardsson, Hair element concentrations in females in one acid and one alkaline area in south Sweden. Ambio **32** (2003) 440-446.

Iron based nano-materials for reductive remediation of pollutants

Paul Duffy, Deirdre Murphy, Laura Soldi, Ronan Cullen and **Paula E. Colavita**

School of Chemistry, Trinity College Dublin, College Green, Dublin 2, Ireland
Corresponding author e-mail: colavitp@tcd.ie

Abstract

Nanosized Zero-valent Iron (nZVI) has been shown to be an effective remediation agent for environmental contaminants. Research has indicated that nZVI can effectively degrade numerous pollutants in groundwater such as organohalides, high valent metals and inorganic anions. Advantages of nZVI include increased reactivity due to a higher surface to volume ratio with respect conventional ZVI and lower costs of implementation. However, Problems remain with practical use of nZVI in the field due to its poor mobility in soil. We report on our work towards developing composites with improved transport properties for the delivery of nZVI through soil matrices.

1 Introduction

In Recent years, the remediation of heavy metals, organohalides and inorganic anions has become an area of much interest and research. This is due to the large amount of sites contaminated with these species due to industry and various other activities. One of the most promising areas of research in this field is the use of Zero-valent iron (ZVI) as a remediant.[1] ZVI has been shown to successfully remediate many of the species mentioned above in a research laboratory environment. ZVI has also been used in permeable reactive barriers to treat contaminated ground water out in the field. ZVI has proven to be successful within these barriers, reducing much of the contaminants in the water. Also, the by-products formed from the ZVI reduction reaction are less toxic and harmful than the starting contaminants.[2],[3],[4]

In general, the reduction process for ZVI is a mechanism which involves active surface sites on the ZVI particle. Research has shown that nano sized ZVI (nZVI) has far superior reactivity towards contaminants than ZVI because of the increased surface to volume ratio for smaller particles. For the same weight of ZVI there are more active surface sites on the nZVI capable of reacting with environmental pollutants. Another advantage of increased reactivity and small size is that in situ remediation is now an effective possibility.[5] In situ remediation is needed for deep aquafiers where permeable reactive barriers are unfeasible due to cost. Also, conventional pump and treat methods are expensive and often lead to the contaminant source itself not being adequately treated. Direct injections of aqueous slurries of nZVI to the contaminant source should in theory be possible for in situ remediation of deep aquafiers.

However, laboratory experiments indicate that nZVI has poor mobility and transport properties in porous soil media.[6] This has been attributed to the severe aggregation of nZVI particles in soil and also from very effective filtration mechanisms by the soil matrix due to the size of the nZVI. It has been reported that mobility can be improved by use of surfactants and other stabilisers which hinder particle aggregation.[7],[8],[9],[10] However, this often comes with a reactivity trade off due to the surfactants causing the surface to be less accessible to contaminants.

We report on our work towards developing a method in order to enhance transport whilst preserving reactivity. We have synthesized carbon microspheres using a method developed by Skrabalak et al.[11],[12] which should in theory be able to minimise filtration of soil via size control. Also, surface functionalities can be tailored onto the carbon to modulate their transport properties via surface charge. These carbon microspheres will act as supports to carry anchored nZVI to contaminated sites. We report preliminary results on the synthesis of iron nanoparticles supported on these carbon microparticles for applications in in situ remediation of contaminants in the subsurface.

2. Materials and Methods

Chemicals. Iron (III) Chloride, sodium borohydride, lithium hydroxide, dichloroacetic acid, Palladium (II) Chloride, Tin (II) Chloride, Ammonium Chloride, Sodium Hypophosphite, Iron Sulphate heptahydrate, Glycine, Potassium Hydroxide and Sodium Hydroxide were purchased from Sigma and used without further purification; absolute ethanol was purchased from the university solvent facility. 4-carboxy benzenediazonium tetrafluoroborate and 4-sulphonate benzenediazonium were synthesized according to previously reported methods.[13],[14]

Carbon Microsphere Synthesis. Carbon microparticles were synthesized according to methods developed by Skrabalak et al.[11] Briefly, a 1.67 MHz piezoelectric crystal placed at the bottom of a flask is used to generate a mist from a solution of the organic salt precursor. Droplets of precursor solution thus generated are in the micrometer range and narrowly dispersed.[15],[16] The mist is carried into a furnace by a flow of inert gas and the organic salt is pyrolysed at 700 °C. Particles were collected in a bubbler containing deionised water and were characterised using a combination of DLS (Zetasizer nano ZS, Malvern Ltd.), nitrogen adsorption and Scanning Electron Microscopy (SEM). Functionalization of carbon microspheres was carried out by immersing the spheres in a solution of the diazonium salt for 24 h; spheres were subsequently washed and centrifuged prior to characterization via infrared spectroscopy and □-potential measurements (Zetasizer nano ZS, Malvern Ltd.).

Carbon/Iron Composite, Iron templating was achieved using the method described by Drovosekov et al.[17] Briefly, Pristine carbon powder is placed in a 0.17 M SnCl2 solution, followed by a 0.14 M acidified PdCl2 solution at 70°C for 1 min each. Carbon microparticles are then placed in a solution consisting of 1.25 M FeSO¬4, 0.2 M Glycine, 2 M NH4Cl and 0.4 M NaH2¬PO2 . The solution is brought to a pH of 10.5 via addition of NH4OH at 90°C in order to initiate Fe0 growth. Carbon Particles were filtered and washed with copious amounts of water after every step.

3. Results and Discussion

3.1 Synthesis of Carbon Microspheres

Carbon microspheres (CM) were synthesized using a home-built ultraspray pyrolysis apparatus (USP). Figures 1a and 1b show SEM images of carbon microspheres pyrolysed from two different precursor solutions. Figure. 1a shows the typical mesoporous structure that is obtained from 1.5 M Lithium Dichloroacetate (LiDCA) precursor solutions. LiDCA carbon microspheres had a BET surface area of 1040 m2/g indicating that the microsphere is highly porous, Figureure 1b shows the open pore structure obtained from carbon microspheres synthesized using 1.5 M precursor solutions of Sodium Dichloroacetate (NaDCA). The images show that carbon microspheres produced from NaDCA have pores

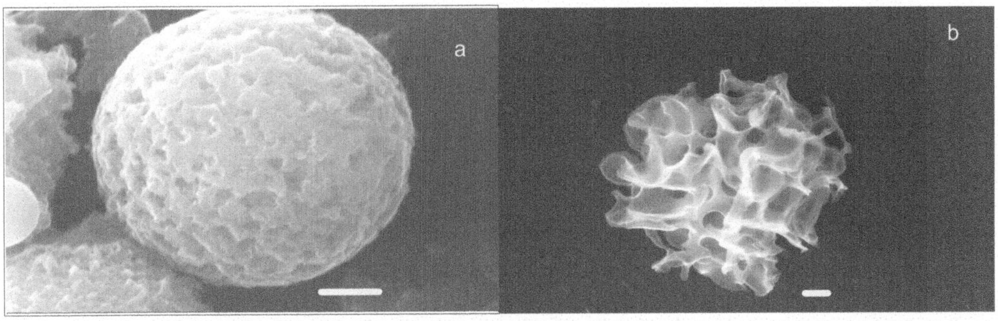

Figure 1: (a) SEM image of CM synthesised from LiDCA precursor. **(b)** SEM image of CM synthesised from NaDCA precursor. Scale bars represent 200 nm.

of larger size than those obtained using LiDCA as a precursor. NaDCA particles have a BET surface area of 470 m2/g, inferior to that of LiDCA; however, the area is still sufficiently high to offer a large number of sites available for both adsorption of contaminants and anchoring of nZVI. Figure 2 shows typical dynamic light scattering size distributions for microparticles obtained using 0.1 M, 1.0 M and 1.5 M NaDCA precursor solutions.[18] The peak diameter was measured at 295 nm, 712 nm and 955 nm for 0.1, 1 and 1.5 M respectively. These results indicate that particle size distribution is controlled by the aerosol generation process.

Surface functional groups and surface charge of carbon particles can be controlled by using diazonium chemistry. Figure 3 shows the infrared transmittance of carbon microparticles before and after functionalization via spontaneous grafting of 4-carboxybenzenediazonium salts. To compare pristine with functionalised, the graphs were baseline corrected and normalised. There appears to be a shoulder peak in the functionalised spectrum at 1695 cm-1 that is not readily present in the pristine carbon spectrum. This is the expected region for C=O stretching peaks of aryl carboxylic acids which absorb in the range 1680-1700 cm-1.[19] This result suggests that aryl diazonium salts successfully graft on the CM surface. Figure 4 shows the zeta potential pH curves for the CM's before and after functionalisation via grafting of

4- diazobenzenesulfonic acid diazonium salts. This data suggests that the surface charge becomes more negative after functionalisation. This result supports the previous FTIR data which implies successful diazonium grafting onto the CM surface. These results indicate that both surface chemistry and inter-particle electrostatic interactions can be controlled via chemical functionalization. The ability to control these properties will prove useful for later mobility and transport studies.

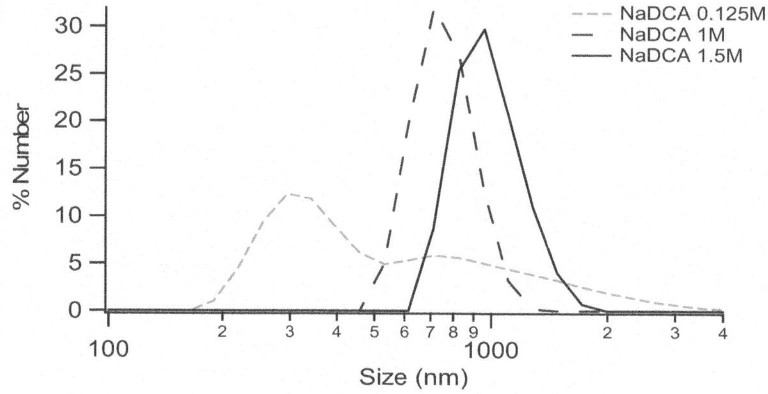

Figure 2: Size distributions obtained using different starting concentrations of NaDCA precursor solution.

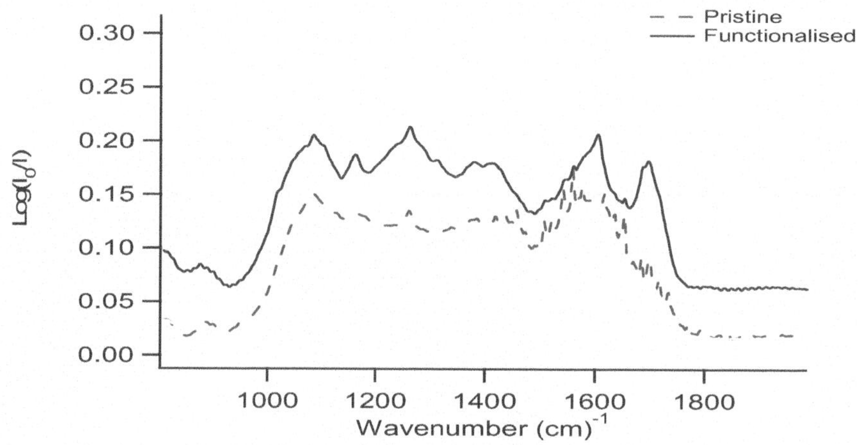

Figure 3: FTIR spectrum comparing pristine CM to CM's which were functionalised with 4-carboxy benzenediazonium tetrafluoroborate.

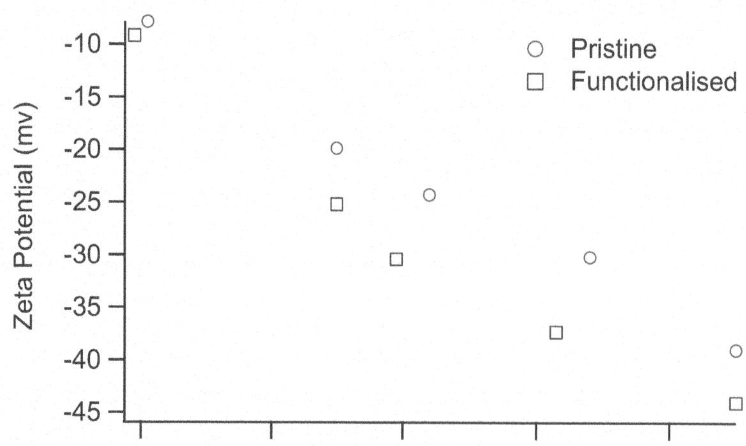

Figure 4: ζ-potential data comparing pristine CM with CM's functionalised with 4-sulfonate benzenediazonium tetrafluoroborate.

3.2 Synthesis of Carbon/Iron Composites

Carbon/Iron composites were synthesised using a surface templating method. First, iron growth was established on a Si wafer in order to test reaction conditions suitable for the forming over the silicon surface. In order to confirm the grey deposit as Fe0, the surface was treated with concentrated HCl to dissolve Fe0 and form Fe2+. Fe2+ was then quantitated via colorimetric methods using 1,10-phenanthroline, which forms a complex with Fe2+ with an absorption maximum at 510 nm.[20] Figure 5 shows the UV-Vis spectrum obtained from this experiment, confirming the presence of Iron on the Si sample. Figure 6a and 6b show the results obtained with this method as applied to a sample of LiDCA and NaDCA particles respectively. A qualitative comparison of these two images indicates that NaDCA carbon is better for producing the supported Iron nano-particles, possibly because LiDCA particles possess extremely small pores which are not accessibly to the iron solution. Figure 6b shows NaDCA microparticles after Fe0 growth for 15 min under the same conditions. Iron grows effectively on the surface of NaDCA and crystal facets of a cubic lattice are clearly seen in microscopy images. We have found that the size of the clusters is very difficult to control by using this methodology and are currently investigating different protocols for the surface templated growth of Fe0.

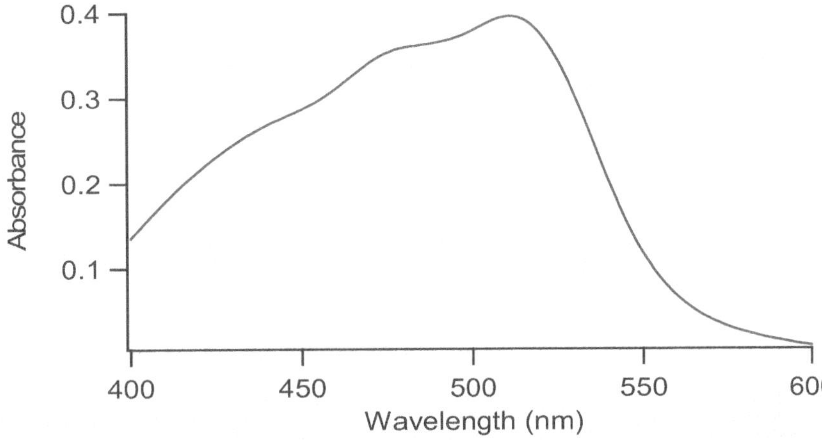

Figure 5: UV-Vis spectrum of complex formed by Fe^{2+} and 1, 10 phenanthroline

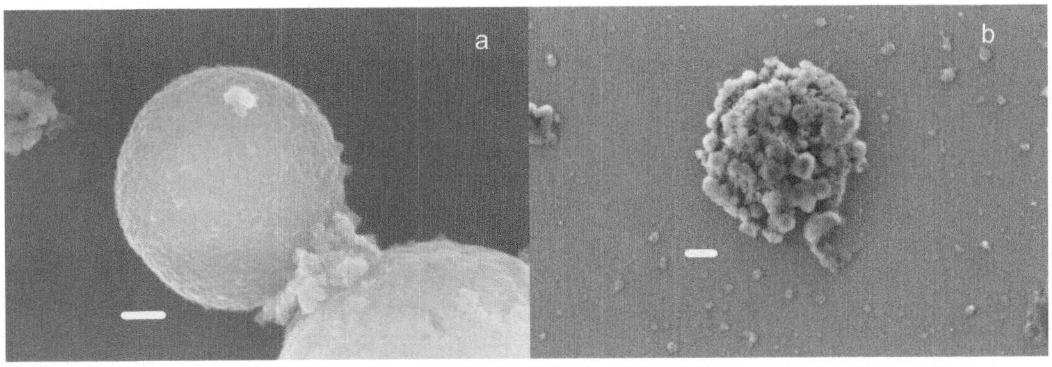

Figure 6: **(a)** SEM image of LiDCA CM after templating; scale bar represents 200 nm. **(b)** SEM image of NaDCA CM after templating; scale bar represents 2000 nm

4. Conclusions

We have successfully synthesized CM's with different structures and porosity. These CM's have a narrow size distribution but more importantly have a size distribution that can be easily controlled by varying precursor parameters as described previously. The surface chemistry and surface charge can be modulated via post-synthesis functionalisation of the carbon surface with Diazonium chemistries. LiDCA microparticles proved to be ineffective as a support for templated Iron growth. However NaDCA was much more effective demonstrating successful templating of Iron particles and clusters at the carbon surface

Acknowlegdments

The authors wish to thank the Environmental Protection Agency, Ireland, for funding and support of this project. DM and RC's work is supported by Science Foundation Ireland under the Research Frontiers Programme, while LS is funded by the Irish Research Council for Science, Education and Technology (IRCSET).

References

(1) Li, X. Q.; Elliott, D. W.; Zhang, W. X. Critical Reviews in Solid State and Materials Sciences 2006, 31, 111.
(2) Zhang, W. X. Journal of Nanoparticle Research 2003, 5, 323.
(3) Narr, J.; Viraraghavan, T.; Jin, Y. C. Fresenius Environmental Bulletin 2007, 16, 320.
(4) Laine, D. F.; Cheng, I. F. Microchemical Journal 2007, 85, 183.
(5) Cantrell, K. J.; Kaplan, D. I.; Wietsma, T. W. Journal of Hazardous Materials 1995, 42, 201.
(6) Phenrat, T.; Saleh, N.; Sirk, K.; Tilton, R. D.; Lowry, G. V. Environmental Science & Technology 2007, 41, 284.
(7) Kanel, S. R.; Nepal, D.; Manning, B.; Choi, H. Journal of Nanoparticle Research 2007, 9, 725.
(8) Saleh, N.; Phenrat, T.; Sirk, K.; Dufour, B.; Ok, J.; Sarbu, T.; Matyiaszewski, K.; Tilton, R. D.; Lowry, G. V. Nano Letters 2005, 5, 2489.
(9) Saleh, N.; Sirk, K.; Liu, Y. Q.; Phenrat, T.; Dufour, B.; Matyjaszewski, K.; Tilton, R. D.; Lowry, G. V. Environmental Engineering Science 2007, 24, 45.
(10) He, F.; Zhao, D. Y. Environmental Science & Technology 2005, 39, 3314.
(11) Skrabalak, S. E.; Suslick, K. S. Journal of the American Chemical Society 2006, 128, 12642.
(12) Skrabalak, S. E.; Suslick, K. S. Journal of Physical Chemistry C 2007, 111, 17807.
(13) Allongue, P.; Delamar, M.; Desbat, B.; Fagebaume, O.; Hitmi, R.; Pinson, J.; Saveant, J. M. Journal of the American Chemical Society 1997, 119, 201.
(14) Hermans, A.; Seipel, A. T.; Miller, C. E.; Wightman, R. M. Langmuir 2006, 22, 1964.
(15) Lang, R. J. Journal of the Acoustical Society of America 1962, 34, 6.
(16) Lozano, A.; Amaveda, H.; Barreras, F.; Jorda, X.; Lozano, M. Journal of Fluids Engineering-Transactions of the Asme 2003, 125, 941.
(17) Drovosekov, A. B.; Ivanov, M. V.; Lubnin, E. N. Protection of Metals 2004, 40, 89.
(18) Berne, B. J. Dynamic Light Scattering with Applications to Chemistry, Biology and Physics; Dover, 2000.
(19) Socrates, G. Infrared and Raman Characteristic Group Frequencies; 3rd ed.; wiley, 2001.
(20) Harvey, A. E.; Smart, J. A.; Amis, E. S. Anal. Chem. 1955, 27, 26.

Removal of lead and chromium (III) by zeolites synthesized from fly ash

Magdalena Zabochnicka-Świątek, Tomasz Doniecki, Artur Błaszczuk and Ewa Okoniewska

Częstochowa University of Technology, Faculty of Environmental Protection and Engineering, Brzeznicka 60a, 42-200 Częstochowa, Poland
Corresponding author e-mail: mzabochnicka@is.pcz.czest.pl

Abstract

The specific composition and structure of zeolites produces unique molecular-sieves, sorption and ion exchange properties which allows for a wide range of applications. Various types of zeolites can be produced form fly ash. The type of fly ash used in the synthesis, kind of the method applied, temperature and solution/fly ash ratio determine the type of zeolite of different structure and efficiency.
Zeolites synthesized from fly ash can be employed in many technologies to remove cations of various metals from solutions by means of adsorption, filtration, ion exchange, coagulation, flotation and sedimentation. The important advantage of zeolites applied to water treatment is its high porosity when comparing to other minerals.

The overall goal of this study was to evaluate the adsorption capacity of zeolites synthesized from fly ash for heavy metals (Pb, Cr) removal from water. The study was carried out under static conditions (batch tests).

The differences in sorption capacity for each of the cation were observed. The obtained results indicate that zeolites synthesized from fly ash provides an economical means of removing these heavy metals from water. The effectiveness of metals removal by zeolites depends on the concentration of ions in solution and type of the metal ions to be removed.

After the process of heavy metal removal zeolites synthesized from fly ash could be regenerated and use many times after the regeneration process. In conclusion, low cost adsorbent, as zeolites synthesized from fly ash could be used in order to minimize processing costs in removal of heavy metals such as lead and chromium from water.

1. Introduction

Coal fly ash is one of the solid wastes produced from the combustion of coal in coal fired power stations. l,5-20% of the coal mass after combustion remains as fly ash and bottom ash. Only half of the fly ash is used as raw material for cement manufacturing and construction while the the other half is disposed of in landfills. As a result, it leads to various environmental problems such as polluting soils and groundwater.

Two major classes of fly ash are specified on the basis of their chemical composition resulting from the type of coal burned; these are designated Class F and Class C. Class F is fly ash normally produced from burning anthracite or bituminous coal, and Class C is normally produced from the burning of subbituminous coal and lignite:

F - is consist of 40÷60% Al_2O_3, 20÷30% SiO_2, Fe_2O_3 and CaO <7% and other compounds, in small quantities,

C - is consist of CaO > 20% and less than F class of: Al_2O_3, SiO_2, Fe_2O_3.

Class C fly ash usually has cementitious properties in addition to pozzolanic properties due to free lime, whereas Class F is rarely cementitious when mixed with water alone [1].

Fly ash is an agglomerate of hollow spheres (Figure.1.) with diameter from 1 to 200 µm that contain silicon and aluminum as major elements, and are crystallo-graphically composed of an amorphous component with some crystals as a quartz, mullite, hematite, and magnetite [2].

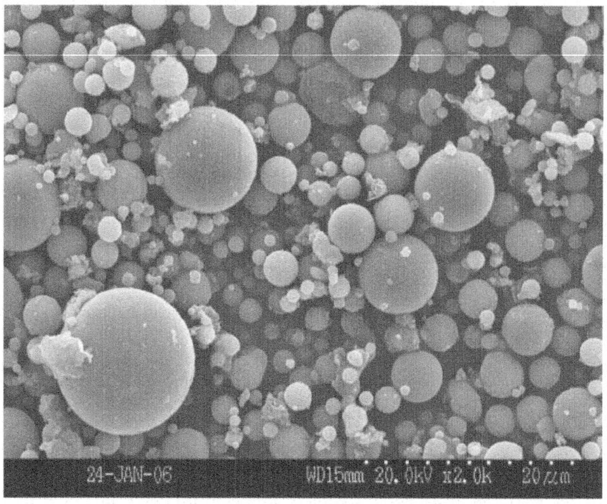

Figure1. SEM photographs of fly ash.

The specific composition and structure of zeolites produces unique molecular-sieves, sorption and ion exchange properties which allows for a wide range of applications. Various types of zeolites can be produced form fly ash [3-5].

The type of fly ash used in the synthesis, kind of the method applied, temperature and solution/fly ash ratio determine the type of zeolite of different structure and efficiency [6,7].

Zeolites synthesized from fly ash can be employed in many technologies to remove cations of various metals from solutions by means of adsorption, filtration, ion exchange, coagulation, flotation and sedimentation. The important advantage of zeolites applied to water treatment is its high porosity when comparing to other minerals.

The overall goal of this study was to evaluate the adsorption capacity of zeolites synthesized from fly ash for heavy metals (Pb, Cr) removal from water. All samples obtained from fly ash were characterized and confirmed by XRD patterns. In addition, controlling crystallization timeThe study was carried out under static conditions (batch tests).

2. Materials and Methods

The coal fly ash was of Polish power plant origin. The fly ash was used as the raw material of zeolite synthesis.

2.1 Synthesis of zeolite

Fly ash was mixed with NaOH solutions in a breaker. The breaker was placed in a water bath and heated at 95^0C for 48h and 72h.

Synthesis of zeolites by hydrothermal treatment of fly ash was prepared as follows - 25g of fly ash and:

- 200mL NaOH solution of 5 mol/L were placed in a 200mL breaker, for crystallization time of 48h: T+NaOH-48,
- 200mL NaOH solution of 5 mol/L were placed in a 200mL breaker, for crystallization time of 72h: T+NaOH-72.

The solid was separated by centrifugation at 4000 rpm for 10 min, washed with distilled water until the pH value was fixed. Then the resulting solid were dried at 105^0C for 12 h in the air, sieved and stored for further use in the adsorption experiments.

Raw fly ash and all samples obtained from fly ash were characterized and confirmed by XRD patterns. XRD pattern was taken on a Bruker D8 Advance X-ray diffraction instrument, the diffraction angle (2θ) in the range 2-65° (Cu Kα radiation), was scanned.

In the table 1 is shown the elementary analysis of fly ash. Results of elementary analysis apply to fly ash from circulating fluidized bed boiler. Solid sample was collected from the elect ostatic precipitator (i.e. ESP) 1st field.

Table 1. Sulphur content, unburned C content and Alkali content in fly ash after 1st section of ESP.

FLY ASH 1st ESP								
S Content [%]	Unburned C Content [%]	SiO_2 Content [%]	Al_2O_3 Content [%]	Fe_2O_3 Content [%]	CaO Content [%]	MgO Content [%]	Na_2O Content [%]	K_2O Content [%]
4,43	1,19	23,09	2,00	8,20	1,05	1,45	14,66	41,76

Inorganic chemicals used in the study, such as NaOH, NaCl, $Pb(NO_3)_2$, $Cr(NO_3)_3$ were all analytical grade reagents.

2.2 Adsorption studies

To study the adsorption capacity of clinoptilolite for heavy metals (Pb, Cr) removal from water 2g of clinoptilolite and 100 mL portions of heavy metal solution (single metal solutions) were placed in a suitable container. The corresponding concentration of each metal (Pb, Cr) was: 0.1 mg/L, 0.5 mg/L, 1.0 mg/L, 5.0 mg/L.

The experiments were performed in triplicates at temperature of 20°C (±5°C). The samples were agitated continuously for 2 hours and the mixtures were equilibrated for 22 hours. Next, final pH was recorded and concentrations of lead and chromium (in the liquid phase) were analyzed using ICP-AES. The solutions were separated by solid phase centrifugation at 4000 rpm for 10 min, before chemical analysis.

Blank solutions without clinoptilolite were prepared in order to examine the possible precipitation of heavy metals.

All the experiments were performed three times and the average experimental relative error was found to be 3.5% for metal concentration measurements.

The sorption capacity of clinoptilolite for Pb and Cr was calculated according to the following formula:

$$A = \frac{(C_0 - C_k)}{m} V \qquad (1)$$

where:

A - the adsorption capacity (mg/g);

C_0 and C_k - the initial and final metal concentration (mg/L);

V - the sample volume (L);

m- the clinoptilolite weight (g).

Metal removal from the solution was determined as follows:

$$Uptake(\%) = \frac{C_0 - C_k}{C_0} 100 \qquad (2)$$

Where C_0 and C_k are the initial and final metal concentrations (mg/L).

3. Results and Discussion

3.1 XRD analysis

The main crystalline phases of raw fly ash are quartz and anhydrite by XRD analysis and trace phases of calcite, hematite and ettringite are also identified (Figure. 2).

According to XRD patterns of all samples obtained, all samples were confirmed by formation of different types of zeolites. Figure 3 illustrates the XRD patterns of five types of zeolites synthesized from fly ash at the condition of 5M NaOH solution during the crystallization time for 48h.

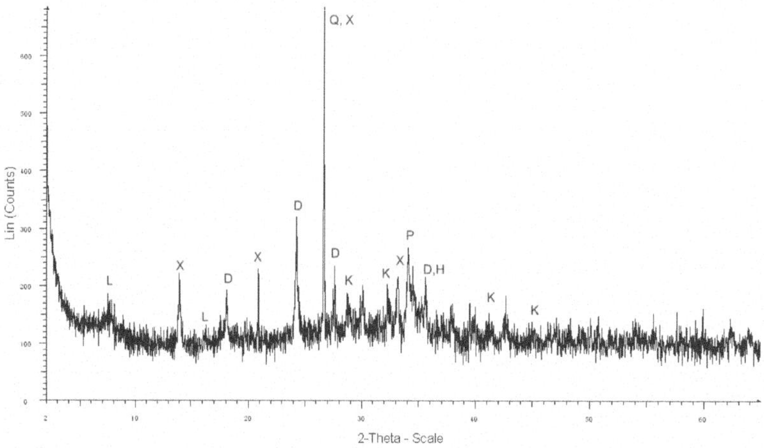

Figure2. XRD pattern of raw fly ash.(E - ettringite, Q - quartz, C - calcite, A - anhydrite, H - hematite)

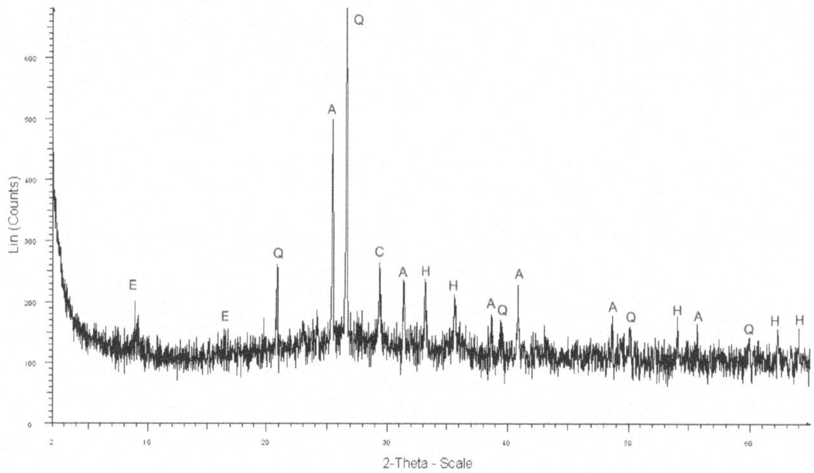

Figure 3: XRD pattern of zeolites samples synthesized from fly ash (synthesis conditions T+NaOH-48) (L-A type zeolite, D - cancrinite, P - portlandite, X - NaP1 type zeolite, Q - quartz, K - katoite, H - hematite)

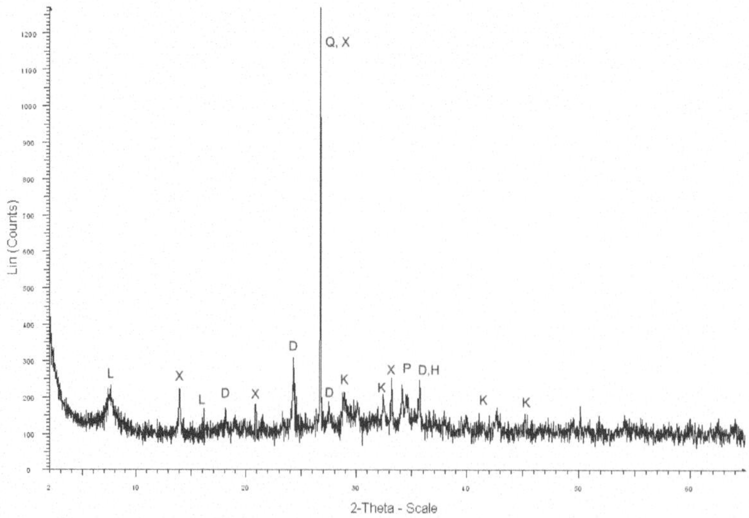

Figure 4: XRD patterns of zeolites synthesized from fly ash at the condition of 5M NaOH solution during the crystallization time for 72h. (For abbreviations see Figure 3).

Four types of zeolites was synthesized from fly ash at the condition of 5M NaOH solution during the crystallization time for 72h (Figure 4).

The obtained results indicate that crystallization time is important parameters that determines compositions of synthesized zeolite during hydrothermal treatment and is in line with the literature [8]. According to the obtained results, increase of crystallization time of 48h and 72h resulted in the different types and quantity of synthesized zeolites. The more quantity of A type zeolite and less quantity of amorphous component was observed.

3.2 Adsorption studies

For the adsorption studies the zeolites synthesized form fly ash at 72h (T+NaOH-72) was chosen. The initial and final pH values of tested samples for adsorption of Pb by the zeolites synthesized form fly ash is presented in Table 1.

The final pH values were within the range of 11.0 to 11.6 for the initial concentrations of lead within the range of 0.1 mg/L and 5.0 mg/L. The initial pH values of all samples were slightly lower than initial pH values.

The initial and final pH values of tested samples for adsorption of Cr(III) by the zeolites synthesized form fly ash is presented in Table 2.

Table 1. The initial and final pH values of tested samples for sorption of Pb.

C_0Pb (II) [mg/l]	T+NaOH-72	
	Initial pH	Final pH
0.1	10.9	11.6
0.5	10.9	11.0
1.0	10.8	11.0
5.0	10.5	11.1

Table 2. The initial and final pH values of tested samples for sorption of Cr(III).

C_0 Cr (III) [mg/L]	T+NaOH-72	
	Initial pH	Final pH
0.1	10.9	11.5
0.5	10.9	11.5
1.0	10.9	11.4
5.0	10.6	11.3

The initial pH values of all samples were slightly lower than initial pH values.

The adsorption capacity of lead and chromium by the zeolites synthesized form fly ash is presented in Figure 5.

The adsorption capacity of the zeolites synthesized form fly ash was slightly influenced by the metal. According to the initial concentration, the differences between the amount of adsorbed metal were observed. Zeolites synthesized form fly ash show the highest adsorption capacity towards lead and chromium for initial concentration of 5 mg/L. The highest adsorption capacities of the zeolites synthesized form fly ash were found towards lead of 0.25 mg/g and chromium 0.246, respectively. Figure 6 presents the uptake (%) of lead and chromium by the zeolites synthesized form fly ash.

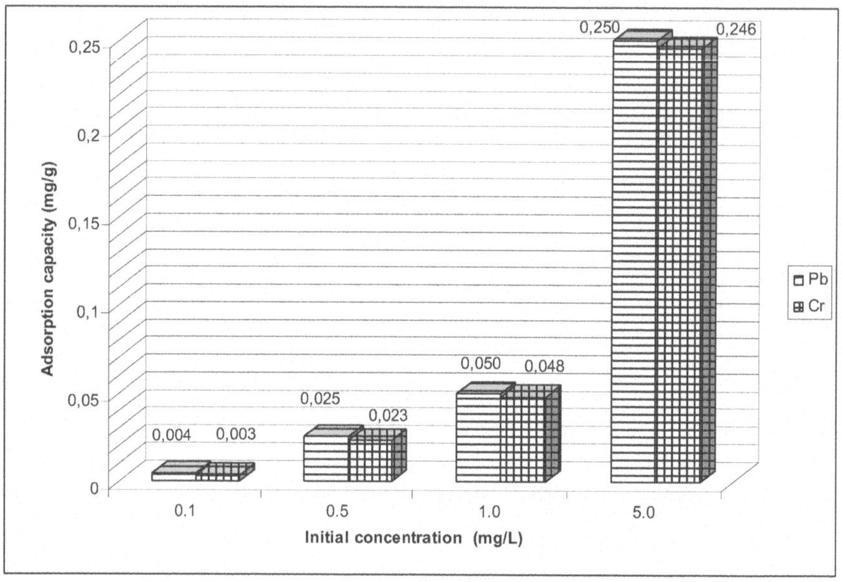

Figure 5: The adsorption capacity of the zeolites synthesized form fly ash (T+NaOH-72h) for lead and chromium.

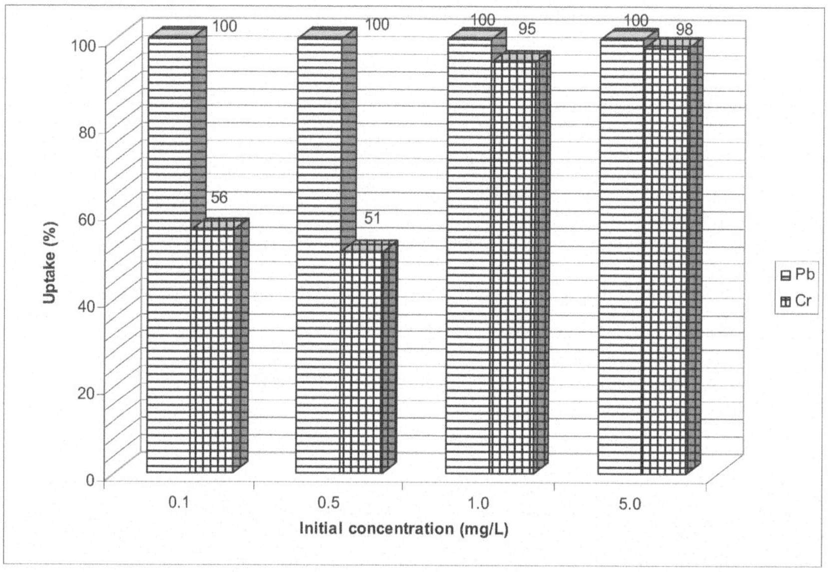

Figure 6: Percentage removal of lead and chromium by the zeolites synthesized form fly ash (T+NaOH-72h).

Based on the conducted investigations, lead was adsorbed with higher efficiency than chromium from the aqueous solutions by the zeolites synthesized form fly ash (T+NaOH-72h). The uptake of lead reached 100% in all concentration. The highest removal level of chromium reached by zeolites synthesized form fly ash (T+NaOH-72h) was 98% in the initial metal concentration of 5.0 mg/L and the lowest of 56% in the metal concentration of 0.1 mg/L.

The obtained results indicate that zeolites synthesized form fly ash are good sorbent material for removal of lead and chromium (III) from aqueous solutions [9, 10].

4. Conclusion

In the present study, the adsorption capacity of zeolite synthesized from fly ash for its lead and chromium (III) retention capacity has been investigated. Zeolite was synthesized by hydrothermal method.

1. Zeolitization of coal fly ash depends on composition of fly ash.
2. The structural formation of zeolites is determined by the crystallization time - controlling crystallization time is necessary during hydrothermal treatment.
3. The differences in sorption capacity for each of the cation were observed.
4. The obtained results indicate that zeolites synthesized from fly ash provides an economical means of removing these heavy metals from aqueous solution.
5. The effectiveness of metals removal by zeolites depends on the concentration of ions in solution and type of the metal ions to be removed.
6. After the process of heavy metal removal zeolites synthesized from fly ash could be regenerated and use many times after the regeneration process.

In conclusion, low cost adsorbent, as zeolites synthesized from fly ash could be used in order to minimize processing costs in removal of heavy metals such as lead and chromium from water.

Acknowledgments
This study was carried out within the frame of the BW 401/206/08 State Committee for Scientific Research (KBN) in Poland.

References

1. Suchecki T. „Zeolity z popiołów lotnych. Otrzymywanie i aplikacje w inżynierii środowiska", wyd. Politechniki Wrocławskiej, Wrocław 2005.
2. Inada M., Eguchi Y., Enomoto N., Hojo J., Synthesis of zeolite from coal fly ashes with different silica-alumina composition, Fuel 84 (2005) 299-304.
3. Xu X., Bao Y., Song C.,Yang W., Liu J., Lin L., Microwave-assisted hydrothermal synthesis of hydroxy-sodalite zeolite membrane, Microporous and Mesoporous Materials 75 (2004) 173-181.
4. Bień J.B., Zabochnicka-Świątek M., 2007, Ion exchange selectivity and adsorption capacity of clinoptilolite, W: Environmental Protection into the Future, Ed. by Nowak W., Bień J.B., Wydawnictwo Politechniki Częstochowskiej, Częstochowa, 383-393.
5. Berkgaut V., Singer A., High capacity of cation exchanger by hydrothermal zeolitization of coal fly ash, Applied Clay Science 10 (1996) 369-378.
6. Zabochnicka-Świątek M., 2007, Czynniki wpływające na pojemność adsorpcyjną i selektywność jonowymienną klinoptylolitu wobec kationów metali ciężkich, Inżynieria i Ochrona Środowiska, Częstochowa 2007, tom 10, nr 1, 27-43.
7. Wu D., Zhang B., Yan L., Kong H., Wang X., Effect of some additives on synthesis of zeolite from coal fly ash, Int. J. Miner. Process. 80 (2006) 266-272.
8. Wang C-F., Li J-S., Wang L-J., Sun X-Y., Influence of NaOH concentrations on synthesis of pure-form zeolite A from fly ash using two-stage method, Journal of Hazardous Materials 155 (2008) 58-64.
9. Somerset V., Petrik L., Iwuohaa E., Alkaline hydrothermal conversion of fly ash precipitates into zeolites 3: The removal of mercury and lead ions from wastewater, Journal of Environmental Management 87 (2008) 125-131.
10. Hui K.S, Chao C.Y.H., Kot S.C., Removal of mixed heavy metal ions in wastewater by zeolite, 4A and residual products from recycled coal fly ash, Journal of Hazardous Materials B127 (2005) 89-101.

Sorption of manganese in the presence of phtalic acid on selected activated Carbons

Ewa Okoniewska and Magdalena Zabochnicka-Świątek

Częstochowa University of Technology, Faculty of Environmental Protection and Engineering, Brzeznicka 60a, 42-200 Częstochowa, Poland

Corresponding E-mail: eokoniewska@is.pcz.czest.pl

Developing from the nineteenth century, the industry takes in more areas of Poland, intensive and polluting them at every stage of its production waste. One of the most dangerous for the environment are organic compounds, formed as industrial products and waste. For such compounds are mainly polycyclic aromatic hydrocarbons, volatile aromatic hydrocarbons, chlorophenols and pesticides. These compounds are toxic by nature, sometimes very strongly, in excessive quantities threaten human health.

For the treatment of wastewater containing complex to degradation of biological or toxic organic substances used almost all physicochemical methods, however, one of the most popular solutions is adsorption using activated carbons, which was recognized by the U.S. Environmental Protection Agency as one of the best available environmental technology.

Activated carbon removes organic compounds, even when they occur in water in small or trace amounts. In addition to good sorption properties for organic substances, also show selective ion exchange properties which makes it can be successfully used to remove metal ions from wastewater

The article presents the results of sorption of manganese in the presence of phtalic acid with two-factor solution at pH 5 and 9.

Section 5

Metals materials and testing & metal leaching

Harmonization of national requirements for metallic materials in contact with drinking water - 4MS approach

Thomas Rapp
Federal Environment Agency, Bad Elster, Germany
Corresponding author e-mail: Thomas.Rapp@uba.de

4MS activities

After the development of the European Acceptance Scheme (EAS) for construction products in contact with drinking water had been postponed or even stopped by the EU Commission the four most active EU Member States (MS) involved in this development started to harmonize their national acceptance procedures. This cooperation of France, Germany, The Netherlands and UK leads to recommendation for the further development of CEN standards, the harmonization of the testing procedures within the 4 MS and the setting of uniform pass-fail criteria. In a second step the voluntary activities by the 4 MS will support the mutual recognition of accepted products. Nonetheless, the 4 MS group provides its results as recommendations to the Commission and strongly supports the implementation of uniform hygienic requirements for materials in contact with drinking water within the Drinking Water Directive (89/83/EC).

Testing procedure for metals in contact with DW

Corrosion of metallic materials may lead to the build-up of layers of corrosion products on the surface of the material. This is a long lasting process and has a strong influence on the further metal release into the drinking water. Additionally, the build-up of these layers and the metal release depend on the water composition. For these reasons the testing for the hygienic fitness of metallic products requires the consideration of different water compositions and a long testing period of at least 26 weeks. However, as the long term metal release is mainly characterized by the material and not by the production process of the final product the metallic materials themselves can be tested for its hygienic fitness. Accepted materials can be listed and the testing for the acceptance of products is then limited to a compliance test of the material composition with the listed materials. For the acceptance of metallic materials CEN developed the test standards EN 15664-1 and EN 15664-2.

Nonetheless the production process may have an influence on the short term (up to 3 to 6 months) metal release e.g. caused by a lead smear films on the surface. The metal release caused by plating processes is also strongly influenced by the production process. Therefore, for some products additional tests will be required. For the limitation of lead films on the surface CEN developed prEN 16057. The nickel release caused by chromium plated products can be determined according to prEN 16058.

Acceptance criteria

Metals in drinking water are derived from a variety of sources. It is therefore necessary to take account of the contribution that other sources, apart from metallic PDW, make to the overall concentrations of metals at consumers' taps by setting a percentage contribution level for each metal. The 4 MS agreed on these percentages as well as on an initial time period of three months tolerating a higher metal release rate during the build-up of corrosion layers.

The extent to which a metallic product contributes to the concentration of a metal in drinking water depends also on its surface area in contact with the drinking water relative to the total surface area of other products in the system. When materials are tested the percentages of the surfaces within a domestic installation for the different products have to be considered. The 4MS defined three product groups with an assumed contact surface of 100%, 10% and 1%, respectively.

Composition List

The interpretation of test results is very complex. Therefore a "Committee of Experts" is required to decide about the acceptance of materials based on the results according to EN 15664-1 and other significant data. The acceptance will then lead to a listing of the accepted materials on a Composition List (Material List). As long as no such European committee exists the 4 MS will rely on their national committees or decision making structures. However, the cooperation of the 4 MS will lead to a common Composition List.

Further details can be found in the 4MS document "Acceptance Of Metallic Materials Used For Products In Contact With Drinking Water". (Online available in due time)

Short period survey of heavy metals concentration in tap water before and after rehabilitation and modernization of water and sewerage services in Baia Mare town

D. Staniloae[1], M Jelea[2], C.Dinu[1], S.M. Jelea[2]

[1] Instrumental Analysis Laboratory, National Research and Development Institute for Industrial Ecology - INCD ECOIND, Romania
[2] North University of Baia Mare, Faculty of Science, 76 Victoriei Street, 430072.Baia Mare, Romania

Corresponding author e-mail: dumitru.staniloae@gmail.com

The Town of Baia Mare has received from 2004 to 2011 an investment of 46 million Euros with ISPA financing for the rehabilitation of the drinking water treatment plant and the rehabilitation and extension of the drinking water networking.

The aim of this work was to survey the heavy metals concentration in drinking water, from the source to the consumer, in the centralized water supply system before and after completion of the investment works. Considering that the work have been completed starting October 2010 (rehabilitation of treatment plant) until June 2011 (rehabilitation and extension of distribution network), the two sampling campaigns have been planned in June 2009 and June 2010.

Different heavy metals were analyzed from samples collected before and after the treatment plant and from 10 final consumers, representatively chosen for all areas served by the water supply system. Sampling was made using the random daily time method in order to have a general aspect on the water quality for a 24 hours interval.

The first campaign as highlighted the particular problem due to the dissolution process of Fe, Mn and Zn in the distribution network. In the second campaign (2010), the values obtained for the same metals (Fe, Mn and Zn) had declines of up to 10 times (Fe).

As a conclusion, the study demonstrates the efficiency of investment it this can be expressed by lowering the concentration of metals in tap water to the consumer.

1. P.G.G. Slaats, E.J.M. Blokler, J.F.M. Versteegh, “Sampling metals at the tap: Analysis of Dutch data over the period 2004-2006”, Metals and related Substances in Drinking Water. COST Action 637.International Conference, European Cooperation in the Field of Scientific and Technical Research, p.61-69, 2007, Turkey.
2. European Commission 1998, Council Directive (98/83/EC, 3 November 1998, concerning quality of water intended for human consumption, Official Journal, 2230/32,23-45, December 1998.
3. World Health Organization and COST 637, Guidance on sampling and monitoring for lead in drinking water, 2009;
4. Law 311/2004, modified Law 458/200, regarding drinking water quality, Official Monitor of Romania, Part I, 382, 2004.

Differences in metal concentrations in water intended for human consumption in the pipe network of the city of Poznań (Poland) in the light of two sampling methods

Józef Górski[1], Marcin Siepak[1], Sławomir Garboś[2] and Dorota Święcicka[2]

[1]Department of Hydrogeology and Water Protection; Adam Mickiewicz University;16 Maków Polnych Str., 61-606 Poznań, Poland

[2]National Institute of Public Heath - National Institute of Hygiene; Department of Environmental Hygiene; 24 Chocimska Str., 00-791 Warsaw, Poland

Corresponding author e-mail: sgarbos@pzh.gov.pl

The article presents the results of a study of metal concentrations (Cd, Cr, Cu, Zn, Ni, Pb, Fe, and Mn) in the tap water of consumers in Poznań (Poland) obtained with the help of two different sampling methods. In the first case, samples were collected daily from 11 randomly selected domestic plumbing systems for a month (October 2008), with one sample taken after overnight stagnation and another during the day, after an exchange of water in the pipes. In the other case, water was sampled from 100 randomly selected plumbing systems using the random daytime sampling (RDT) method in two weeks of May 2010. The study was conducted on the water distribution system of the city of Poznań supplied from an artificial recharge of groundwater (Dębina well field) and from wells drawing groundwater through bank infiltration (Mosina-Krajkowo well field).

The determinations of Cd, Cr, Cu, Zn, Ni, Fe and Mn in 2008 were performed by atomic absorption spectrometry with flame atomisation (F-AAS) using a Varian apparatus (SpectrAA280Z, Varian, Australia). Pb was determined by atomic absorption spectrometry with graphite furnace atomization (GF-AAS) using a Varian apparatus (SpectrAA280Z, Varian, Australia). In 2010, Cd, Ni and Pb levels were determined using inductively coupled plasma mass spectrometry (XSeries II CCT spectrometer, Thermo Electron Corporation, UK). For Cu, Zn, Fe and Mn, inductively coupled plasma optical emission spectrometry with CID detector was used (IRIS Advantage Duo ER/S spectrometer, Thermo Jarrell Ash, USA).

The analysis of the results shows domestic plumbing to have a significant effect on the content of metals in water, which is most readily visible in samples taken after overnight stagnation. This especially concerns Cu and Zn, since high Cu levels were recorded in pipes made of copper, and Zn levels, in those made of galvanized steel. The concentration figures were also distinctly higher after overnight stagnation for Ni, Cr and Pb, and less so for Cd, Fe and Mn. In turn, the results for samples collected from flushed plumbing are largely similar to those obtained using the RDT method. This especially concerns Zn, Cu and Pb.

The research was financed from the 2009-2010 research fund as project No. 398/N COST/2009/0 of the Ministry of Science and Higher Education.

Galvanic impacts of partial lead service line replacement on lead leaching into drinking water

Simoni Triantafyllidou and Marc Edwards
Virginia Tech
Blacksburg, VA 24061, USA

Corresponding author E-mail: striant@vt.edu

Abstract

Due to jurisdiction issues, partial lead service line replacement (and not full) is widely implemented in the United States (US), in order to alleviate lead-in-water problems. A portion of the lead service line is replaced with copper, and the dissimilar pipe materials are then connected to restore drinking water service. This practice creates an electrochemical or galvanic cell, which can accelerate corrosion of the lead pipe by galvanic action. The adverse effects of such connections in the context of lead leaching were verified in experiments of simulated lead service line replacement. Galvanic connections between lead pipe and copper pipe increased lead release, compared to lead pipe alone. The extent of galvanic corrosion was dependent on drinking water quality, and specifically on the Chloride to Sulfate Mass Ratio (CSMR) of the water. Higher galvanic currents between lead and copper were measured when the CSMR was high, mechanistically explaining the trends in lead release. Consideration of galvanic corrosion long-term impacts after partial lead service line replacements is deemed important, on the basis of the results presented herein.

1. Introduction

Lead (Pb) is widely recognized as one of the most pervasive environmental health threats in the United States (US). Water consumption contributes to an estimated 10-20% of the general population's total lead exposure [1], but can occasionally be the dominant source of exposure [1, 2]. The harmful health effects from lead exposure through drinking water have been recognized since the 1850s. In that era drinking water contamination by lead pipes was the main source of human-ingested lead, causing infant mortality, neurological effects, and digestive problems [3].

Partial Lead Service Line Replacements. Lead service lines were the standard in many US cities through the 1950s, and were even occasionally installed up to the ban of lead pipe in 1986. Old lead service lines can still be significant contributors to lead-in-water hazards. Lead in US drinking water is currently regulated under the Lead and Copper Rule (LCR), which may require replacement of utility-owned lead service lines, if the LCR lead action limit of 15 ug/L is exceeded.

If the lead service line extends onto the homeowner's property, the utility is only required to replace the portion of pipe that it owns, leaving behind the customer-owned portion of lead pipe (Figure 1, left). Although numbers vary dramatically from city to city or even from home to home, a national survey [4] indicated that the total length of service lines in the US averages 55-68 feet, with 25-27 feet (i.e. 40-45%) being under the utility's jurisdiction. The cost of replacing the customer portion can add up to several thousand dollars, and few customers voluntarily replace their portion of the lead service line [5].

The practice of only replacing the utility portion of lead pipe, referred to as a "partial lead service line replacement", has been known to increase the concentration of lead in drinking water. The increased lead can arise from a variety of mechanisms and can possibly be short-term (days to weeks) or even long-term (months to years) in duration. Short-term problems occur from disturbing the lead rust (i.e., scale) that has accumulated on the pipe over decades of use, and/or from creating metallic lead particles when the pipe is cut [6]. In the US these short-term mechanisms from cutting and scale disturbance were documented in laboratory experiments [7] and in field studies undertaken by several utilities [4].

Longer-term problems might arise from creation of a new electrochemical or galvanic cell, between the old lead pipe and the newly installed copper pipe (Figure 1, right). Britton and Richards [8] documented a case in Glasgow where more than four months were required for lead levels to drop. In the US Swertfeger *et al.* [5], who measured lead levels in water of homes after partial lead service line replacements, stated that even after 1 year of sampling, replacement did not show an improvement over keeping a complete line in place. On the other hand, potential long -term problems after partial replacements were described as "likely inconsequential" in one case study [9], and the logic behind this practice is that at some indefinite future time it will provide benefits. A consensus opinion has therefore not yet been reached on the long-term implications, either adverse or beneficial, of partial lead service line replacements.

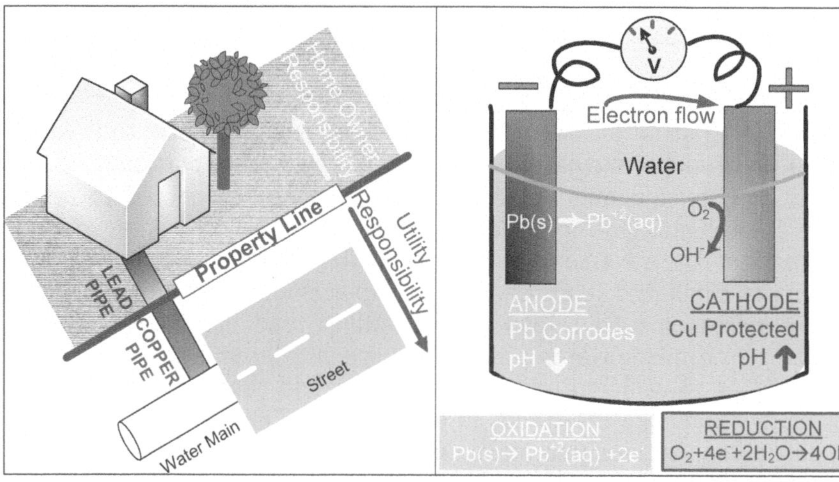

Figure 1: Typical plumbing configuration after partial lead service line replacement, where copper pipe is directly connected to lead pipe (left). Conceptualization of galvanic corrosion due to direct electrical connection of copper pipe to lead pipe (right).

The Effect of Chloride-to-Sulfate Mass Ratio (CSMR) on Galvanic Corrosion of Lead. Oliphant [10] and Gregory [11] first showed that the water chemistry controls the magnitude of galvanic corrosion between lead and copper, with one critical factor being the relative concentration of chloride to sulfate. Gregory [11] developed the concept of chloride-to-sulfate mass ratio (CSMR) to explain this effect. To illustrate, for water containing 15 mg/L Cl^- and 30 mg/L SO_4^{-2}, the resulting CSMR is 0.5. Gregory [11] determined that CSMRs above 0.5 increased galvanic corrosion of lead solder connected to copper pipe, as evidenced by increased galvanic voltage measurements.

Edwards *et al.* [12] later determined that, for a sub-set of surveyed US water utilities studied in-depth, 100 percent of utilities with CSMR below 0.58 met the LCR lead action level of 15 ug/L, whereas only 36 percent of utilities with CSMR above 0.58 were in compliance. The identified critical CSMR of 0.58 cited for causing lead compliance problems was remarkably similar to the 0.5 threshold identified as causing galvanic corrosion of lead in the preceding English studies. Other laboratory experiments [13, 14, 15] as well as anecdotal evidence from specific US water utilities [14, 16] supported the notion that lead release was impacted by higher CSMR.

Clark and Edwards [17] provided a mechanistic explanation for the success of the empirical CSMR in explaining lead contamination of potable water. They examined the solution chemistry of Pb^{+2} in the presence of chloride and sulfate. They conducted experiments at relatively low pH (≈ 3.0-5.0), since previous microelectrode measurements [14] had shown that local pH at the lead surface of galvanic connections to copper can drop substantially. Pb^{+2} formed sulfate solids which were relatively insoluble even at pH of 3.0. On the contrary, Pb^{+2} formed soluble complexes with chloride, which could significantly increase the solubility of lead. Since increased lead solubility can translate to lead contamination of drinking water, these data illustrated how a high CSMR can indeed worsen lead problems at low pH.

2. Materials and methods

The experimental apparatus was constructed to track lead leaching from simulated partial lead service line replacements. The test rigs consisted of a new copper pipe section that was electrically connected to new lead pipe, with a total rig length of three feet (Figure 2). The lead portion and the copper portion of each rig were separated by an insulating spacer, and could be externally connected via grounding strap wires instead (Figure 2). If the wires were disconnected, direct galvanic corrosion between lead and copper was not possible.

The portion of the pipe that is lead and the portion that is copper were systematically varied - as could occur in partial replacements with different percentage of consumer ownership of the service line. Specifically, 100% lead pipe (simulating a lead service line before replacement), 100% copper pipe (simulating full replacement), and four increments in between (17% copper pipe, 50% copper pipe, 67% copper pipe, and 83% copper pipe to simulate partial replacements) were tested.

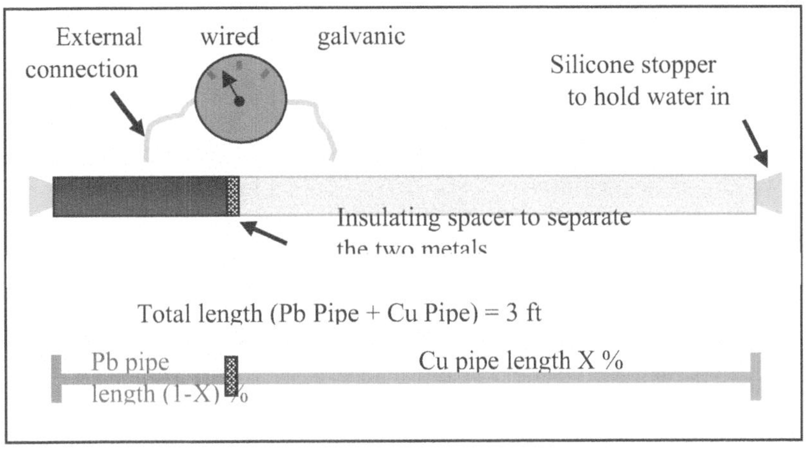

Figure 2: Generalized schematic of experimental setup. This design assessed the contribution of galvanic corrosion to lead in water, as it would occur after partial lead service line replacement.

Three distinct phases of experimentation were undertaken:

- **Phase 1.** During weeks 1-11, all rigs were exposed to synthetic tap water with a low chloride to sulfate mass ratio (CSMR) of 0.2 ("low CSMR water"). This water also had an alkalinity of 15 mg/L as $CaCO_3$, monochloramine disinfectant dosed at 4.0 mg/L as Cl_2, ionic strength of 4.6 mmol/L (by addition of salts to mimic other tap water constituents) and pH of 8.0 (Table 1).
- **Phase 2.** After baseline results were established in the non-aggressive water, the test water was switched to an aggressive synthetic tap water with a high CSMR of 16 during weeks 12-25 ("high CSMR water"). All other water parameters such as alkalinity, monochloramine, ionic strength and pH were kept the same as in the "low CSMR water" of Phase 1 (Table 1).
- **Phase 3.** For weeks 26-29 the rigs continued to be exposed to the "high CSMR water" as in Phase 2, but without direct galvanic corrosion between the lead and copper pipe due to removal of the connecting strap wires.

Table 1: Key characteristics of the two synthetic waters utilized in the experiment.

Types	Cl^- (mg/L)	SO_4^{-2} (mg/L)	CSMR	Alkalinity (mg/L as Cl_2)	NH_2Cl (mg/L as Cl_2)	Ionic Strength (mmol/L)	pH
Low CSMR Water	22	112	0.2	15	4.0	4.6	8.0
High CSMR Water	129	8.0	16	15	4.0	4.4	8.0

Throughout the experiment (i.e. in all three phases), water was completely changed inside the pipes three times per week, using a "dump and fill" protocol. The contribution of galvanic connection (or lack thereof) to lead release was assessed by measuring total lead concentration in water and galvanic current magnitude:

- One composite water sample was collected at the end of each week from each rig, by pouring the three water samples of that week into the same container. Total lead was quantified in these unfiltered composite water samples after digestion with 2% nitric acid, using an inductively coupled plasma mass spectrophotometer (ICP-MS).
- Galvanic current between the dissimilar metals (for Phases 1 and 2 when the external wires were connected) was measured with a hand-held multi-meter. The currents were measured by connecting the multi-meter in-line for 15 seconds after disconnecting the wire between the two metals.

3. Results and discussion

With the exception of weeks 1-3, when lead release had not yet stabilized, results were otherwise synthesized by averaging the lead data for each experimental phase.

Effect of Galvanic Corrosion and CSMR on Lead Release. All simulated partial replacements (17%, 50%, 67% and 83% of Pb replaced by Cu) released the same or more lead to the water than did the rig consisting of pure lead (i.e. 0% Cu) (Figure 3). This was true for all three experimental phases, and results were statistically significant at the 95% confidence level (error bars plotted, Figure 3). Even though these results appear to be counter-intuitive, they are not surprising considering early knowledge on galvanic corrosion. That is, the 0% Cu rig, consisting only of lead pipe, has a higher Pb surface area in contact with the water compared to all other conditions. This condition would thus be expected to release more lead to the water. However, when lead pipe is connected to copper pipe, the effect of galvanic corrosion in enhancing lead release is so strong, that even a smaller lead surface area exposed to the water results in worse lead contamination of the water.

High CSMR released much more lead to the water compared to low CSMR, when the wires were connected. In fact, when comparing high CSMR (Phase 2) to low CSMR water (Phase 1), lead release increased by 5 times (in the case of 17% Cu) to as much as 12 times (in the case of 83% Cu) (Figure 3). These differences were statistically significant at the 95% confidence level (Error bars plotted, Figure 3). Disconnecting the wires under high CSMR water (Phase 3) decreased lead release by 4-6 times in all Cu:Pb galvanic couples (Figure 3). This demonstrates the direct role of galvanic corrosion in sustaining high lead concentrations in water, when lead and copper pipe are electrically connected.

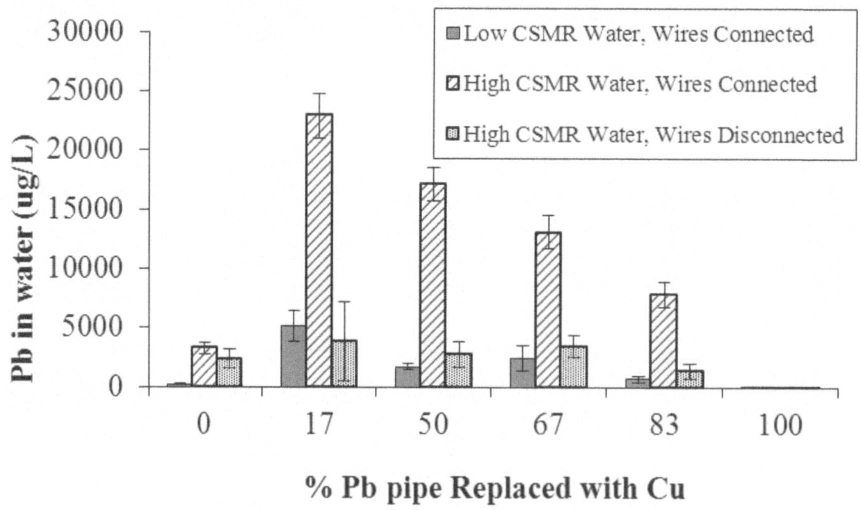

Figure 3: Lead leaching versus extent of lead replacement by copper. The error bars denote 95% confidence intervals. CSMR: Chloride-to-Sulfate Mass Ratio.

Mechanistic Insights via Galvanic Current Measurements. Measurement of the galvanic current between the lead and copper portion of the rigs provided mechanistic insights on the observed lead leaching trends. Galvanic current measurements were taken during Phase 1 and Phase 2 of the experiment, when galvanic corrosion was activated. Since the wires were disconnected during Phase 3, thereby blocking electron flow between the two dissimilar metals, no current was flowing and thus no such measurements were conducted during that phase.

Higher currents were measured when high CSMR water was fed to the rigs (Phase 2), compared to when low CSMR water was fed to the rigs (Phase 1) (Figure 4). This is consistent with the lead leaching results (Figure 3). Under the high CSMR water condition, the highest current of 87 uA was measured for the 17% Cu rig, followed by the 50% Cu, 67% Cu, and 83% Cu rigs (Figure 4). These differences were statistically significant at the 95% confidence level (Error bars plotted, Figure 4). The ranking of the rigs with respect to the magnitude of the measured galvanic currents is consistent with that based on the lead leaching results. For instance, the 17% Cu rig had the highest measured current (Figure 4), and it also resulted in the highest lead-in-water concentrations (Figure 3).

Under the low CSMR condition, the highest current of 52 uA was measured for the 17% Cu rig, followed by the 50% Cu, 67%, and 83% Cu rig. These differences were statistically significant at the 95% confidence level (Error bars plotted, Figure 4). This ranking in terms of galvanic current measurements under the low CSMR condition did not always agree with the ranking based on the lead leaching results (Figure 3).

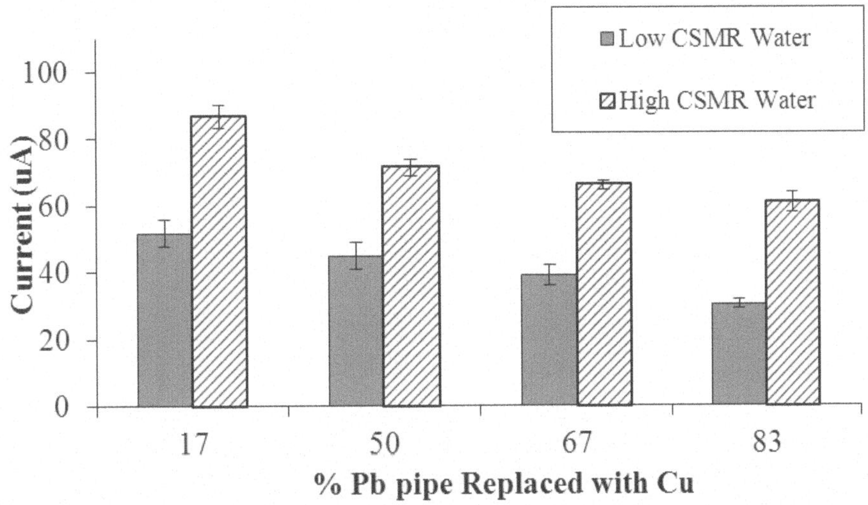

Figure 4: Galvanic current versus extent of lead replacement by copper. The error bars denote 95% confidence intervals. CSMR: Chloride-to-Sulfate Mass Ratio.

4. Conclusions

Under controlled experiments of simulated partial lead service line replacements that lasted for more than seven months:

- Galvanic connections between copper pipe and lead pipe increased lead release, compared to lead pipe alone.
- Inactivation of galvanic corrosion between lead and copper resulted in a 4-6 times decrease in lead release, under an "aggressive" water condition of high CSMR.
- The two test waters, one with low CSMR of 0.2 and one with high CSRM of 16, represented extremes in enhancing lead release by galvanic corrosion. That is, high CSMR released 5-12 times more lead to the water than did low CSMR.
- High sustained galvanic currents between lead and copper, resulting in corrosion of the lead, were measured when the CSMR was high. When the CSMR was low, galvanic currents were also low, consistent with corresponding low lead leaching results.

On the basis of a literature review and of these initial results, the future desirability of partial lead service line replacements should be carefully considered. That is, depending on drinking water chemistry, galvanic corrosion might significantly contribute to lead leaching even in the long term. As a result, the practice of partial lead service line replacements may actually worsen lead contamination of potable water, defeating their purpose. In such cases, alternative remedial strategies would need to be considered. More work is needed, in order to quantify the relevant contribution of galvanic corrosion to lead release, compared to other mechanisms such as normal dissolution, deposition corrosion, particle detachment, and lead retention in pipe scale. Future work should also utilize more realistic flow regimes and flow rates, compared to the worst-case stagnant conditions of this study.

Acknowledgments

The authors acknowledge the financial support of the Water Research Foundation (formerly known as AwwaRF) under project # 4088 PCR. The Water Research Foundation maintains copyright of this material as part of the report "Contribution of Galvanic Corrosion to Lead (Pb) in Water after Partial Lead Service Line Replacements". Opinions and findings expressed herein are those of the authors and do not necessarily reflect the views of the Water Research Foundation.

References

[1] Levin, R.; Brown, M.J.; Michael, E.; Kashtock, M.E.; David, E.; Jacobs, D.E.; Elizabeth, A.; Whelan, E.A.; Rodman, J.; Schock, M.R.; Padilla, A. and Sinks, T., 2008. *Environ. Health Perspect.*, 116:1285-1293.

[2] Edwards, M.; Triantafyllidou, S. and Best, D., 2009. *Environ. Sci. Technol.*, 43(5):1618-1623.

[3] Troesken, W., 2006. Cambridge, MA: MIT Press.

[4] *Water Research Foundation*, 2009. *Report 91229*. Prepared by A. Sandvig, P. Kwan, G. Kirmeyer, B. Maynard, D. Mast, R. Trussell, S. Trussell, A. Cantor, and A. Prescott.

[5] Swertfeger, J.; Hartman, D.J; Shrive, C.; Metz, D. H. and DeMarco, J., 2006. *Proceedings of the 2006 Annual AWWA Conference*. San Antonio, TX.

[6] Schock, M.R.; Wagner, I. and Oliphant, R.J., 1996. Internal Corrosion of Water Distribution Systems. AWWA Research Foundation/DVGW-Technologiezentrum, Denver, CO (Second Edition).

[7] Boyd, G.; Shetty, P.; Sandvig, A., and G. Pierson., 2004. *Jour. Envir. Engrg.*, 130(10):1188-1197.

[8] Britton, A. and *Richards,* W.N., *1981. J. Inst. Water Eng. Scient.*, 35:349-364.

[9] Boyd, G. R.; Dewis, K. M.; Korshin, G. V.; Reiber, S. H.; Schock, M. R.; Sandvig, A. M. and Giani, R., 2008. *Jour. AWWA*, 100(11): 75-87.

[10] Oliphant, R.J., 1983. Water Research Center Engineering, Swindon, External Report 125-E.

[11] Gregory, R., 1985. Water Research Center Engineering, Swindon, Interim Report 392-S.

[12] Edwards, M.; Jacobs, S. and Dodrill, D., 1999. *Jour. AWWA* 91(5): 66-77.

[13] Nguyen, C.; Triantafyllidou, S.; Hu, J. and Edwards, M., 2009. *Proceedings of the 2009 Annual AWWA Conference*. San Diego, CA.

[14] Edwards M., and Triantafyllidou, S., 2007. *Jour. AWWA* 99(7):96-109.

[15] Dudi, A., 2004. Master's Thesis, Department of Civil and Environmental Engineering, Virginia Tech.

[16] Kelkar, U.; Schulz, C.; DeKam, J.; Roseberry, L.; Levengood, T. and Little, G., 1998. *Proceedings of the AWWA Annual National Conference*, Dallas, TX.

[17] Clark, B. and Edwards, M., 2007. Virginia Water Resources Research Center, Special Report No. SR42-2008.

Metal and organic release from construction products in contact with drinking water disinfected with Sodium Hypochlorite

E. Veschetti, V. Melini, L. Achene, L. Lucentini and M. Ottaviani

Section of Inland waters, Department of Environment and primary prevention, Istituto Superiore di Sanità, Rome, Italy

Corresponding author e-mail: enrico.veschetti@iss.it

According to the Council Directive 98/83/EC on the quality of water intended for human consumption, Member States shall ensure that no substances or materials used in the preparation or distribution of water remain in water in concentrations higher than is necessary for the purpose of their use and do not reduce the protection of human health. To this end, the Directive sets parametric values for a number of metals and organics that can be released into water by construction products in contact with drinking water (CPDW) and can pose a threat to human health. It is well known that migration from CPDW into water can be affected by many factors such as surface properties of materials, design of distribution system, flow regime as well as chemical and physical characteristics of water. In particular, disinfectant residues may promote a significant release of chemical species by altering redox equilibriums or reacting with material surface.

In this study the effect of chlorine on the migration of metals and organics into water from CPDW was evaluated. Eight among the most commonly used construction products were selected: galvanized steel, corroded galvanized steel, stainless steel AISI 304, stainless steel AISI 316, cast iron, copper, polyethylene and polyvinylchloride. Single test pieces of the above-mentioned materials were preliminary degreased with organic solvents as acetone and pentane. Then, they were soaked with acetic acid at 5% for 30 min to remove unavoidable coatings left after manufacture from their surface and to make them active. After dipping into test water for 5 min, every examined piece underwent a migration trial performed in a new aliquot of test water containing 1 mg/L of sodium hypochlorite at $30\pm1\,^{\circ}C$. The liquid level was set so that the ratio between the piece surface and the water volume was approximately 1 dm^{-1}. Disinfectant degradation was monitored over 48 hours and aliquots of test solution were periodically collected and stabilized with nitric acid. The concentration of metals possibly released by construction products were analyzed by optical emission spectrometry with inductively coupled plasma (ICP-OES). Additional samples were collected to perform analyses of not-purgeable organic carbon (NPOC). The investigation was carried out in two different aqueous matrices both distributed in Rome (Italy) for human consumption: groundwater and surface water collected by Peschiera river and Bracciano lake, respectively. The migration tests were repeated at a 2.5-mg/L starting concentration of the disinfectant. The outcomes of experimental tests were compares with metal and organic migration from materials in contact with the corresponding non-disinfected waters.

Dezincification of brass fittings - effects of metal solvency control measures

Larry L. Russell[1] and Brian T. Croll[2]

[1]REED International LTD. USA
[2]Consultant, United Kingdom
Corresponding author e-mail: REEDINTERNLTD@aol.com

Selective dissolution of zinc from brass fittings can occur when the chloride (mg/l) to alkalinity (mg/l as calcium carbonate) ratio exceeds 0.5. This can be rapid and at pH above 8.3 can lead to precipitated zinc salts blocking pipes (meringue). This phenomenon can present itself in a variety of brass materials, but is focused on the use of high Zinc yellow brasses. Recently, these impacts have been observed as a result of the use of plastic PEX type tubing that is connected with brass fittings. The authors are working on projects in Europe and throughout the United States and as far west as Hawaii.

Dezincifying conditions are made worse by treatment processes, such as softening and during nitrate removal using ion exchange. Treatment for metal solvency control can also produce dezincification and recent experience in the USA will be presented and discussed. Additionally, the water quality is impacted by the dissolved metals that are introduced into the water. Water purveyors have taken the approach of claiming immunity from liability in the occurrence of dezincification.

There are methods for modifying the brass alloy to minimize dezincification, such as adding Arsenic or using copper rich brasses (red brass), but to date these have not been adopted in the domestic water supply market due to water quality concerns (arsenic) and cost (red brass). Lawsuits in the United States alone are approaching 1 billion Euros in damages due to failed brass fittings.

This paper will present information on the status of the use of yellow brasses in domestic water supply systems and on means of controlling dezincification in domestic plumbing systems. Additionally, the water treatment aspects of this phenomenon will be reviewed in detail. This is a problem that is occurring worldwide and it involves a large number of European, American and Chinese companies who supply these parts for use throughout the world. It is a problem destined to visit all of the water users in areas with the water quality described in the opening of this abstract.

Concentration of heavy metals on surface of filter materials and in backwash water

Alina Pruss, Joanna Jeż-Walkowiak and Marek M. Sozański

Institute of Environmental Engineering, Poznan University of Technology, Piotrowo 3a, 60-965 Poznan, POLAND

Corresponding author e-mail: alina.pruss@put.poznan.pl

Abstract

Filtration is the main technological process realized at Water Treatment Plant (WTP) in Poland. Main target of rapid filters is elimination of iron and manganese from groundwater. In this paper, rapid filter exploitation at three WTP in Poznan agglomeration was analyzed.

Incoming water, except iron and manganese, contained also small concentration of other metals. These metals were eliminated from water with different effectiveness by the filtration process, mainly by adsorption on iron hydroxide flocks.

To prove, that metals eliminated from water by filtration process are durable built in covering of filter material grains, the covering layer were analyzed to find a iron, manganese, chromium, nickel, copper, arsenic, selenium and lead in it.

Samples of filter media (sand) were taken from selected filter of each WTP.

Additionally, the backwash water samples were taken directly during filter backwashing and concentration of heavy metals was determined. The concentration of heavy metal was analyzed by ICP-MS methods.

Results show different heavy metals concentration at the filter material covering and backwash water. Concentration of iron and manganese were dominating both, on surface of filter material and in backwash water.

1. Introduction

City Poznan as well as communes adjacent to him is being supplied by the Poznan Waterworks System with water (PWS). This system at present is providing over 755 000 residents of Poznan and nearby communes with water: of Luboń, of Puszczykowo, of Mosina, of Swarzędz, of Czerwonak, of Brodnica, of Suchy Las, of Kórnik, of Murowana Goślina [Chomicki and other 2008].

The Poznan Waterworks System is being powered mainly from 3 water intakes. Over 51 % water for Poznan is delivering Water Treatment Plant in Mosina (intake Mosina- Krajkowo). Water Treatment Plant Wisniowa (Debina intake) is providing system 37 % waters, however Water Treatment Plant Gruszczyn (Gruszczyn - Promienko intake) only 7 % waters [K Wilmański., Lasocka-Gomuła 2004]

The scheme of the Poznan Waterworks System was presented in Figure 1.

The technology of treating water in 3 analyzed stations is based on processes of aerating, the filtration and the disinfections. In the process of the filtration carried out in gravitational filters above all iron and the manganese are being removed, main polluting treated waters. Medium parameters of the quality of inlet and outlet water from rapid filters on exploited for the WTP were described in the table No. 1.

2. Materials and Methods

Samples of filter material (quartz sand) were taken to determine the concentration of heavy metals deposit in every filter. Samples were dried off in the drier to fixed mass, and then 10 g was charged and 2 ml of the concentrated hydrochloric acid was flooded about the concentration about 35 % and with solution of the nitric acid (1 + 1) in the amount of 20 ml. This way samples made out were put in the water-powered bath and they were cooking to the moment of parrying given earlier acidities. In the more distant stage 50 ml of the distilled water was added to samples and after chilling filters were filtered through earlier weighed and dried off in temperature 105oC. Separated in this way filter the ml in amount 10 they handed over for analysis with a view to indicating the concentration of heavy metals, and the deposit with the filter again was dried off and they weighed.

Samples of used backwash water taken during backwashing were indicated concentration of heavy metals. Heavy metals were determined in the certified AQUANET laboratory in Poznan with ICP-MS technique[norm PN-EN ISO 17924]

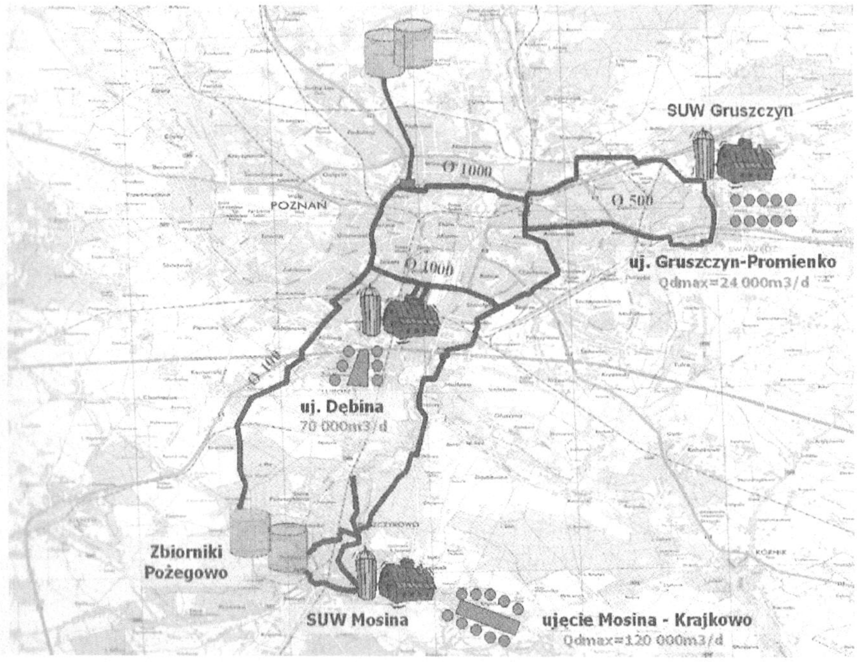

Figure 1. Poznan Waterworks System [K Wilmański, Lasocka-Gomuła I., 2004].

Table 1. Concentration of heavy metals in inlet and outlet waters - averages from 2008 [Witkowicz, 2009].

Metal		WTP Wisniowa		WTP Mosina		WTPGruszczyn	
		inlet	outlet	inlet	outlet	inlet	outlet
Iron	mg/l	0.319	0.029	2.030	0.070	3.189	0.016
Manganese	mg/l	0.338	0.006	0.538	0.011	0.130	<0.005
Nickel	mg/l	0.003	0.001	0.002	0.001	0.000	0.000
Copper	mg/l	0.001	0.016	0.001	0.002	0.000	0.002
Arsenic	mg/l	0.002	0.002	0.001	0.001	0.001	0.000
Chromium	mg/l	0.000	0.000	0.000	0.000	0.000	0.000
Selenium	mg/l	0.000	0.000	0.000	0.000	0.000	0.000
Cadmium	mg/l	0.000	0.000	0.000	0.000	0.000	0.000
Lead	mg/l	0.000	0.000	0.000	0.000	0.000	0.000
Mercury	mg/l	0.000	0.000	0.000	0.000	0.000	0.000

Filters inlet waters didn't contain chromium, selenium, cadmium, lead and mercury in 2008.

3. Results and Discussion

In table 2 get masses were described of the filter material covering downloaded from analyzed WTP.

Table 2. The filter material covering.

WTP	Gruszczyn depth 1.0 m	Wisniowa surface	Mosina depth 1.0 m
The filter material covering [g]	0.6	2.75	0.6

They observed, that biggest the filter material covering taken of the surface layer of the sand filter of WTP Wisniowa was characterized. Undoubtedly it correlated with producing filter material of the coating covered with oxides of the manganese in the upper party and long-lasting stopping iron. Mass the filter material covering taken at depth were 1 m much lower, were an about 0.6 g. Concentration of metals in the filter material covering were described in table 3 and in figure No. 2 and 3.

Table 3. Concentration of metals in the sand filter material covering.

Metal		WTP Gruszczyn depth 1,0 m	WTP Wiśniowa surface	WTP Mosina depth 1,0 m
Chromium	mg/g	0.367	0.026	0.250
Manganese		5267	6436	3017
Iron		6300	1880	567
Nickel		0.417	10.545	2.167
Copper		0.483	0.982	0.800
Arsenic		3.800	0.200	0.600
Selenium		0.010	0.001	0.007
Cadmium		0.008	0.056	0.010
Lead		0.517	0.040	1.833
Mercury		0.137	0.065	0.040

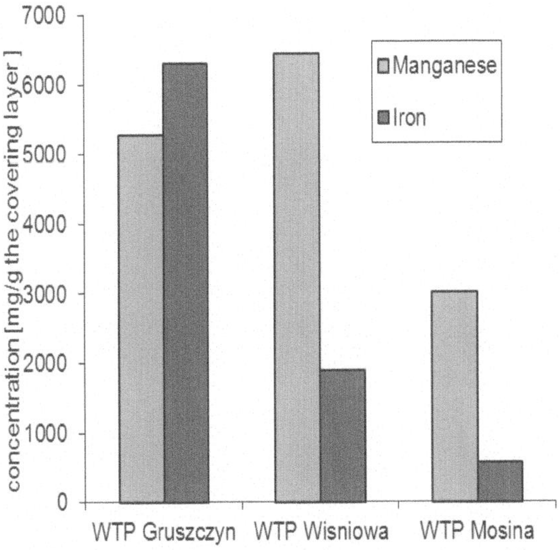

Figure 2. Concentration of iron and the manganese in the sand filter material covering

Conducted analysis showed the high concentration of iron and the manganese in everyone analyzed the filter material coverings. However the concentration of cadmium and the selenium was little what is confirming deficiency in these elements in waters taken away. Moreover on the filter material covering fold the WTP Mosina and Gruszczyn appeared arsenic. He most probably stayed adsorbed on hydroxides of iron covering quartz sand. Concentration of nickel in the filter material covering in the WTP Wisniowa is high, much higher than in other analyzed plants. It is probably effect of the quality of water inlet to filters, which contained this metal.

Results of analyses of backwash waters were described in the table No. 5 and on the figure No. 4 and 5.

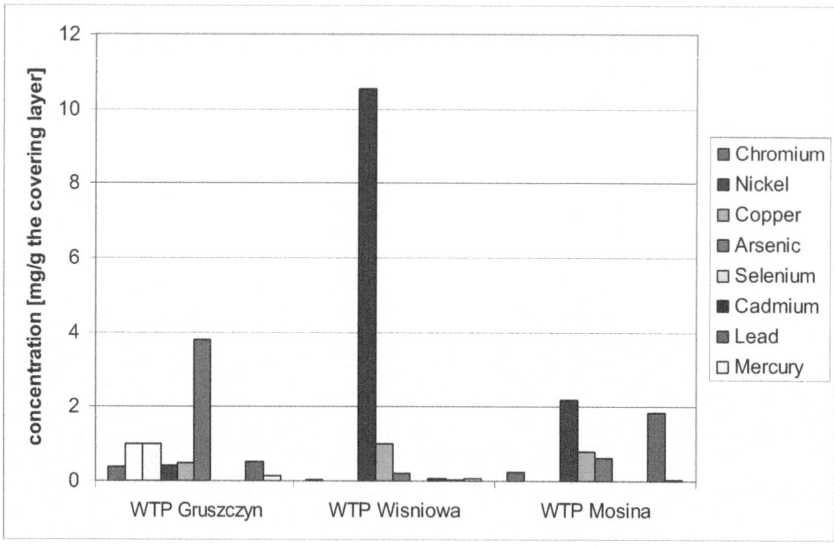

Figure 3. Concentration of chosen metals in the sand filter material covering

Table 5. Concentration of heavy metals in backwash waters.

Metal		WTP Gruszczyn	WTP Wiśniowa	WTP Mosina
Chromium	mg/l	<0.01	<0.01	0.01
Manganese		1.3	3.8	6.3
Iron		1100	95	680
Nickel		<0.02	<0.02	0.02
Copper		<0.03	<0.03	0.05
Arsenic		0.14	0.14	0.27
Selenium		<0.01	<0.01	<0.01
Cadmium		<0.002	<0.002	<0.002
Lead		<0.01	<0.01	<0.01
Mercury		<0.0005	<0.0005	<0.0005

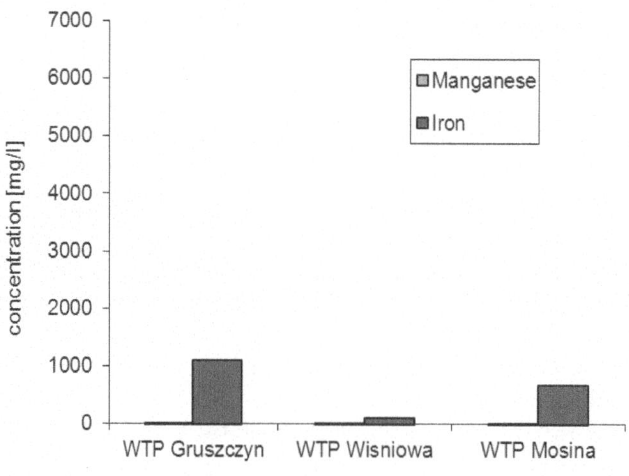

Figure 5. Concentration of iron and the manganese in backwash waters.

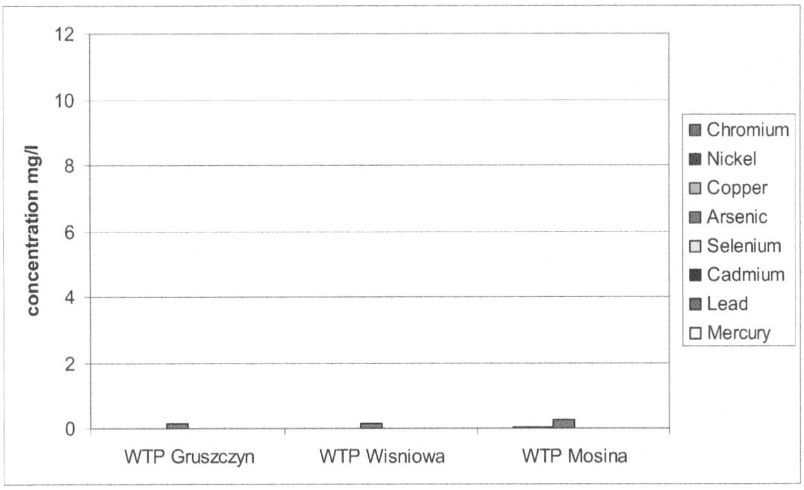

Figure 6. Concentration of chosen metals in backwash waters.

In backwash waters were detected above all iron, the manganese and the arsenic. It can attest to the fact that these metals were stopped (settled) on the surface of filter material into the transitory way and backwash the filter enabled to dismiss them.

4. Conclusions

During the process of the filtration of water carried out on three Water Treatment Plants for the city of Poznan above all iron and manganese and some heavy metals that are inlet to rapid filters are being removed from water.

Analysis of the filter material covering shows that all of metals indicated in inlet water are accumulated in covering layer which determine effectiveness of filtration process.

During backwashing only iron, manganese and arsenic were removed from filter material covering and indicated in used backwash water.

Acknowledgments

Authors would like to thank to AQUANET COMPANY for availability of water quality dates and possibility of additional sampling and analysis.

References

[1] Chomicki I., Bartosik A.: Doświadczenia z funkcjonowania infiltracyjnego ujęcia wody Dębina w Poznaniu i wstępna koncepcja jego modernizacji." VIII International conference "Municipal and rural water supply and water quality", Poland, Poznań 2008 r.

[2] Chomicki I., Graczyk A., Kijko D.: „Konflikt największego ujęcia wody dla aglomeracji poznańskiej z obszarem NATURA 2000." VIII International conference "Municipal and rural water supply and water quality", Poland, Poznań 2008 r.

[3] COST ACTION 637: METEAU - Metals and Related Substances in Drinking Water, http://www.meteau.org.

[4] Directive 98/118/EC of the European Parliament and of the Council of 12 December 2006 on the protection of groundwater against pollution and deterioration. Official journal L 372/19, 2006 r.

[5] DWD, 98, Council directive 98/83/EC on the quality of water intended for human consumption. Official Journal L 330, 05/12/1998 p. 0032 - 0054

[6] Huck P.M., Sozański M.M., Biological Filtration for membrane pre-treatment and other applications: toward the development of a practically-oriented performance parameter, Journal of Water Supply: Research and Technology - AQUA, IWA Publishing, 57.4, 2008.

[7] J. Jeż-Walkowiak, A. Pruss, M.M. Sozański: Development of iron and manganese removal from groundwater in rapid filters with chalcedonit bed. Cost Action 637, 3rd International Conference, Ioannina, Greece, 21-23 October 2009,

[8] Jeż-Walkowiak J., Sozański M.M, Weber Ł.: Iron and manganese removal in filtration process through chalcedonit sand, Polish Journal of Environmental Studies, volume 16, nr 2a, part II, 2007.

[9] Jeż-Walkowiak J., Pruss A., Puk E., Sozański M.M., Weber Ł.: Research on arsenic sorption on selected filtration materials. 2nd International conference "Metals and Related Substances in Drinking Water " (METEAU) Cost Action 637, Lisbon, Portugal, October 29-31, 2008

[10] Konieczny K., „Treatment of waters polluted with organic substances" Second National Congress of Environmental Engineering, Lublin 2005 r.

[11] Lacey M., Filtration: venerable and versatile workhorse, JAWWA, December 2001

[12] Norma PN-EN ISO 17294-1 „Jakość wody. Zastosowanie spektrometrii mas z plazmą wzbudzoną indukcyjnie (ICP-MS). Część 1: Wytyczne ogólne." Polski Komitet Normalizacyjny, Warszawa 2007 r.

[13] Norma PN-EN ISO 17294-1 „Jakość wody. Zastosowanie spektrometrii mas z plazmą wzbudzoną indukcyjnie (ICP-MS). Część 2: Oznaczanie 62 pierwiastków." Polski Komitet Normalizacyjny, Warszawa 2006 r.

[14] Proceedings of Int. Conf. METEAU - Metals and Related Substances in Drinking Water. COST Action 637. Antalya, Turkey October 2007

[15] Pruss A. Jeż-Walkowiak J., Sozański M.M., Dymaczewski Z., Michalkiewicz M.: Elimination of heavy metals from water of Warta river by infiltration and filtration process. 2nd International conference "Metals and Related Substances in Drinking Water " (METEAU) Cost Action 637, Lisbon, Portugal, October 29-31, 2008

[16] Pruss A. Krzemieniewska E: Influence of shut down sand filter at Poznań Water Treatment Plant "Wiśniowa" on biological activity of filter bed. /W: IX International conference "Municipal and rural water supply and water quality", Poland, Poznań 2008 r.Poznań, 2010 r. Red. M.M. Sozański., Z. Dymaczewski, J.Jeż-Walkowiak [Organiz.]: PZITS - Oddz. Wlkp., Canadian Society for Civil Eng., Politechn. Pozn. - Inst. Inż. Środ. [i in.]. - Kołobrzeg 21 -23 czerwca 2010

[17] Pruss A., Jeż-Walkowiak J., Dymaczewski Z., Michałkiewicz M., „Usuwanie metali ciężkich z wody z rzeki Warty w procesach infiltracji filtracji." IV National Congress of Environmental Engineering, Lublin 2009 r.

[18] Rozporządzenie Ministra Zdrowia z dnia 29 marca 2007 r. w sprawie jakości wody przeznaczonej do spożycia przez ludzi. Dz. U. 2007 r. nr 61 poz. 417.

[19] Świątczak J., Skotak K., Bratkowski J., Witczak S., Postawa A.: „ Metale i substancje towarzyszące w wodach przeznaczonych do spożycia w Polsce." VIII International conference "Municipal and rural water supply and water quality", Poland, Poznań 2008 r.

[20] Wilmański K., Lasocka-Gomułka I.: „Modernizacja procesu uzdatniani wody pitnej dla aglomeracji poznańskiej." VI International conference "Municipal and rural water supply and water quality", Poland, Poznań 2004 r.

[21] Witkowicz Karol, Efektywność usuwania metali ciężkich z wody w procesach technologicznych realizowanych na Stacjach Uzdatniania Wody dla miasta Poznania. Praca magisterska, Politechnika Poznańska, Instytut Inżynierii Środowiska, Poznań 2009, promotor: dr inż. Alina Pruss

[22] World Health Organization "Guidelines for drinking - water quality. First addendum to third edition." 2006 r.

[23] World Health Organization WHO, 2004, Guidelines for drinking water quality, 3rd edition, Geneva, 2004.

[24] www.aquanet.pl

The influence of dissolved natural organic matter on the stability of Arsenic species in groundwater

E. Veschetti, L. Achene, P. Pettine, E. Ferretti and M. Ottaviani

Section of Inland waters, Department of Environment and primary prevention, Istituto Superiore di Sanità, Rome, Italy

Corresponding author e-mail: enrico.veschetti@iss.it

Arsenic bioavailability, toxicity and mobility depend on its speciation. In aquatic environments arsenic is mainly found in two oxidation states and three oxianion forms, i.e., H_3AsO_3, $H_2AsO_4^-$ and $HAsO_4^{2-}$. The stability of As(III) and As(V) has been reported to be dependent on water pH, redox potential, microbial activity, concentration of iron, manganese and natural organic matter (NOM). NOM is ubiquitous in aquatic and terrestrial systems. NOM particles, such as fulvic acid (FA) and humic acid (HA) particles, bind very strongly to (hydr)oxide minerals limiting arsenic adsorption on mineral surface. In addition to the competition effects for adsorption, NOM may influence arsenic distribution via some other mechanisms. For instance, degradation and oxidation of NOM may be coupled with reduction of arsenate to arsenite. More recently, few studies have also postulated that arsenic can form organic complexes with the dissolved fraction of natural organic matter (DOM). In spite of this recent evidence, quantitative data on As-DOM interaction are still missing.

Aim of this study was to investigate the effect of DOM on temporal stability of inorganic arsenic species in aqueous matrices taking into account the influence of pH, redox potential, ionic strength, Fe and Mn concentrations. Batch experiments were performed using arsenate and arsenite-containing solutions spiked with variable quantities of DOM extracted from aliquots of lake water. During the contact time, all the reaction systems were incubated at a preset temperature in the range 8-25°C. At first As(III) and As(V) stability was studied in synthetic water solutions containing inorganic salts, then the same experiments were repeated using real samples collected from an Italian volcanic aquifer. Speciation analyses to evaluate the ratio between the two inorganic species were carried out with solid phase extraction followed by electrothermal atomic absorption spectrometry. The complexation of arsenic with DOM was examined with gel filtration chromatography.

Quality control of arsenic determination in drinking water with ICP-MS: Krakow Tap Survey 2010

K. Wątor and E. Kmiecik

AGH University of Science and Technology, Kraków, Poland

Corresponding author e-mail: ewa.kmiecik@agh.edu.pl

Abstract

In Poland in 2010 screening tap survey is performed in selected cities as a part of European research project "Metals and related substances in drinking water in Poland" (Project No. 28.28.140.7013). Also in Kraków drinking water samples from taps were collected and analyzed using ICP-MS. Some possible sources of uncertainty in the arsenic determination were examined. The first one was impact of nitric acid used to sampling preservation. The results show that nitric acid is not significant source of uncertainty. Also using bottles gives not important part of uncertainty. The analysis of control samples was also performed. The mean value of arsenic concentration in laboratory blank samples was 0.242 µg/L, in field blank samples 0.371 µg/L, whereas parametric value for this element according to Drinking Water Directive [1] is 10 µg/L. Total relative expanded uncertainty for arsenic in drinking water in Cracow is bigger than 50% (it should be considered when we compare the results to the parametric value), but measurement relative uncertainty equals only 6.43%.

1. Materials and Methods

In Poland in 2010 screening tap survey is performed in selected 10 cities as a part of European research project "Metals and related substances in drinking water in Poland" (Project No. 28.28.140.7013). Also in Kraków drinking water samples from taps were collected. All taken samples derived from one water intake, Raba River. During sampling campaign 101 routine samples and 25 control samples were collected (14 duplicate and 11 blank samples) in may and june of 2010. Control samples are necessary for QA/QC purposes. Both routine and control samples were collected in the same way, using the same sampling protocol. All samples were analyzed using ICP-MS method in certified hydrogeochemical laboratory of Hydrogeology and Engineering Geology Department at the University of Science and Technology in Kraków (Certificate of Polish Centre for Accreditation No AB 1050). ICP-MS method is the good one for the arsenic concentration analysis in water [3].

The hydrogeochemical laboratory of Hydrogeology and Engineering Geology Department has implemented internal quality control/quality assurance system. ICP-MS method was validated and certified laboratory limit of detection for arsenic is 1 µg/L. But laboratory validated arsenic concentration from 0.1 to 1000 µg/L.

Both accuracy and precision are on acceptable level in validated range of measurements.
The laboratory compares its data and quality control system by participating in interlaboratory studies with other certified laboratories and by analysis of certified reference material – traceability. The accuracy of analysis of arsenic in certified natural water is very high for this laboratory and amount for about 99%.

2. Results

Some possible sources of uncertainty in the arsenic determination were examined.

2.1 Nitric acid

The first one was impact of nitric acid used to sampling preservation (Figure 1). The mean value of arsenic determination in deionized water was 0.236 µg/L, while in deionized water with addition of nitric acid was 0.242 µg/L. Nitric acid gives not important portion of uncertainty because results lower than LOD were obtained.

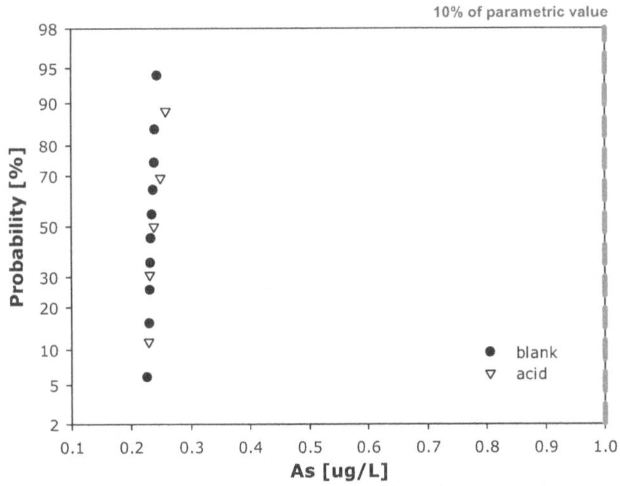

Figure 1. Concentration of arsenic[µg/L] in deionized water (blank) and in deionized water with addition of nitric acid (acid).

2.2 Bottles

The impact of using bottles on arsenic determination was also considered. In this case deionised water in four different types of bottles (different material, size, and/or producer) was analyzed. Analyses were performed four times for each bottle during five months. The results are shown in Figure 2.

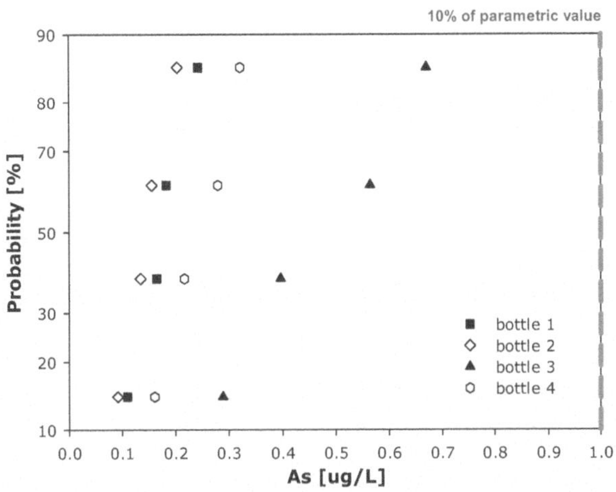

Figure 2. Concentration of arsenic in deionized water in four different types of bottles [µg/L].
The results show that type of bottles is not significant source of uncertainty (all results are lower tha certified limit of detection and lower than 10% of parametric values).2.3 Control samples

The mean value of arsenic concentration in blank samples was 0.236 µg/L, in field blank samples – 0.371 µg/L and vary from 0.212 µg/L to 0.623 µg/L, whereas parametric value for this element according to Drinking Water Directive is 10 µg/L. Certified laboratory limit of detection for arsenic is 1 µg/L – it is required 10% of parametric value for arsenic in drinking water. Laboratory measures also lower concentrations with good precision and accuracy (even on the level of 0.1 µg/L). Field blank samples gives also possibility to estimate practical detection limits for analyzed elements [8].

Researches prove that sampling process could be important source of uncertainty influencing final result and general quality of results achieved during water quality monitoring [2, 4-7, 10]. Duplicate samples collected during screening tap survey in Kraków were analyzed using simplified version of unbalanced duplicate method with only one analysis per sample for calculating between-target and measurement variance and uncertainty.

ROB2 software was used to determinate between-target and measurement (analytical + sampling) variance and uncertainties. Results are shown in Figure 4 and Table 1.

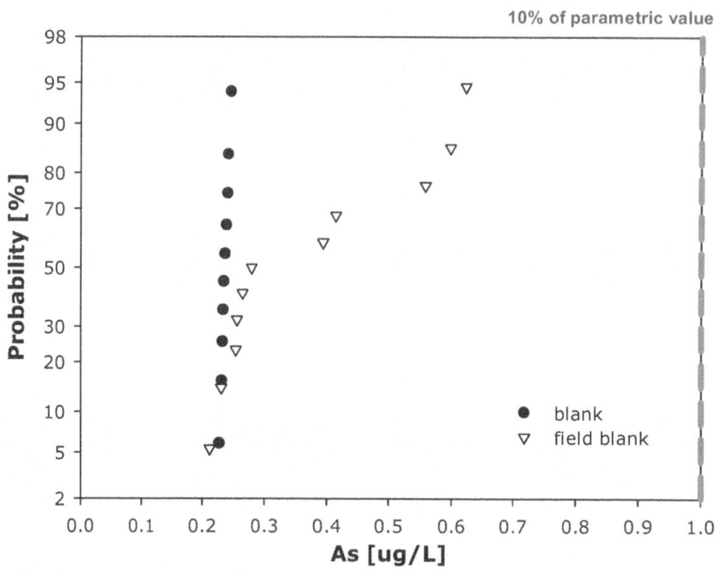

Figure 3. Concentration of arsenic in blank and field blank samples [μg/L].

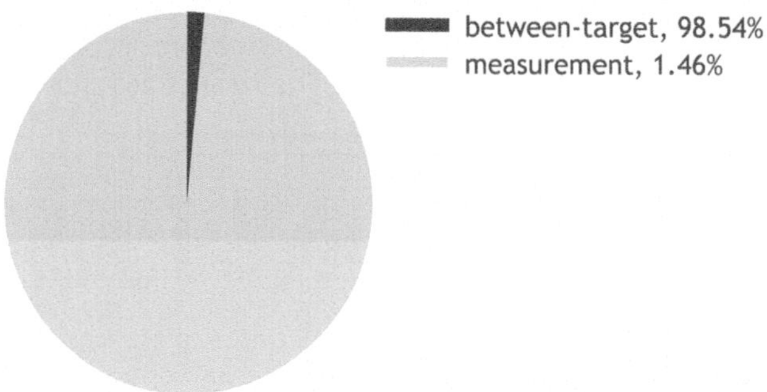

Figure 4. Percentage variances – between target and measurement.

Table 1. Uncertainty results (standard – u, expanded – U and relative – U'; coverage factor k=2) for arsenic concentration in drinking water in Kraków.

Parameter	Value
x_{mean} [μg/L]	0.778
u_{total} [μg/L]	0.207
U_{total} [μg/L]	0.414
U'_{total} [%]	**53.21**
u_{meas} [μg/L]	0.025
U_{meas} [μg/L]	0.05
U'_{meas} [%]	**6.43**
$U_{between-target}$ [μg/L]	0.206
$U_{between-target}$ [μg/L]	0.412
$U'_{between-target}$ [%]	52.96

Total expanded uncertainty for arsenic in examined drinking water in Kra¬ków is equal 0.414 µg/L and the total relative uncertainty is bigger than 50% (it should be considered when we compare the results to the para¬metric value), but measurement relative uncertainty equals only 6.43%.

4. Conclusions

During screening tap survey in Poland in 2010 water samples from Krakow's taps were collected (101 routine, 14 duplicate and 11 blank samples) using the same sampling protocol (RDT – random daytime). All samples were analyzed using ICP-MS method in the certified hydrogeochemical laboratory of Hydrogeology and Engineering Geology Department at the University of Science and Technology in Krakow.

The mean value of arsenic concentration in blank samples was 0.242 µg/L, in field blank samples – 0.371 µg/L and vary from 0.212 µg/L to 0.623 µg/L, whereas parametric value for this element according to Drinking Water Directive is 10 µg/L. Certified laboratory limit of detection for arsenic is 1 µg/L.

Some possible sources of uncertainty in the arsenic determination were examined. Neither using bottles nor nitric acid used to sampling presser¬vation have not significant influence on arsenic concentration in samples. The mean arsenic concentration values for deionized water in four tested bottles and with addition of nitric acid are not significantly different than mean arsenic concentration in used deionized water. Also sampling and analysis gives not big contribution to total uncertainty – only 6.45%.

Acknowledgments

The study was partially supported by AGH-UST 18.18.140.605.

References

[1] DWD, Drinking Water Directive, Council Directive 98/83/EC on the quality of water intended for human consumption, 1998.

[2] Eurachem, Estimation of measurement uncertainty arising from sampling, 2007.

[3] Kalevi K., Gustafsson J., Analytical aspects concerning to set threshold values for substances in groundwater, BRIDGE FP6, Deliverable 7: State-of-the-art knowledge on behaviour and effects of natural and anthropogenic groundwater pollutants relevant for the determination of groundwater threshold values. Final reference report, 2006.

[4] Kmiecik E., Assessing uncertainty associated with sampling of groundwater: Raba river basin monitoring network (South Poland), New developments in measurement uncertainty in chemical analysis, Symposium at BAM, Berlin 15-16 April 2008.

[5] Kmiecik E., Drzymała M., Uncertainty associated with the assessment of trends in groundwater quality (Krolewski spring, Krakow, Poland), New developments in measurement uncertainty in chemical analysis: Symposium at BAM, Berlin 15-16 April 2008.

[6] Kmiecik E., Drzymała M., Podgórni K., Uncertainty associated with groundwater sampling (Królewski spring, Kraków, Poland). Chemometria: metody i zastosowania, Komisja Chemometrii i Metrologii Chemicznej. Komitet Chemii Analitycznej PAN, 2009.

[7] Kmiecik E., Podgórni K., Estimation of sampler influence on uncertainty associated with sampling in groundwater monitoring, Biul. PIG no 436(9/1) 2009.

[8] Postawa A., Kmiecik E., Implementation of limit of detection (LOD) and practical limit of detection (PLOD) values for the assessment of uncertainty involved in sampling and analytical processes during drinking water quality monitoring, COST ACTION 637 METEAU: metals and related substances in drinking water : 3rd international conference: Ioannina, Greece, 21-23 October 2009.

[9] Ramsey M.H., Thompson M., Hale M., Objective evaluation of precision requirements for geochemical analysis using robust analysis of variance, J. Geochem. Explor 44(1992).

[10] Witczak S., Bronders J., Kania J., Kmiecik E., Różański K., Szczepańska J., BRIDGE FP6. Deliverable 16: Summary Guidance and Reccomendations on Sampling, Measuring and Quality Assurance, Final reference report, 2006.

High fluoride concentrations in surface water - example from a catchment in SE Sweden

Tobias Berger, Mats Åström, Pasi Peltola and Henrik Drake

Geochemistry Research Group, Linnaeus University, SE-391 82 Kalmar, Sweden

Corresponding author e-mail: tobias.berger@lnu.se

We studied the fluoride occurrence and its relationship to local geological properties in a small catchment (2700ha) located at the Baltic Sea coast in southeast Sweden. The aim was to investigate the possible impact of a granite intrusion (1.4 ga old Götemar granite) on the fluoride (F^-) concentrations in a stream, both on a spatial and temporal scale. This type of anorogenic granite is younger than the surrounding bedrock types and typically recognized by its richness in fluorine. The catchment is dominated by exposed bedrock (51.2%) and a thin till cover (30.3%) and 86 % is covered by coniferous and mixed forest. The intrusion is situated in the lower reaches just north of the main stem. Since the stream, which is perennial, is located in the boreal zone (N 57°) it is recognized by strong discharge fluctuations due mainly to snow melt during spring. Time series of surface water chemistry and discharge have been analyzed and combined with targeted sampling within the catchment. On a continental scale European stream waters (n=808, 25 countries) have concentrations (95 percentile) below 0.36 mg l^{-1} [1]. In the stream highlighted in this study (Kärrsvik) this is the case for the upper reaches of the catchment with maximum and median concentrations of 0.79 mg l^{-1} and 0.37 mg l^{-1} respectively. However, towards the stream outlet the F^- concentrations increase 1.6 to 4.7 times (during equal discharge conditions). The highest concentration measured in the lower reaches was remarkably high for a surface water with 4.16 mg l^{-1} (median 1,13 mg l^{-1}) . In comparison, the WHO guideline value is 1.5mg l^{-1} for drinking water[2]. The results describe a spatial and temporal behavior of F^- that confirms the hypothesis of the Götemar granite as a source for elevated fluoride concentrations in the surface water of the catchment. The mechanism is weathering of glacial deposits, partially consisting of Götemar granite, and greisen fractures (which are strongly connected to the intrusion and, as well, rich in fluorite). This knowledge can be of significant importance in areas where overburden waters frequently exceed the maximum limit of fluoride, as occurs in parts of Sweden.

References

[1] http://www.gtk.fi/publ/foregsatlas/

[2] http://www.who.int/water_sanitation_health/dwq/fulltext.pdf

Leaching of nickel and the other elements from kettle by domestic using

Vladimira Nemcova[1], Jana Kantorová [1], Frantisek Kozisek[2, 3], and Daniel Weyessa Gari [3]

[1] Institute of Public Health, CZ-70200 Ostrava, Czech Republic Republic
[2] Department of General Hygiene, Third Faculty of Medicine, Charles University in Prague
[3] National Institutes of Public Health, Department of Environmental Health, CZ-10042 Prague, Czech

Corresponding author e-mail: vladimira.nemcova@zuova.cz

This contribution is focused on the concentration of metals mainly nickel dissolved to water from kettles. The limit value of Ni for drinking water mentioned in the DWD No. 98/83/ES. set maximum permissible limit (MPL) 20μg/l. In general, the Ni concentration in Czech Republic's drinking water is not high. In recent years attention to the release (dissolution) of Ni from water bacteria was given, where significant presences especially in first part of the sample after long (eg. night or week end) stagnation, furthermore the Ni presentation in hot waters where the requirement is the same as for drinking water. The third area is that the possibility of Ni dissolution from kittle referred in the foreign literatures.

Kettles are very necessary equipments in household, water stagnation in these equipments and follow-up boiling is usual household phenomena. Tap water with high mineral concentration and in second phase with low mineral concentration was selected as dissolving medium. The scheme of this experiment gives attention on the consumer's practices. Aside from Ni concentration, the metals such as Pb, Cd, Cr, Zn, Fe, and Ca are also monitored.

The poster presents the results of this study.

Monitoring of metals concentrations in water intended for human consumption sampled from the area of Warsaw performed by ICP-MS and ICP-OES techniques

Dorota Święcicka, Sławomir Garboś and Jakub Bratkowski
National Institute of Public Health - National Institute of Hygiene, Department of Environmental Hygiene, 24 Chocimska Str., 00-791 Warsaw, Poland

Corresponding author e-mail: dswiecka@pzh.gov.pl

The materials applied for construction of the water supply installations (including pipes and pipe fittings) and taps may be responsible for considerable increase of observed concentrations of several metals in drinking water such as: chromium, nickel, copper, lead and zinc.

Monitoring of metals concentrations in water intended for human consumption sampled from the area of Warsaw was performed within DWM/N176/COST/2008 project financed by Polish Ministry of Science and Higher Education. Several metals which are listed in Directive 98/83/EC (Al, As, Cd, Cr, Cu, Fe, Mn, Ni, Pb) and additionally Co and Zn were determined in 100 tap water samples collected from the area of Warsaw.

The part of Warsaw supplied in drinking water by Central Water Supply System was chosen as control area. This area was split into approx. 100 sampling squares. Thus one sample was collected from the area of 0.09 km^2 (square with dimensions of 300 m $\times$ 300 m).

National monitoring of drinking water quality performed in Poland is fully based on FFS (*Fully Flushed Samples*) method. Therefore it does not include of monitoring level of releasing metals from plumbing installations and taps present at the point of sampling. Therefore in our work *Random Day Time* (RDT) monitoring based on taking 1 L of water directly from the tap used for consumption water drawing at a time randomly chosen within the day during normal office hours (there were no water abstraction, flushing, cleaning of the tap prior to the sampling) was applied for collection of tap water samples. In order to characterize the quality of water in the supply zone the main constituents of drinking water, Na, Mg, Ca, Cl^-, NO_3^-, PO_4^- and SO_4^{2-} were determined in representative of 10 % samples from this area. Additionally pH and temperature were determined in all collected samples.

For the determination of Al, As, Cd, Ni and Pb inductively coupled plasma mass spectrometry was applied (XSeries II CCT spectrometer, Thermo Electron Corporation, UK) while for the determination of rest of metals simultaneous inductively coupled plasma optical emission spectrometry with CID detector was used (IRIS Advantage Duo ER/S spectrometer, Thermo Jarrell Ash, USA). For the determination of anions high performance ion chromatography with conductometric detection was applied (ICS-2500 chromatographic system, Dionex, USA).

Stagnation time determined during sampling was crucial parameter influenced metals concentrations observed in sampled tap waters. In analyzed drinking water samples exceeding maximum admissible concentrations levels of Ni, Fe and Pb was observed.

Short period survey of metals and related substances in Racibórz town tap water, Poland

S. Jakóbczyk, H. Rubin, A. Kowalczyk and K. Rubin

Department of Hydrogeology and Engineering Geology, University of Silesia, Sosnowiec, Poland

Corresponding author e-mail: sabina.jakobczyk@us.edu.pl

Almost 60 000 inhabitants of Racibórz (Poland) are supplied with groundwater from sandy and gravely aquifer of Pleistocene and Miocene age, which is abstracted by three well fields: Gamowska, Strzybnik and Bogumińska. Groundwater extraction in this area has been proceeded for over 100 years. In the 80's groundwater withdrawal amounted to about 18 000 m^3/d, and the water table was lowered of about 15 - 18 m. For the last 15 years water extraction has been decreased twice what resulted in rise of the groundwater table. At the same time the increased concentration of some metals in groundwater has been observed: Ni - up to 0.16 mg/L, Fe - up to 10 mg/L, Mn - up to 1 mg/L. The increased concentrations of these metals were caused mainly by geochemical processes induced by groundwater level fluctuations. Raw water for consumption is purified by aeration and filtration through the quartz-sand bed with the addition of anthracite and manganese dioxide.

In March 2010, within the confines of COST action 637 recognition of occurrence of metals and related substances in water sampled from consumer's taps in NW part of Racibórz was conducted. Groundwater samples from the Gamowska and Strzybnik well fields were also collected.

Consumer's tap water was sampled in 100 randomly chosen points within regular grid divided into elements of dimensions 200 m x 200 m. One-liter samples were taken at a random time of a working day directly from the tap in a property without previous flushing (Random Daytime Sample). For the quality control (QA/QC) 11 doubled and 11 field blank samples were collected. Using ICP-MS method all samples were analyzed for Al, As, Cd, Cr, Cu, Fe, Mn, Ni, Pb and Zn. Results for consumer's tap water show exceeded concentrations for Ni (one sample), Pb (one sample) and Fe (four samples) with respect to Polish regulations and European directive. Groundwater sampling revealed in all wells exceeded concentrations for Fe (max. 2.05 mg/L) and Mn (max. 0.229 mg/L) and in one well for Pb (0.015 mg/L). Comparison of maximum concentrations of analyzed metals for groundwater (raw water) and water after purification shows its significant decline what results in very low concentrations of aforementioned metals (no exceeded values). Yet, comparing maximum concentrations of metals for purified water and the tap water one may observe increased values for all analyzed constituents, especially for Fe, Zn and Cu in the tap water.

Section 6

Source waters

Geogenic arsenic in groundwaters and soils - re-evaluating exposure routes and risk assessment

D.A. Polya[1], D.Mondal[1,2,3], B.Ganguli[4], A.K.Giri[2], M.Banerjee[2], S.Khattak[1,5], N.Phawadee[1,6] and C.Sovann[1,7]

[1] SEAES, University of Manchester, M13 9PL, UK
[2] Molecular and Human Genetics Division, IICB, Kolkata-700 032, INDIA
[3] now at LSHTM, Keppel Street, London WC1E 7HT, UK
[4] Dept. Statistics, University of Calcutta, Kolkata - 700 019, INDIA
[5] NCE in Geology, University of Peshawar, 25120, NWFP, PAKISTAN
[6] CEDS, National University of Laos, Dongdok Campus, Vientiane, LAO PDR
[7] Royal University of Phnom Penh, Phnom Penh, CAMBODIA

Corresponding E-mail: david.polya@manchester.ac.uk

Abstract

We present here data from the AquaTRAIN, CALIBRE and PRAMA networks and also from recent work from other groups that demonstrate that: (i) the 10 µg/L guideline for arsenic in drinking water may not be as protective of human health as for other chemicals; (ii) rice is a major exposure route for many individuals, including in the European Union, and that re-assessment of the arsenic-in-food regulations within the European Union is required; and (iii) exposure to arsenic through drinking water and rice may result in genetic and other damage in individuals, that are otherwise externally asymptomatic, at least in the earlier stages of the development of cancers and other detrimental sequela, some of which have latency periods of decades. There is a clear and present need for more critically determining the human health and socio-economic impacts of current levels of human exposure to arsenic within the European Union and elsewhere. The relative merits of regulatory/remediatory strategies need to explicitly take into account substitution of risks.

1. Introduction

Regulation and remediation of hazardous chemicals in drinking waters (including groundwaters used for that purpose) and soils has been very largely, though not exclusively, driven by studies (speciation, biogeochemistry, hydrology/hydrogeology, exposure, remediation, human health impacts) of anthropogenic chemicals and the application of conservative safety factors. Uncertainties and misperceptions regarding exposure routes, dose-response relationships and consequent human health risks associated with geogenic chemicals, notably arsenic, have contributed to drinking water and food regulations in both the European Union and elsewhere they are not as demonstrably protective of public health as those for many chemicals of known anthropogenic origin [1, 2, 3]. Both EFSA [4] and FAO/WHO [5] have recently raised similar concerns. Interestingly just a decade ago, reviews by the NRC [6,7] raised similar concerns over the previous WHO guide value of 50 µg/L for drinking water, leading to the tightening to the current WHO provisional guide value of 10 µg/L [8], which is widely but not universally adopted.

We present here data from the AquaTRAIN, CALIBRE and PRAMA networks and also from recent work from other groups that demonstrate that: (i) the 10 µg/L for arsenic in drinking water may not be as protective of human health as for other chemicals; (ii) rice is a major exposure route for many individuals living in the European Union and that re-assessment of the arsenic-in-food regulations within the European Union is required; and (iii) exposure to arsenic through drinking water and rice may result in genetic and other damage in individuals, that are otherwise externally asymptomatic, at least in the earlier stages of the development of cancers and other detrimental sequela, some of which have latency periods of decades [6,7]

2. Risks arising from WHO provisional guide value of 10 ug/L

2.1 Arsenic-attributable health risks

Chronic exposure to arsenic at concentrations equivalent to as low as 300 µg/L has been unequivocally linked to a wide variety of detrimental cancer and non-cancer health end-points [3,6,7,9,10,11]. Non-cancer end-points include development of highly visible skin hyperpigmentation and keratoses, as well as hypertension, ischaemic heart disease and diabetes, although there are considerable uncertainties in dose-response data [12,13,14,15], not least of all because of wide variety of dietary, genetic and environmental confounding factors [16, see also references in 3,6,7] Such exposure also contributes to the development of cancers of the skin, bladder, liver and lung - the latter being considered by some as the most significant [11].

2.2 Lung cancer risks attributable to arsenic in drinking water

Smith [17] estimated the lifetime cancer risks per million population attributable to chronic exposure to arsenic in drinking water at 500, 50 and 10 µg/L to be approximately 100,000, 10,000 and 2,000 respectively. These estimates are considerably higher than the values typically utilised by the USEPA as the upper bound for acceptable risks for individual carcinogens in drinking water. [18,19, see Table 1]

Table 1. Comparison of model lifetime cancer risks from exposure to arsenic with those typically used to establish USEPA MCLs (maximum contaminant levels).

Source / Daily exposure	Carcinogen	Risk[a] / 10^6
Well water with 500 µg/l arsenic (2 litre)[b]	Arsenic (1000 µg)	100,000 [c]
Well water with 50 µg/l arsenic (2 litre)[b]	Arsenic (100 µg)	10,000 [c]
Well water with 10 µg/l arsenic (2 litre)[b] [d]	Arsenic (20 µg)	2,000 [c]
USEPA Typical upper range of acceptable cancer risk [e]	10^{-4} lifetime risk	100
USEPA Typical upper range of acceptable cancer risk [e]	10^{-6} lifetime risk	1

[a] based on lifetime exposure; [b] USEPA default value; [c] based on data of Smith [11]; [d] note the lack of any safety factor even at 10 µg/L; [e] Whilst the USEPA do not currently prescribe a single value for acceptable lifetime cancer risk, USEPA [18] states that "for regulating chemical carcinogens, ... MCLs are set as close to the MCLG [maximum contaminant level goal] as is technically and economically feasible, but also with an acceptable cancer risk range of 10^{-4} to 10^{-6}", Cross [19] note that "EPA drinking water MCLs for carcinogens are generally set from about 1 x 10^{-4} to 1 x 10^{-6} theoretical upper-bound lifetime cancer risk".

2.3 Uncertainties

The estimates of Smith [11] are broadly based upon a linear extrapolation of strong epidemiological data obtained for populations chronically exposed to arsenic in drinking water with concentrations greater than 100 µg/L and assuming that there is no threshold concentration for its carcinogenic impact.

There is not a consensus regarding the how dose-response relationships from arsenic and arsenic-attributable cancers, including lung cancer, should be extrapolated to arsenic concentrations in drinking water near the WHO provisional guide value [1,6,7]. In the absence of such consensus it may be concluded that:

(i) the WHO provisional guide value for arsenic in drinking water is not as demonstrably protective of human health as are the values for other chemical components

(ii) further research work is needed to resolve the uncertainties

It is further noted that exposure to arsenic from rice and other foodstuffs may have led, in some studies, to some underestimation of the health effects arising from arsenic in drinking water because the exposure of "unexposed" groups to arsenic may have been underestimated - this is analogous (though not necessarily the same magnitude of effect) to that subsequently noted for early classic studies by Wynder & Graham and Doll & Hill on the impact of smoking of health because of the relatively high proportion of smokers in the hospital-based reference cohorts [20].

3. Rice is a major route of arsenic exposure

3.1 Case study - Indian Sub-continent

Some of world's areas most highly impacted by geogenic arsenic are in the Indian sub-continent [21,22,23] where the highest arsenic exposed populations are predominantly exposed through consumption of arsenic-bearing drinking water. Where such waters contain, say, 1000 µg/L As, not atypical of many highly impacted areas [21,22], exposure through rice was typically less than 5 % of total exposure [24] - remediation efforts therefore reasonably focussed on drinking water, and rice became generally perceived as an inconsequential exposure route, not least of all because only the inorganic (i-As) content is generally considered to be toxic/carcinogenic (although see [17] for discussion of MMA toxicity).

However, as remediation efforts became effective, the relative importance of rice as an exposure route has increased [25,26,27,28]. Given the high bioavailability of i-As [29], such studies have highlighted that arsenic exposures from rice (e.g. with as little as 100 µg/kg As of which 50 % was inorganic - see Figure 1) may exceed recommended maximum tolerable weekly intakes arising from exposure to drinking water with arsenic concentrations at the WHO provisional guide value [25,26,27,28].

3.2 Europe

There has been a perception that only rice from areas impacted by high arsenic groundwaters may be high in arsenic and hence populations in regions such as the USA and Europe are not particularly at risk. This is demonstrably not correct. Indeed, Zavala [30] reports that the mean total arsenic in rice grown in Europe and USA (198 µg/kg) is higher than that for Asia (70 µg/kg) (although the health impacts of this can be ameliorated by differences in the percentage of i-As in rice). Meharg [31,1] clearly demonstrates not only the importance of exposure through rice to certain groups in Europe but also the significant consequent human health risks.

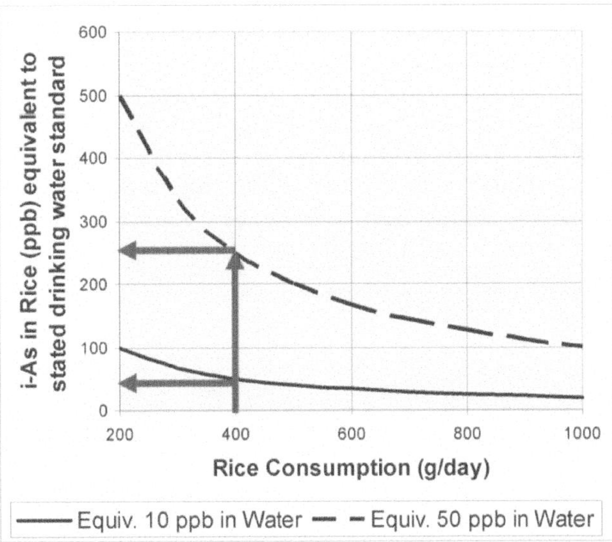

Figure 1. i-As (inorganic arsenic) concentration in rice equivalent to the old (50 ppb (µg/L)) and more recent (10 ppb (µg/L)) WHO provisional guide values in drinking water, assuming consumption of 2 L of drinking water per day.

4. Externally asymptomatic citizens may also be at risk

The widespread and often highly visible arsenic-attributable hyper-pigmentation and keratosis has led to the perception that externally asymptomatic people may not be at risk of arsenic attributable diseases. Since there is increasing evidence that good nutrition and certain genetic polymorphisms may be protective [32], the development of such a perception was not altogether unreasonable. Nevertheless studies of genetic damage in symptomatic and asymptomatic groups both chronically exposed to high(>

300 µg/L) As drinking water (see Figure 2) show that both groups have substantially higher genetic damage, as evidenced by micronuclei frequency, than an unexposed group [33] - thus asymptomatic citizens may also be suffering genetic damage from arsenic exposure.

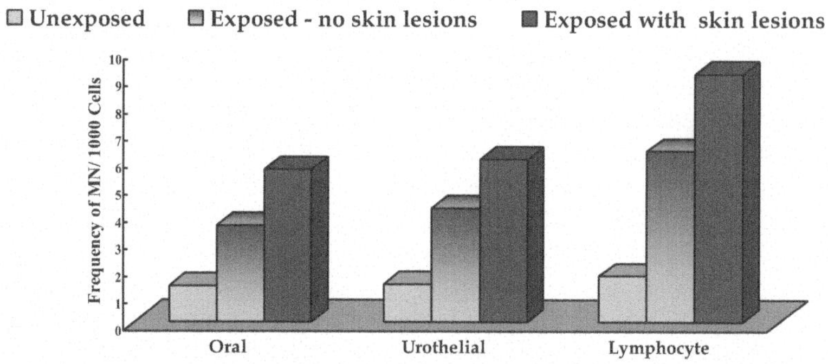

Figure 2. Frequency of micronuclei damage in cohorts

(i) unexposed; (ii) exposed to high arsenic drinking water but with no skin lesions; (iii) exposed to high arsenic drinking water and with visible skin lesions. For each and every one of the three cell types investigate, exposed populations exhibited significantly higher frequencies of micronuclei damage than unexposed population, irrespective of whether or not skin lesions were visible. Data from Basu [33].

Within the European Union, hyperpigmentation is very rare in the historically highly exposed population in arsenic-impacted regions of the Pannonian Basin [34], yet excess cancer mortality attributable to arsenic exposure are estimated to be as high as 10 % [35].

5. Discussion and Conclusions

There is a clear and present need for more critically determining the human health and socio-economic impacts of current levels of human exposure to arsenic within the European Union and elsewhere.

The use of known biomarkers of exposure [36], adsorption [37], metabolism [38] and early and late biological effects [39,40] as well as the development of novel biomarkers may be of considerable assistance in identifying at-risk groups within the population as a whole, as well as reducing the uncertainties of does-response relationships for key arsenic attributable detrimental health end-points.

As shown in Figure 3, the re-evaluation of geogenic arsenic exposure routes and human health risks lead to consideration of revised health targets and of drinking water and other safety plans.

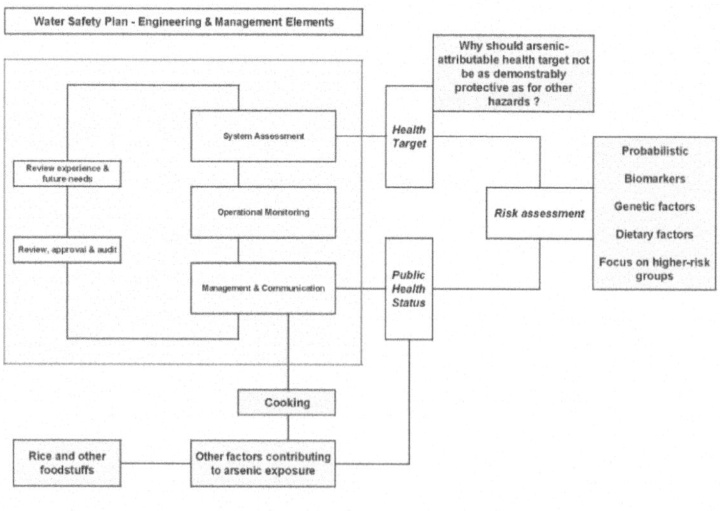

Figure 3. Water safety plan (modified after [41]) showing relationship to public health status, risk assessment and health targets, and to re-evaluation of exposure routes and assessment of geogenic arsenic attributable risks discussed in this study.

Revising regulatory values will ultimately not only take into accounts revised risk assessments, but also consider what are perceived to be acceptable upper bounds of risk [42] and weigh these against economic costs and benefits [43,44]. GIS and other tools for spatial mapping and interpolation of hazard, exposure and environmental and genetic confounding factors impacting dose-response relationships will also be important to facilitate the adoption and implementation of regulatory policies suitable for particular regions and/or population groups [45]. Lastly, the relative merits of regulatory/remediatory strategies need to explicitly take into account substitution of risks.

Acknowledgments

This is a contribution, to the 4th International Conference COST Action 637, Kristianstad, Sweden, October 13-15, 2010, of the AquaTRAIN MRTN (Contract No. MRTN-CT-2006-035420) funded under the European Commission 6th Framework Programme (2002-2006) Marie Curie Actions, Human Resources & Mobility Activity Area - Research Training Networks, the EU Asia-Link CALIBRE Project (Contract No. KH/AsiaLink/04 142966) and the UKIERI PRAMA (Contract No. SA07/09) Project. The views expressed do not necessarily reflect those of any of the funders including the European Community, which is not liable for any use that may be made of the information contained herein. DAP thanks Ingegerd Rosborg, Prosun Bhattacharya and the organisers for their kind invitation and the opportunity to present this work in Kristianstad. DM acknowledges the receipt of a Dorothy Hodgkins Postgraduate Award. SK was funded by a NCE Geology University of Peshawar Faculty Development Scholarship.

References

[1] Meharg, A.A., Raab, A., *Environmental Science & Technology*, **2010**, *44*, 4395-9

[2] Polya, D.A., Berg, M., Gault, A.G. and Takahashi, Y., *Applied Geochemistry*, **2008**, *23*, 2968-2976

[3] Polya, D.A., Mondal, D. And Giri, A.K., in Preedy and Watson (Eds) *Handbook of Disease Burdens and Quality of Life Measures*, **2009**, pp 702-728. Springer-Verlag, ISBN: 978-0-387-78665-0

[4] European Food Standards Authority, *Scientific Opinion on Arsenic in Food*, **2009**, EFSA-Q-2008-425

[5] FAO/WHO Joint Expert Committee on Food Additives, **2010**,TRS 958-JECFA 72.

[6] National Research Council, *Arsenic in Drinking water*, **1999**, National Academy Press, Washington, DC.

[7] National Research Council, *Arsenic in Drinking water*, **2001**, Update. National Academy Press, Washington, DC.

[8] WHO *Environmental Health Criteria 224. Arsenic and Arsenic Compounds*, **2001**, World Health Organization. Geneva.

[9] Brown J., *Public Health Goal for Arsenic in Drinking Water*, **2004** ,Office of Environmental Health Hazard Assessment, California EPA.

[10] USEPA, *Issue Paper: Inorganic Arsenic Cancer Slope Factor*, Final Draft, July 22, **2005**

[11] Smith, A.H., Lingas, E., Rahman, M. *Bulletin of the World Health Organization*, **2000**, 78, 1093-1103.

[12] Navas-Acien A., Sharrett, A.R., Silbergeld, E.K., Schwartz, B.S., Nachman, K.E., Burke, T.A., Guallar, E., *American Journal of Epidemiology*, **2005**, 162, 1037-1049.

[13] Navas-Acien A., Silbergeld, E.K., Streeter, R.A., Clark, J.M., Burke, T.A., Guallar, E., *Environmental Health Perspectives*, **2006**, 114, 641-648.

[14] Mondal, D., Adamson, G.C.D., Nickson, R., Polya, D.A., *Applied Geochemistry*, **2008**, 23, 2998-3008

[15] Adamson, G.C.D., Polya, D.A., *Journal of Environmental Science and Health Part A*, **2007**, 42, 1909-1917.

[16] Vahter, M.E., *J Nutrition*, **2007**, 137, 2798-2804

[17] Smith, A.H., Steinmaus, C.M., *Annu. Rev. Public Health*, **2008**, 30,107-122; Smith, A.H., Royal Geographical Society Meeting, London, August **2007**, http://www.geog.cam.ac.uk/research/projects/arsenic/symposium/S1.4_A_Smith_et_al.pdf

[18] US Environmental Protection Agency, *Use of Microbial Risk Assessment in Setting US Drinking Water Standards*, **2002**, EPA 814/S-92-001.

[19] Cross, F.B., Byrd, D.M., Lave, L.B., *Admin. Law Review*, **1991**, 61.

[20] Stolley, P.D., Lasky, T., *Investigating disease patterns*, **1995**, Scientific American Library.

[21] Smedley P., Kinniburgh D., *Applied Geochemistry*, **2002**, 17, 517- 568.

[22] Charlet, L., Polya, D.A., *Elements*, **2006**, 2, 91-96.

[23] Mazumder DNG, Haque R, Ghosh N, De BK, Santra A, Chakraborty D, and Smith AH (1998) Int. J. Epidemiol. 27: 871-877
[24] Khattak, S.A., Polya, D.A., Samanta, M., abstract, in Jean, J.-S., *Arsenic 2010*, NCKU, Taiwan, May **2010**.
[25] Juhasz, A. L., Smith, E., Weber, J., Rees, M., Rofe, A., Kuchel, T., Sansom, L., Naidu R. *Environmental Health Perspectives*, **2006**, 114, 1826-1831.
[26] Williams, P.N., Islam, M.R., Adomako, E.E., Raab, A., Hossain, S.A., Zhu, Y.G., Feldmann,J., Meharg ,A.A., *Environmental Science & Technology*, **2006**,40, 4903-8
[27] Mondal, D. and Polya, D.A., *Applied Geochemistry*, **2008**, 23, 2986-2997.
[28] Mondal, D., Banerjee, M., Kundu, M., Banerjee, N., Bhattacharya, U., Giri, A.K., Ganguli, B., Sen Roy, S., Polya, D.A., *Environmental Geochemistry and Health*, **2010**, 32, 463-477.
[29] Kile, M.L., Andres Houseman, E., Breton,C.V., Smith,T., Quamruzzaman, Q., Rahman,M., Mahiuddin,G., Christiani, D.C., *Environmental Health Perspectives*, **2007**, 115, 890-893.
[30] Zavala, Y.J., Duxbury, J.M., *Environmental Science & Technology*, **2008**, 42, 3856-3860.
[31] Meharg, A.A., Williams, P.N., Adiomako, E., Lawgali, Y.Y., Deacon, C., Villada, A., Campbell, R.C.J., Sun, G., Zhu, Y.-G., Feldmann, J., Raab, A., Zhao, F.-J., Islam, R., Hossain, S., Yanai, J., *Environmental Science & Technology*, **2009**, 43, 1612-1617.
[32] Banerjee, M., Sarkar, J., Das, J.K., Mukherjee, A., Sarkar, A.K., Mondal, L. , Giri, A.K. *Carcinogenesis*, **2007**, 28: 672-676.
[33] Basu, A., Ghosh, P., Das, J.K., Banerjee, A., Ray, K., Giri, A.K., *Cancer, Epidemiology, Biomarkers and Prevention*, **2004**, 13, 820-827
[34] Lindberg, A.L., Goessler, W., Gurzau, E., Koppova, K., Rudnai, P., Kumar, R., Fletcher, T., Leonardi, G., Slotova, K., Gheorghiu, E., Vahter, M., *Journal of Environmental Monitoring*, **2006**, 8, 203-208.
[35] Tony Fletcher, London School of Hygiene and Tropical Medicine, pers. comm..
[36] Gault, A.G., Rowland, H.A.L., Charnock, J.M., Wogelius, R.A., Gomez-Morilla, I., Vong, S., Samreth, S., Sampson, M.L. and Polya, D.A. *Science of the Total Environment*, **2008**, 393, 168-176.
[37] Button, M., Watts, M.J., Cave M.R., Harrington C.F., Jenkin, G.T., *Environmental Geochemistry and Health*, **2009**, 31, 273-282.
[38] Brima, E.I., Harrington, C.F., Jenkins, J.O., Gault, A.G., Polya, D.A. and Haris, P.I., *Applied Toxicology and Pharmacology*, **2006**, 216, 122-130.
[39] Bonassi, S., Au, W.W., *Mutation Research*, **2002**, 511, 73-86.
[40] Bonassi, S., Hagmar, L., Stromberg, U., Montagud, A.H., Tinnerberg, H., Forni, A., Helkklla, P., Wanders, S., Wilhardt, P., Hansteen, I.-L., Knudsen, L.E., Norppa, H. For the European Study Group on Cytogenetic Biomarkers and Health, *Cancer Research*, **2000**, 60, 1619-1625.
[41] Medema G., Ashbolt, N. *MICRORISK QMRA : its value for risk management*, **2006**, European Commission FP5 Report (Contract EVK1-CT-2002-00123)
[42] Kelly, K.E., *The Myth of 10^{-6} as a definition of acceptable risk*, Delta Toxicology Inc., updated from **1991,** 84th Ann. Meeting, AWMA, Vancouver, Canada.
[43] US Environmental Protection Agency, *Arsenic in Drinking Water Rule Economic Analysis,* **2000**, EPA 815/R-00-026.
[44] Smith A.H., Smith, M.M.H. *Toxicology*, **2004**, 198, 39-44.
[45] Rodriguez-Lado, L., Polya, D.A., Winkel, L., Berg, M. and Hegan, A. *Applied Geochemistry*, **2008**, 23, 3010-3018.

Arsenic distribution in surface and groundwater in the central bolivian highland

M. Ormachea[1,2 *], P. Bhattacharya[2] and O. Ramos[1,2]

[1] Laboratorio de Hidrogeoquímica, Instituto de Investigaciones Químicas (IIQ), Universidad Mayor de San Andrés,Casilla 303 La Paz, Bolivia

[2] KTH-Groundwater Arsenic Research Group, Department of Land and Water Resources Engineering, Royal Institute of Technology (KTH), SE 100 44 Stockholm, Sweden

Corresponding author e-mail: maurormache@gmail.com

Abstract

This study deals with the chemical quality of water samples taken from manually constructed wells with depths between two to nine meters. Almost all well-waters are used for consumption as drinking and irrigation water, especially during dry season. The wells are located around the Poopó Lake, situated in the central part of the Bolivian Altiplano (BA). The wells are mostly shallow and ontaminated by arsenic (As) and other trace metals from natural and anthropogenic sources. The north east side of the lake is a semiarid area where strong mining activities are carried out since last century. The west south side of the lake is an arid area where agricultural and cattle activities are carried out. Due the mining and geothermal sources, rivers, soils and some wells in the semiarid area are polluted by trace metals. Few rivers in the arid area are seasonally used for irrigation and become scarce or disappear before reaching the lake and many wells become dry as well. Detailed hydrochemical analyses of the well waters around the Poopó Lake reveal elevated As concentrations in almost all wells in the region.

1. Introduction

In Bolivia, there are no comprehensive studies on the contamination of groundwater from geogenic sources or mining activities, nor their impact on the population. Several mining areas in the Bolivian Altiplano (BA) have been exploited for five centuries from the colonial period to the present time for silver and gold deposits generally associated with polymetallic sulfides comprising Fe, Cu, Zn, Pb and Co (SERGEOMIN, 1999) as well as other trace elements like Li and B associated with the saline lakes in the BA. Earlier studies in the BA by Garcia (2006) have identified that mining and smelting activities have resulted in an extensive contamination of the rivers, groundwater and sediments adjoining the mining areas around the Poopó basin by the toxic metals and arsenic (As) through atmospheric deposition as well as acid mine drainage. This has also posed a considerable risk to the agro-industrial products, especially vegetables like potato, onion, carrots and others. The presence of As has been documented in groundwater of the Poopó basin with concentration levels above the WHO drinking water guideline (10 µg/L). Similar elevated concentrations are also documented in the surface water and the sediments in the region (Quintanilla *et al.* 2009).

Poopó Lake is located in the middle of the Bolivian altiplano at an altitude of 3686 m asl. The lake has a maximum length of 90 km and a maximum width of 53 km. The Poopó Lake (Figure. 1) has an area that varies from 2650 km^2 to 4200 km^2 on a seasonal basis (Quintanilla 1994). This paper deals with the assessment of surface and groundwater quality in the shallow wells and their relationship to the distribution of As and other trace metals around the Poopó basin at the county of Oruro in the BA. The groundwater quality is severely impacted by local geology and the historical and present mining and smelting activities around the major cities. Open shallow wells were sampled as they are commonly used for drinking water and for irrigation purposes.

2. Materials and methods

A total of 32 water samples were collected from manually constructed wells which are placed at shallow depths between 2-9 m bgs. The geographical location of the wells was recorded using a hand-held global positions system (GIS) GARMIN 12. Samples were collected following standard protocol for water sampling (Bhattacharya *et al.*, 2002). The pH, electrical conductivity, temperature and redox potential (Eh) were determined in the field and alkalinity was measured *in situ* using a HACH digital titrator. Phosphate was determined *in situ* using a portable HACH spectrophotometer at 540 nm wavelength.

Analyses of anions and cations were carried out at the hydrogeochemistry laboratory of the Chemical Research Institute (IIQ), La Paz, Bolivia. The major anions, Cl^-, NO_3^- and SO_4^{2-} were analyzed using an Alltech Model ion - chromatograph with an ion pack column. The major cations, Na^+, K^+, Ca^{2+}, and Mg^{2+}, were analyzed in the filtered acidified water samples using Perkin Elmer AAnalyst100 flame atomic absorption spectrometer. Total trace metals and As concentrations were analysed by Inductively Couple Plasma Atomic Emission Spectrometry (ICP-OES) at Stockholm University, Sweden

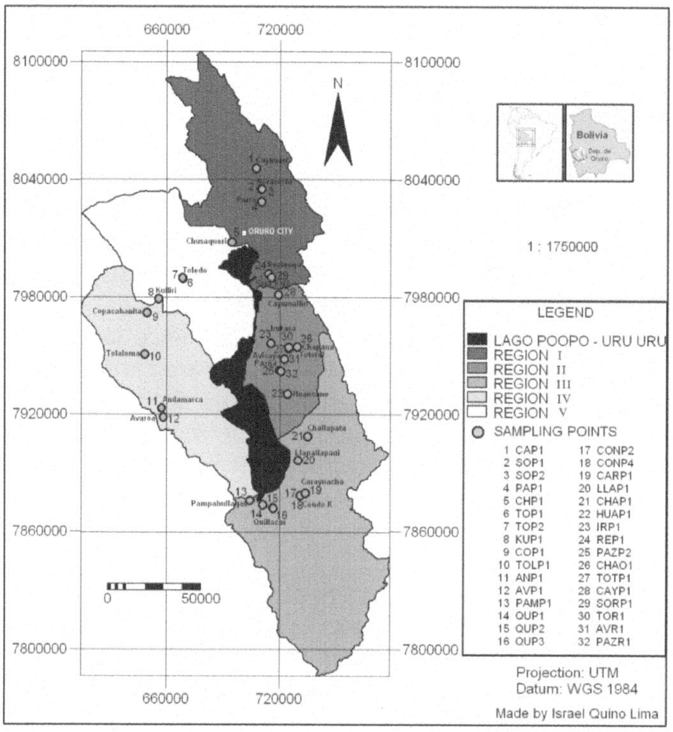

Figure 1. Inset map showing the location of the Bolivian Altiplano and the study areas

3. Results

The pH values were circum-neutral ranged from 3.10 to 7.90 with an average of 6.83. The redox potential ranges from +141 to +417 mV (average: +199 mV). The major ion and selected trace element chemistry of the groundwaters is presented in Figure. 2. The analysed water samples indicated a diversity of water types: 21.4% Na-Ca-Mg-SO_4-HCO_3; 14.3% Na-Ca-HCO_3, 10.7% Ca-HC_3-SO_4; 7.1% Na-Ca-Cl; 7.1% Na-HCO_3-SO_4.

Dissolved As concentration in the groundwater ranges from below detection limit to 242 μg L^{-1} and averaged 63 μg L^{-1} (n=32). The south west region presents the highest levels of natural As from 116.8 μg L^{-1} in Pampahullagas (south) to 242 μg L^{-1} in Toledo (west). Three wells located north east of the basin present As below detection limit (5 μg L^{-1}). More than 78% of the wells exceeded the WHO guideline (10 μg L^{-1}). Arsenic in groundwater is attributed to the oxidation of sulphide minerals; but volcanic ash in the area also might be a source of natural As in the groundwater.

The distribution of heavy metals shows close relation to the local mining activities. The highest concentrations of zinc, iron and manganese are found in well samples affected by mining activities. Zinc concentrations (0.007-205.3 mg L^{-1}, average: 14.5 mg L^{-1}. Among the redox sensitive elements, Fe and Mn showed wide variability in the ranges of 0.01-35.2 mg L^{-1}; (average 1.3 mg L^{-1}) and 0.01-19.6 mg L^{-1} (average 2.0 mg L^{-1}), respectively. These water samples also showed elevated concentration of Al (0.014-62.2 mg L^{-1}; average 4.6 mg L^{-1}) and Si ranging from 7.3 to 26.1 mg L^{-1} with an average of 15.0 mg L^{-1}.

Among the trace elements, boron shows high concentration and is possibly related to the presence of non metallic minerals and geothermal activities in the area. Boron concentration ranges from 0.3 to 6.1 mg L^{-1} and averaged 1.4 mg L^{-1}.

The geochemical program Visual Minteq and MINTEQ database was used to calculate the speciation of As and the equilibrium concentrations of As species. Speciation modelling indicates that the predominant species are: $HAsO_4^{2-}$ (80%), $H_2AsO_4^-$ (17%), others (3%).

Further studies in the area will be carried out to develop a conceptual model of the genesis, mobilization and transport of As in the region and potential health impacts on the local population

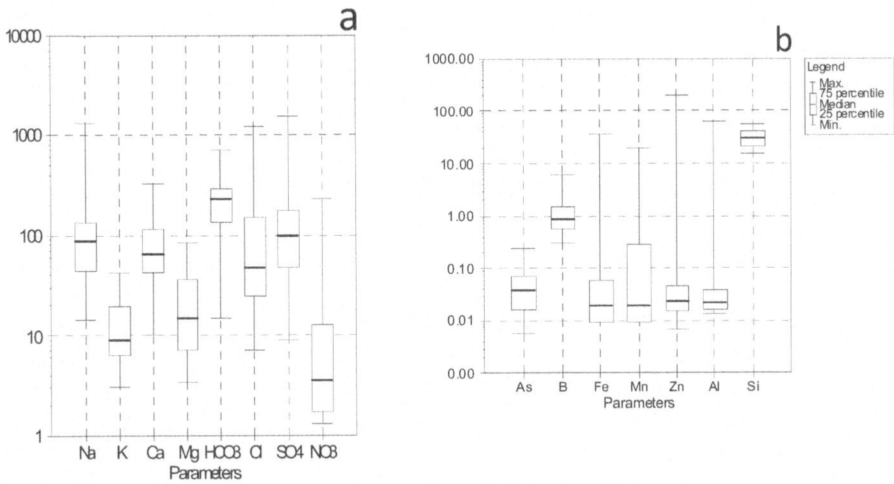

Figure 2. Box and Whisker plot for the distribution of a) major ions and b) selected trace elements in the groundwater samples.

4. Discussion and conclusions

Most of the rivers and wells located near mines are polluted by high concentrations of iron, zinc, manganese and lead this shows a close relation between polluted waters and industrial activities.

Arsenic distribution is very randomly in all the study area, but to the south is possible to identify natural contamination from geogenic sources, most of the wells in this area are used for consumption as drinking water where arsenic concentration is more than 20 times the safe values from OMS (10 µg L^{-1})

Acknowledgments

The financial support of this project by the Swedish International Development Cooperation Agency (Sida Contribution:7500707606) is gratefully acknowledged..

References

Bhattacharya, P., Jacks, G., Ahmed, K.M., Routh, J., Khan, A.A., 2002. Arsenic in groundwater of the Bengal delta plain aquifers in Bangladesh. Bull. Environ. Cont. Toxicol. **69:** 538-545.

Garcia M., 2006. Transport of arsenic and heavy metals to lake Poopo, Bolivia.

Gustafsson, J. P., (2008) Visual MINTEQ v 2.60.

http://www.lwr.kth.se/English/Oursoftware/vminteq/index.htm.

Quintanilla, J. 1985. Estrategia de estudio del sistema fluviolacustre del Altiplano. Ecología en Bolivia N°7: 65-74, La Paz.

Quintanilla, J. 1994. Evaluación Hidroquímica de la cuenca de los lagos Uru Uru y Poopó. IIQ-UMSA,

Quintanilla, J., Ramos Ramos, O.E., Ormachea, M., García, M.E., Medina, H., Thunvik, R. & Bhattacharya, P. 2009. Arsenic contamination, speciation and environmental consequences in the Bolivian plateau. In: *Natural Arsenic in Groundwater of Latin America -— Occurrence, health impact and remediation.* J. Bundschuh, M. Armienta, P. Birkle, P. Bhattacharya, et al. (eds.) CRC Press/Balkema, Leiden, pp. 91-100.

SERGEOMIN, 1999. Inventariación de recursos naturales renovables y no renovables del departamento de Oruro. Boletín del Servicio Nacional de Geología y Minería N° 24, 44 pp.

Genesis of arsenic enriched groundwater and relationship with bedrock geology in northern Sweden

P. Bhattacharya[1], **G. Jacks**[1], M. Svensson[2] and M. von Brömssen[1,2]

[1] *KTH International Groundwater Arsenic Research Group, Department of Land and Water Resources Engineering, KTH, SE-100 44 Stockholm Sweden*
[2] *Department of Soil and Water Environment, Ramböll Sweden, Box 4205, SE-102 65 Stockholm, Sweden*

Corresponding author e-mail: gunnjack@kth.se

Abstract

A growing concern over incidents of widespread human exposure to arsenic (As) from groundwater sources has been noticed during the past three decades. Väasterbotten county in northern Sweden hosts a large number of sulphide ore deposits and a number of gold deposits are recently discovered. Both are accompanied by elevated arsenic contents. Proterozoic metasediments sandwiched in the bedrock and mixed into the till contains elevated amounts of arsenic as well. During the present study about 80 groundwater samples were collected from dug wells, bore-wells and springs in the Skellefte field in Västerbotten County in northern Sweden. Data from community environmental offices were also collected and included in the study. Arsenic concentrations were elevated in borewells and wetland springs while none of the dug wells had arsenic contents above 10 mg/l. The highest content seen in borewells was 300 mg/l and in wetland springs 100 mg/l. The As(III)/As(tot) varied largely in borewells while it was mostly above 0.8 in wetland springs indicating more reducing contents in the latter. The use of a redox classification indicated that two nechanisms were involved in the mobilisation of he arsenic, oxidation of sulphides and reduction of ferric oxyhydroxides. In some cases the borewells showed a mixed pattern, indicating inflow from different environments.

1. Introduction

Arsenic in groundwater is an emerging threat, discovered almost globally during the last three decades. The reason for the recent discovery of that threat from an old poison is that the chronic toxicity was underestimated [1] and that the detection of arsenic at low levels previously required special analytical methods not frequently used. The mobilisation of arsenic into groundwater is highly dependant on redox conditions. Mainly three mechanisms are responsible for most occurrences of arsenic in groundwater: 1) oxidation of sulphides, 2) reduction of ferric oxyhydroxides releasing adsorbed arsenic and 3) high pH favouring desorption of arsenic from ferric and aluminium phases [2, 3]. In the studied area there are conditions that would enable the first two mechanisms to be active.

The aim of the work has been to assess the risk of arsenic mobilisation into groundwater and its relation to the geologic environment. An extensive presentation of data and results are available in [4].

2. Site, Materials and Methods

2.1 Sampling sites

The Skellefte field is underlain by Paleoproterozoic rocks of ages between 1800 to 1900 Ma [5]. The rock sequence is formed in a marine environment and comprises two volcanic sequences overlain by sediments and another younger volcanic sequence. Granitoids are cutting through the volcanics and sediments and the whole pack is subject to tectonic movements. The massive sulphide ores are situated high up in the second volcanic sequence, just below to metamorphosed sediments. A newly discovered line of gold deposits is another source of arsenic [6]. The sulphide orebodies contain various arsenic contents up to as much as 7 % in the Boliden oreboby [7]. The metamorphosed sediments underlying about 4 000 km^2 are partly fined grained and contain about 1 % sulphur and around 100 mg/kg As [8,9].

The bedrock is overlain by till and glacifluvial sediments. The till is a mixture of the rocks found to the northwest of a site as the ice-direction was from NW to SE which is mirrored in the content of arsenic [10, 11]. The till is podzolic with pH in the order of 5-6. Abundant wetlands in the form of peat bogs are present.

2.2 Sampling and analytical methods

Samples were filtered in the field through 0.2 □m filters and pH of the samples was determined in the field. In some samples we also measured Eh although that is difficult and results are reliable only when the Fe^{2+}/Fe^{3+} couple govern the redox-potential [12]. Further arsenic speciation was done in the field by the use of cartridges (MetaSoft Centre, PA, USA). The site of sampling was determined by a GPS-navigator and the coordinates registered were imported in ArcGIS and plotted on bedrock and quaternary geology maps to enable correlation between the geology and the water chemistry at the respective sites.

Alkalinity was determined by titration with 0.02 M HCl using a Radiometer ABC 80 autoburette and a Radiometer PHM 82 pH-meter. Major anions were analysed by ion chromatography on a Dionex DX-120 analyser.

Major cations and trace elements were determined by ICP-OES (Varian Vista-PRO) at Stockholm University.

3. Results and Discussion

As mentioned two mechanisms of mobilisation are expected to be present in this area both redox-dependent. This motivated a distribution of the samples into redox-classes using a classification from Swed. Environmental Protection Agency [13] (Table 1).

Table 1. Redox classes as defined by Swed. Environ. Prot. Agency [12].

Redoxcharacteristics	Fe mg/l	Mn mg/l	SO_4 mg/l
Aerated, oxic water (I)	<0.1	<0.05	>2
Moderately oxic (II)	<0.1	>0.05	>2
Anaerobic (III)	>0.1	>0.05	>2
Strongly anaerobic (IV)	>0.1	>0.05	<2
Mixed water type 1 (V:1)	<0.1	All values	<2
Mixed water type 2 (V:2)	>0.1	<0.05	All values

The result of the redox classification is seen in Figure. 1. Dug well are generally having aerated water not favourable for the mobilisation of arsenic mirrored in the fact that none of the dug wells had > 10 □g/l of As. The wetland springs have on the contrary reduced water with high contents of iron and manganese and arsenic contents were found up to 100 □g/l [14]. The borewells were equally distributed in all classes and they also contained mixed water types showing that the inflow to borewells could come in several sections from quite different environments (Figure. 2).

The As(III)/As(tot) ratio varies largely for borewells as they have water from both reducing and oxidising environments (Table 2). The ratios for the wetland springs show predominantly As(III) as would be expected in a reducing environment. A redox parameter which is not included in the redox classification is nitrate, indicating oxidising conditions. It is observed that elevated nitrate is a good indicator of arsenic safe groundwater (Figure. 3).

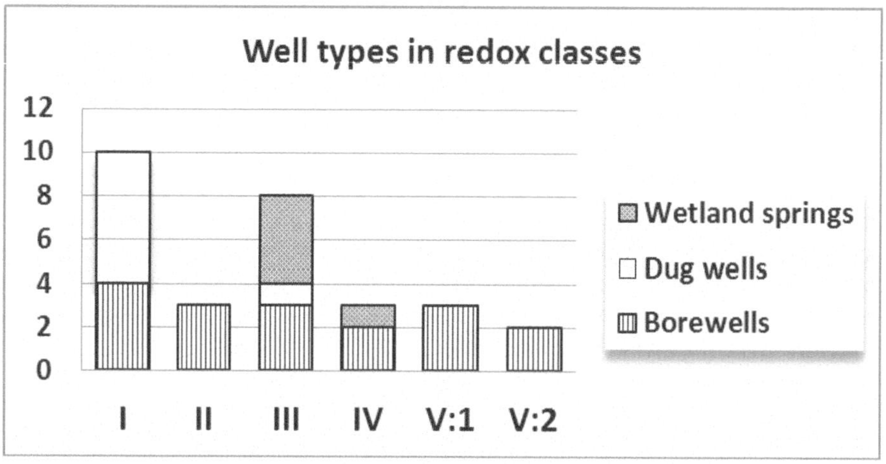

Figure 1. Number of wells in the different redox classes.

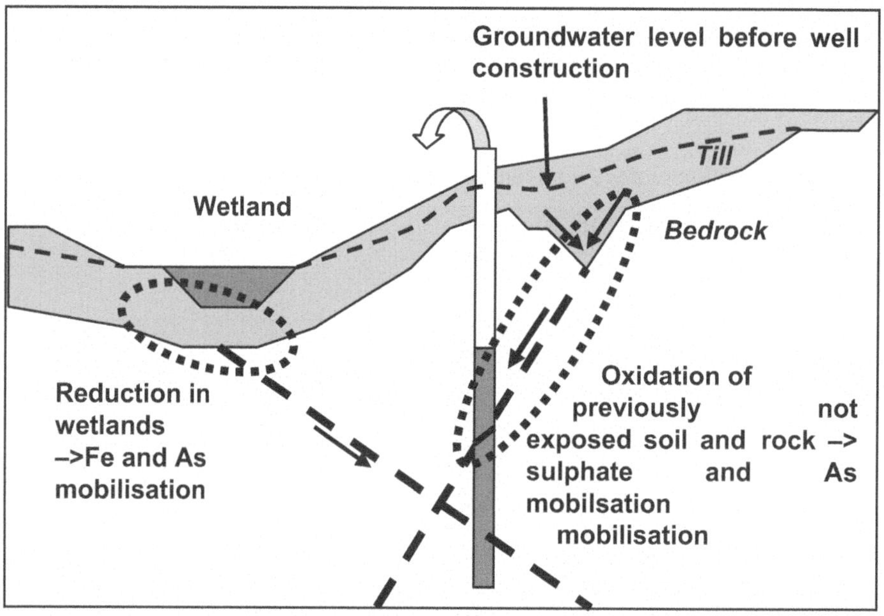

Figure 2. Borewell drawing water from two different redox environments.

An effort was done to assess which of the major rock types present in the investigated area that was showing elevated arsenic concentrations (Table 3). The volcanic rocks turned out to give the highest arsenic concentrations in groundwater. There was a large variation in each of the rock groups. Peculiarly enough the sedimentary rock which do include metamorphic sulphide-containing members showed generally low contents of arsenic in groundwater. However, these rocks also include arsenic sediments without sulphides. It should also be remembered that the till cover constitutes a mixture of different parent materials derived from upstream in the glacial ice direction.

Table 2. Arsenic speciation in a selection of the wells with elevated arsenic concentrations.

Sample ID	Source*	As(tot) µg/L	As(III) µg/L	As(III)/As(tot)
VB-5	BW	293	318.9	1.09
VB-22	BW	12.9	0.25	0.02
VB-56	BW	6.5	.8	0.12
VB-57	BW	24.2	24.3	1.00
VB-59	BW	6.1	1.2	0.20
VB-60	BW	178.4	18.1	0.1
VB-61	BW	201.2	49.2	0.24
VB-85	WS	7,1	6,.5	0.93
VB-86	WS	31.8	11.9	0.37
VB-87	WS	41.6	35.2	0.85
VB-88	WS	63.9	52.2	0.82
VB-89	WS	70.0	66.7	0.95

* BW = Tube well; WS = Wetland spring.

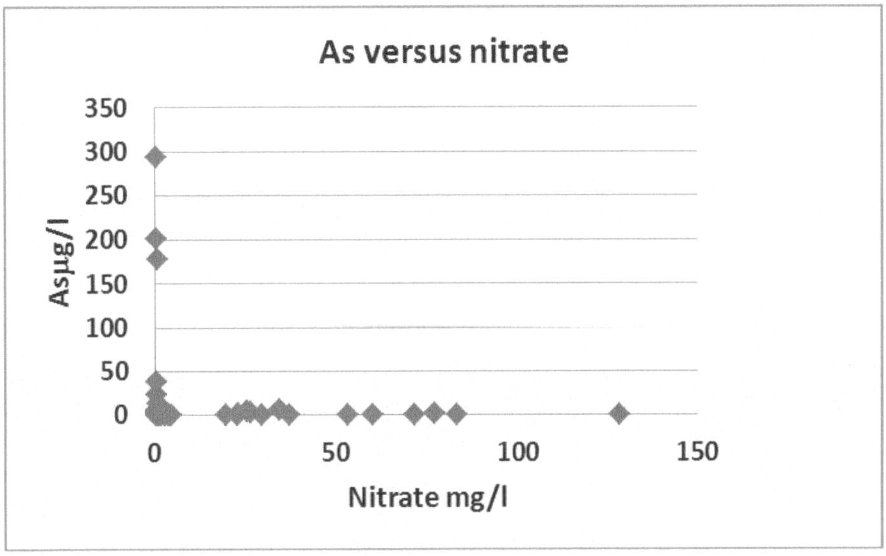

Figure 3. Arsenic versus nitrate I nall wells

Table 3. Arsenic in borewells from different rock environments.

Bedrock group	Number of wells	Min µg/L	Max µg/L	Mean µg/L	Median µg/L
Acid-intermediate volcanic rock	12	0,05	300	52	3
Alkaline volcanic rock	8	0,1	270	46.8	7.9
Acid-intermediate plutonic rock	30	0,05	201,2	13.4	2.1
Alkaline plutonic rock	0				
Sedimentary rock	10	0,25	22	3,2	0,55

4. Conclusions

A number of the borewells have elevated arsenic concentrations and a few of them as high as close to 300 µg/l. Two mechanisms are responsible for the mobilisation of arsenic, oxidising conditions mobilising arsenic from sulphides and reducing conditions leading to reduction of ferric oxyhydroxides and mobilisation of adsorbed arsenic. Dug wells are found to be safe presumably due to oxidising groundwater in which ferric oxyhydroxides are stable and a good sink for any arsenic mobilised [15, 16]. It is recommended that arsenic should be analysed especially in borewells and in wells with elevated iron concentrations. Volcanic rocks tended to show the highest mean levels of arsenic while sedimentary rocks had low contents. However some or the arsenic may come from till which contains at places sulphic metamorphic sediments.

Acknowledgments

This investigation was carried out by support from Swedish Geological Survey.

References

[1] Vahter, M., Concha, G., **2001**, *Pharmacol Toxocol,* 89, 1-5.

[2] Bhattacharya, P., Welch, A. H., Stollenwerk, K. G., Laughlin, M. J., Bundschuh, J., Panaullah, G., **2007**, *Sci Tot Environ*, 379, 109-120.

[3] Bhattacharya, P., Claesson, M., Bundschuh, J., Sracek, O., Fagerberg, J., Jacks, G., Martin, R. **2006**, *Sci Tot Environ*, 358(1-3), 97-120.

[4] Bhattacharya, P., Jacks, G., von Brömssen, M., Svensson, M. 2010, Arsenic in Swedish Groundwater. KTH. Land & Water Resources Engineering, 25 pp + appendices.

[5] Billström, K., Weihed, P., **1996**, *Econ Geol*, 90(6), 1054-1072.

[6] Bark, G., Weihed, P., **2007**, *Ore Geology Reviews* 32(1-2), 431-451.

[7] Ödman, O. H., **1941**, *Swed. Geol Survey, SGU* Ser C 438.

[8] Svensson, U., **1980**, *Swed. Geol. Survey* Ser C 764. 79 pp.

[9] Dumas, H., **1985**, *Luleå Technical Univ., Report.* 54 pp.

[10] Andersson, M., Lax, K., **2000**, *Swed. Geol. Survey* Ser GK 2.

[11] Lax, K. & Selinus, O., **2005**, *Geochemistry, Exploration Environmental Analysis* 5, 337-346.

[12] Back, P-E., **2001**, *Vatten* 2001-2, 153-160.

[13] SEPA (Swedish Environmental Protection agency), **2000**, *Environmental Quality Critera – Groundwater.* 140 pp.

[14] Jacks, G., Mörth, M., Sjekovec, Z., **2011**, *Int. Applied Geochem. Symposium*, 22-26 Aug., Rovaniemi, Finland.

[15] Pierce, M. L., Moore, C. B., **1982**, *Water Res* 16, 1247-1253.

[16] Gustafsson, J-P., Bhattacharya, P., **2007**, *Trace Metals and Other Contaminants in the Environment* 9, 159-206.

Nickel in groundwater - a case study from northern Sweden

G Jacks[1] and D. Fredlander[2]
[1] Department of Land and Water Resources Engineering, KTH, SE-100 44 Stockholm Sweden
[2] UMEVA, SE-901 84 Umeå, Sweden
Corresponding author e-mail: gunnjack@kth.se

Abstract
A large groundwater plant in Northern Sweden has experienced problems with elevated nickel concentrations in some of the 20 water wells that are used for water supply. There are two possible sources for the nickel, some sulphide deposits present in the direction of the glacial ice movement and glacial sulphidic clays. The elevated nickel contents appeared after the shifting of the basins for artificial recharge to new sites. No elevated nickel contents is found in the glacifluvial deposits and neither in the ferric oxyhydroxides in the B-horizons of the podzolic soils. Empetrum nigrum is known to pick up nickel from soils but the content of a composite sample is not elevated. A clay layer is sandwiched in the glacifluvial deposits and there is a pronounced relation between low groundwater levels and elevated nickel in the wells with elevated nickel. Thus the most likely source is the sulphidic glacial clays.

1. Introduction

Nickel is a mobile metal with a low affinity to organic matter in the same order as zinc [1]. Nickel is found to be a groundwater pollutant in connection to mines [2, 3] and also in drainage from acid sulphate soils [4, 5, 6]. Toxicity of nickel has been observed as industrial exposure through inhalation [7]. The Swedish Food Board considers that nickel in excess of 20 □g/l could exacerbate nickel contact allergy. It is still debated whether nickel is essential for higher organisms but microorganisms do have Ni-dependant enzymes like Helicobacter pylori [8].

In a large groundwater plant certain wells have been found to contain elevated amounts of nickel in excess of the current Swedish limit of 20 ug/l [9]. Only a fraction of the 20 wells have been found to contain elevated levels of nickel and there is also a sizeable seasonal variation in the wells from below 20 □g/l to about 50 □g/l.

The aim of present analysis was to identify likely sources of the nickel. Two possible sources have been brought forward, mineralsations in the bedrock brought by the glacial processes and glacial sulphidic fine sediments sandwiched into the glacifluvium.

2. Site, Materials and methods

2.1 Sampling site

The site is a large groundwater aquifer used for water supply. The groundwater is recharged in ponds by surface water from a river and is extracted after about 600 m passage through the aquifer in a line of 20 wells. The aquifer is a glacifluvial fill in a valley with a thickness up to a maxiumum of about 60 m [10]. The lower portion is an esker type of formation, in portion of it overlain by finer sediments in parts containing sulphides. The uppermost portions are glacial outwash deposits consisting of sand and gravel.

2.2 Sampling and analytical methods

The analytical results are taken from the records of the waterworks and by a sampling of a selection of the 20 wells. In connection to that samples for S-isotope analysis were taken but the results are as yet not available.

In addition to water sampling, some sample of the podzolic soil was taken especially to see whether the B-horizon contained elevated amounts of nickel. Also a composite sample of Empetrum nigrum was taken as this specie is known to pick up nickel from contaminated soils [11].

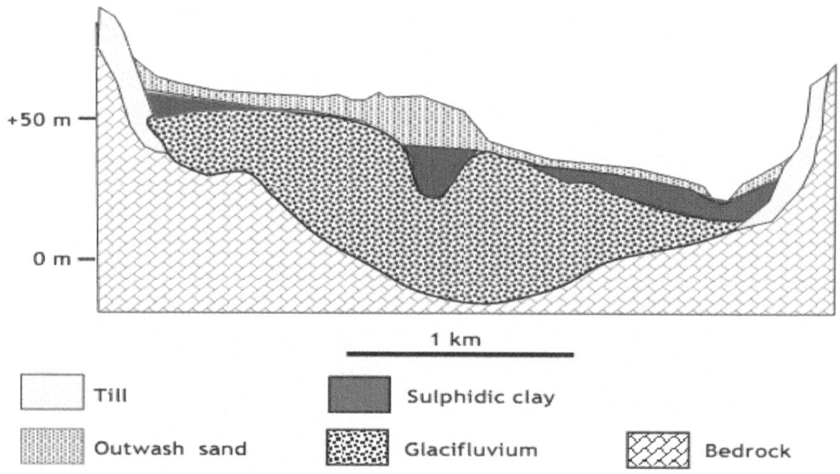

Figure 1. Cross section of the glacifluvial valley fill at the waterworks.

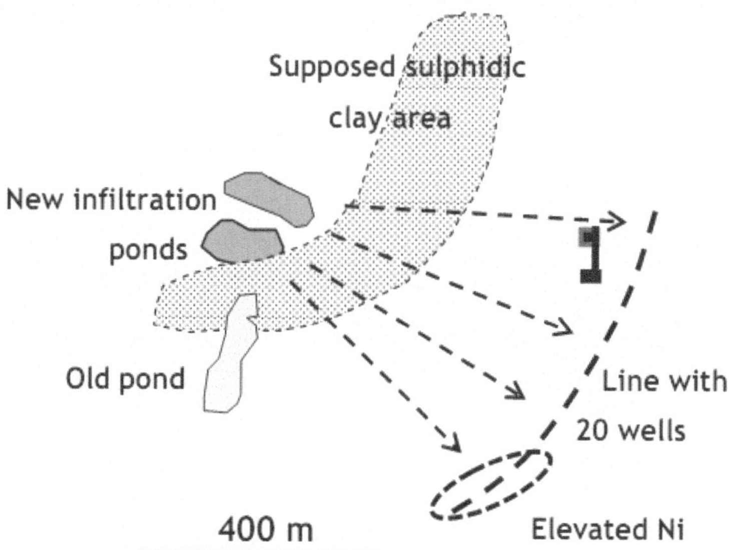

Figure 2. Map sketch of the recharge area and well-field.

3. Results and discussion

The first hypothesis that the nickel is derived from sulphides or oxidized sulphides implies that especially the B-horizons in the podzolic soils could be a source of the nickel. The artificial recharge would tend to raise the groundwater levels so that especially manganese oxides could be subject to reduction and divalent nmanganses mobilized along with the built in nickel [12, 13]. However, a composite sample from the B-horizon in the close to the recharge basins (Figure. 2) contained only 3,3 mg/kg of Ni. Neither did a composite sample of Empetrum nigrum, known to pick up nickel to some extent from nickel contaminated soils, contain above background values for nickel.

There is a relation between groundwater level and nick

el concentration in groundwater in the sense that nickel concentrations increase when the groundwater levels falls (Figure. 3). This indicates that the sulphidic clay-silt layer sandwiched in between the true esker deposits and the glacial outwash (Figure. 1 and Figure. 2) is oxidized and releases nickel at low groundwater levels. The elevated nickel concentrations were observed when the recharge ponds were shifted NE-wards and a bit upstream of the probable extent of the sulphidic clays (Figure. 2) which further supports the hypothesis that these are the source of the nickel.

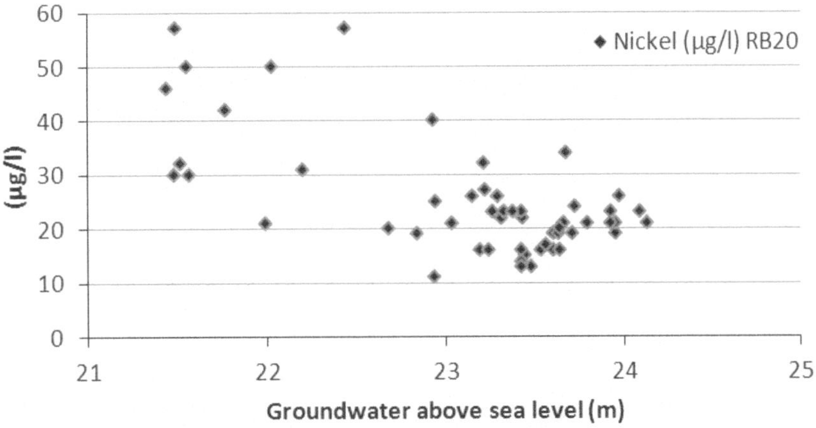

Figure 3. Nickel in µg/l versus groundwater level above the sea level in well 20, the most affected on in the well-field.

4. Conclusions

Two sources of the elevated nickel concentrations in some of the extraction wells at the water works have been hypothetised. There is no indication that the sulphide mineralisations in the bedrock upstream in the glacial ice direction are disseminated sources in the glacifluvial aquifer material. All evidence points to that sulphidic clays sandwiched in the glacifluvium are the source. When the groundwater levels fall exposing the clay to oxidation the nickel levels increase in some of the wells. The content of nickel is not a serious problem as only a few wells are affected and that most of the nickel is removed during the water treatment and discarded with the iron precipitates.

Acknowledgments

This investigation was carried out by support from J. Gust. Richert Foundation.

References

[1] Ashworth, D. J., Alloway, B. J., 2008, Communications in Soil Science and Plant Analysis 39, 538-550.
[2] Herbert, Jr R., 2006, Journal of Geochemical Exploration 90, 197-214.
[3] Heikinen, P. M., Räisänen, M L., 2008, Journal of Geochemical Exploration, 97, 1-20.
[4] Österholm, P., Åström, M., 2002, Applied Geochemistry, 17, 1209-1218.
[5] Sohlenius, G., Öborn, I., 2004, Geochemistry and partitioning in acid sulphate soils in Sweden and Finland before and after oxudation. Geoderma 122, 167-175.
[6] Lax, K., 2005, Agricultural and Food Science, 14, 83-97.
[7] Denkhaus, E., Salnikow, K., 2002, Critical Reviews in Oncology Hematology 42, 35-56.
[8] Havtin, P. R., Delves, H. T., Nevell, D. G., 1991, FEMS Microbiology Letters 77, 51-54.
[9] Swedish Food Board, 2001, Drinking Water Criteria, SLVFS 2001:30.
[10] Jacobsson, M-L, 2001, Hydrogeological investigation anf groundwater modelling, M Sc thesis, Dept. of Geology, Gotherburg Univ. 53 pp.
[11] Uhlig, C., Salemaa, M., Vanha-Majamaa, I., Derome, J., 2001. Environmental Pollution, 112. 425-442.
[12] Plekhanova, o, 2003, Eurasian Soil Science, 36, 1183-1190.
[13] Du Laing, G., Meers, E., Dewispelaere, M., Rinklebe, J., Vandecasteeele, B., Verloo, M. G., Tack, M. G., 2009, Water, Air & Soil Pollution, 202, 353-367.

Arsenic in the different environmental compartments of Switzerland: an updated inventory

Hans-Rudolf Pfeifer[1], Mohammad Hassouna[1] and Nadia Plata[2]
[1] IMG-Centre d'Analyse Minérale, Faculté de Géosciences et de l'Environnement, Université de Lausanne, CH-1015 Lausanne, Switzerland
[2] EPTES Sàrl, Rue de la Madeleine 28, CH-1800 Vevey, Switzerland
Corresponding E-mail: Hans-Rudolf.Pfeifer@unil.ch

Abstract
Switzerland has three main areas with elevated natural arsenic concentrations : 1) the northern part, where a number of thermal mineral springs are located, 2) the Jura mountains with iron- rich limestone and clays and 3) the Alps, where arsenic-bearing ore deposits and silicate rock aquifers are found. In addition in the Alps, there are also isolated arsenic-bearing thermal and mineral springs. A complete survey of all public drinking water supplies carried out between 1997 and 2002 showed that about 20'000 people lived in areas with arsenic between 10 and 50 mg/L in spring waters and a few hundred depended on waters with As between 50 and 180 mg/L. In the meanwhile, most communities have access to drinking water < 2 mg/L. In most cases the waters were well oxygenated and the arsenic was in its pentavalent form (arsenate). In flooded soils rich in organic matter (forest, wetlands), with reducing conditions and elevated dissolved iron, trivalent arsenite predominated. The origin of these naturally contaminated waters is in As-bearing rocks and soils, in which the As is most often located in sulfides (pyrite, arsenopyrite) and Fe-oxyhydroxides. They either occur as dm-m-sized veins or disseminated in areas of several hundred meters. Only very little contamination can be attributed to waste materials, such as mine dumps or old industrial waste repositories. Plants growing on As-rich soils usually contain less than 5 mg/kg As. Monitoring data for mosses suggests that dust particles rich in As can locally contribute to a week air pollution. The only available study on the relation of As- concentrations in drinking water and cancer incidence did not give significant results.

1. Introduction

In Switzerland, like in other European countries, up to 1970, due to industrial activities, such as mining, smelting and glass manufacturing and its use as medical drug for humans, pesticide, wood preservative and growth promoter for animals, considerable amounts of arsenic had been introduced to different environmental compartments, especially in soils and waste repositories [1]. Between 1970 and 1990 various efforts were undertaken, to reduce anthropogenic input and local contamination by arsenic, especially the build-up of high levels in soils. However, the environmental monitoring was centered on toxic trace metals, such as mercury, cadmium and lead.

The discovery of natural arsenic in Swiss soils and waters started in 1989, when a Canadian mining company proposed to start an exploration campaign on the site of the former gold mine of Costa-Astano in southern Switzerland. In order to know more about the local groundwater composition, the state authorities of the canton Ticino analyzed As in the waters and soils around the mine and found slightly elevated values between 5 and 12 mg/L As in the surface waters (pers. comm. M.Jäggli) and a up to 200 mg/kg of As in soils adjacent to mine waste [2]. A study on the possible contamination by mercury on the same site confirmed a halo of elevated As-concentrations of about 200 m around the former mining site [3], [4], [5]. In Fall 1996, a microbiological contamination forced the authorities to survey the drinking water of the adjacent village of Astano (200 inhabitants) in the same area. By accident, concentrations of up to 80 mg/L As were discovered in one of the two drinking water reservoirs. This lead to a systematic survey of the drinking water of the whole Ticino area [6].

Also alarmed by the contemporaneous discovery of high As-concentrations in West-Bengal, Bangladesh [7], [8], between 1997 and 2002, all Swiss drinking and mineral waters have been checked for elevated As-concentrations. This systematic survey revealed other areas with elevated As-contents in drinking water comparable to those of southern Switzerland, touching about 20'000 people [9]. In the meanwhile, most spring waters with concentrations > 10 mg/L have been taken off the distribution systems of drinking water, however without officially lowering the limiting value of 50 mg/L.

This paper intends to present an update of existing data about Switzerland since about 2001 for the different environmental compartments, i.e. rocks and minerals, waste materials, soils, waters, wetlands and air. Figure. 1 and Table 1 give an overview of the different regions, the origin of the arsenic and typical concentrations found.

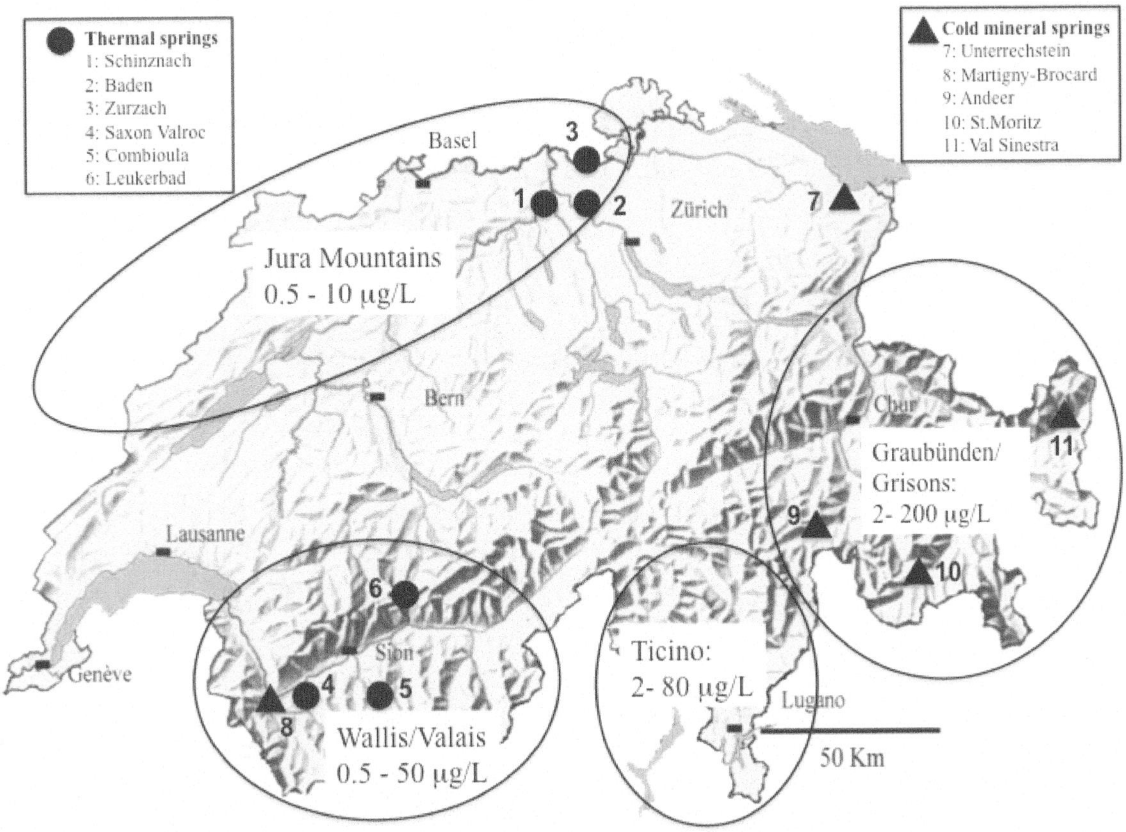

Figure. 1. Areas with elevated concentrations of arsenic in Switzerland. In several areas there are thermal or cold mineral springs with As-contents above 10 mg/L used for balneological purposes. The main areas with spring waters are marked with large open circles and the marked ranges refer to drinking waters used up to 2002.

2. Regional variation

Regions with elevated As contents in the environment comprise the thermal spring area of Northern Switzerland, the Jura Mountains, and the Alps with the areas of Wallis/Valais, Ticino and Grisons/Graubünden (Figure. 1).

In northwestern Switzerland, in the Jura mountains As-bearing ferruginous limestones and ferruginous red clays are at the origin of elevated As-concentrations in soils, however all known drinking waters are below 2 mg/L [33]. In Northern Switzerland some up to 45°C warm thermal mineral springs with an origin in the deep seated crystalline basement occur [26]: Schinzach (25 mg/L), Baden (38 mg/L, Zurzach (123 mg/L, Figure. 1). In southwestern Switzerland, in the Wallis/Valais area, there are several As-bearing (mainly sulfide) formerly mined ore deposits and areas with aquifers, exhibiting dispersed elevated concentrations of the same minerals [10]. Springs used for the supply of drinking water in several districts contain As-concentrations between 5 and 50 mg/L and in 2004 about 14'000 people in 18 communities drank water with >10 mg/L As (table 1 and [11], [12], [13]). There are also some thermal mineral springs used for balneological purposes with elevated As-concentrations [26]: Saxon Valroc (15mg/L), Combioula (24mg/L) and Leukerbad (27mg/L).

In southern Switzerland, in the Ticino area, apart of a few formerly mined Fe- and As-bearing gold mines, many areas with aquifers with dispersed As-bearing minerals exist, which served as drinking water resource (table 1). In total, in 1996 about 5000 people in 12 communities depended on water containing between 10 and 80 mg/L As[6] .

Table 1. The different areas in Switzerland with elevated As in the environment and their features

AREA Region No. of people touched (in 2001)	Origin of As	Typical range of As rocks (r) soils (s) plants (p) waste (w)	Range of As in spring water and surface water	References
WALLIS/VALAIS Collonges Dorénaz Vernayaz Finhaut Bagnes St.Niklaus Eisten Blatten Oberwald 14'000 persons	- Ore deposits and local veins rich in pyrite and arsenopyrite - A few occurrences of more rare As-rich minerals such as gersdorffite, skutterutite, As-bearing sulfosalts	r: 20- 1000 mg/kg (ore: max. 46%) s: 20- 1600 mg/kg p: 0.1 - 3.6 mg/kg(ferns) w: up to 40% (former mines)	0 - 50 mg/l	[5],[10], [11], [12],[13], [14],[15], [16],[17], [18],[19]
TICINO Morcote Malcantone Gambarogno Valcolla Val Isone 4000 persons	- Ore deposits and local veins rich in pyrite, pyrrhotite, arsenopyrite, goethite and scorodite - Glacial deposits rich in Fe-oxyhydroxides, clays and adsorbed As - Silicate rocks with dispersed pyrite, arsenopyrite and or allanite	r: 20- 90'000 mg/kg (ore max. 46%) s: 20- 2000 mg/kg p: 0.2 - 10 mg/kg w: up to 40%	0 - 80 mg/l	[3], [5], [6], [20], [21], [22], [23], [24], [25]
GRAUBUENDEN/ GRISONS Upper Engadine Lower Engadine (Val Sinestra) Val Poschiavo 1000 persons	- Silicate rocks with dispersed pyrite, arsenopyrite, hematite, goethite, rarely veins or local ore deposits - Cold CO_2-rich mineral springs	r: 20- 500 mg/kg (ore max. 46%) s: no data, except peat: 100-1400 mg/kg p: no data w: no data	0 - 180 mg/l (Val Sinestra: 3000 mg/L !)	[5], [26], [27], [28], [29], [30], [31]
JURA MOUNT. Délémont Weissenstein Probably 0 persons	- Fe-bearing limestones - Fe-rich goethite-bearing clays (Bolus/sidérolitique)	r: 20- 800 mg/kg s: 20-150 mg/kg p: 0.1- 0.4 mg/kg w: no data	0 - 5 mg/l	[32], [33], [34], [35], [36], [37]

The highest As-concentrations in spring waters occur in southeastern Switzerland: in the Engadine and Poschiavo Valleys drinking waters with up to 180 mg/L have been discovered in 2000. Maximum hundred people were exposed to waters above 10 mg/L. The aquifers resemble those in southern Switzerland, however, springs with very contrasting As-concentrations occur within a few tens of meters in the same area, indicating vein-type occurring of As-bearing minerals. There are also some cold springs with very contrasting As contents (Figure. 1; [26]: Andeer (12 mg/L), St. Moritz (50 mg/L) and Val Sinestra (up to 3000 mg/L), the latter two being also very rich in CO_2.

3. Rocks and minerals

Aside from the ferruginous carbonate and clay rocks of the Jura mountains and a few other rare sedimentary occurrences [5], most arsenic bearing rocks are silicate rocks which contain As-bearing minerals, either dispersed in small amounts throughout a volume of a few hundred meters in extension, or concentrated in dm- to m-wide vein-type features of several tens of meters often forming exploitable ore deposits. Most of these rocks are of granitic origin and carboniferous in age (around 300 mio y), more rarely basaltic of various age. Permian red beds and lower Triassic of the Black Forest massive underlying northern Switzerland can also contain arsenic [38], [39] and are at the origin of the As of the thermal springs in northern Switzerland. Another group of sediments enriched in As are of glacial and fluvioglacial origin (till and gravel) in Southern Switzerland [22]. Most As-rich rocks contain typically between 500 and 100'000 mg/kg.

In terms of primary minerals, sulfide type As-bearing minerals, such as arsenopyrite (FeAsS, 46 % As), pyrite (FeS_2, up to 5% As), pyrrhotite (FeS, up to 3% As) are the most widespread [5]. Occasionally loellingite ($FeAs_2$, 72% As), skutterutide (Co,Fe,Ni)As 2-3 (max. 80% As), gersdorffite (NiAsS,), beudanite ($PbFe_3(AsO_4)(OH)_6$, 12% As) and As-bearing sulfosalts, e.g. tennantite ($(Cu,Fe)_{12}As_4S_{13}$, 12% As) occur. Secondary minerals issued from high and low temperature alteration include scorodite ($FeAsO_2 \cdot 2H_2O$, 32% As), hematite (Fe_2O_3, up to 2.5% As), goethite (FeOOH, up to 2% As), allanite ($(Ce,Ca,Y,La)_2(Al,Fe^{+3})_3(SiO_4)_3(OH)$, 3% As).

4. Waste materials

Most As-bearing materials in Switzerland are natural (rocks, untouched or mined ore deposits, soils), only a few As-bearing industrial wastes are known: coke remaining from coal gas production factories (max. 120 mg/kg As, [40]) and waste from glass production (up to 2600 mg/kg As). Percolation waters from several waste repositories are described in [41].

5. Soils

Soils are usually enriched in As with respect to the bed rock by a factor between 1.5 and 2. Typical normal agricultural soils in Switzerland exhibit a mean concentration of 10 mg/kg [42], [43]. Forest soils in areas with elevated As-contents in the rocks contain between 20 and 800 mg/kg As [22], [23]. The highest concentrations are typically found in organic A and B horizons, often also rich in amorphous Fe-oxyhydroxides [23]. Agricultural fields in the same area are often depleted in As (< 5 mg/kg), which can be explained by the desorbing effect of phosphate-ions of fertilizers and manure on arsenate sorbed on soil particles [23].

6. Waters

The typical regional variations in As-concentrations have already been discussed above. 80% of the waters (small lakes, creeks and groundwater) are well oxygenated and contain the As in its pentavalent oxyanion form (arsenate), together with low total iron concentrations. For these, the local presence of rocks and soils with elevated As-concentrations and often pH values above 7.5 seem to be at the origin of the observed values in the waters. Cold and warm mineral springs (often with high total dissolved ions and CO_2) contain between 20 and 50% trivalent As (arsenite) and some tens of mg/L of iron. In water flooded forest soils and especially in wetlands arsenic concentrations can reach 1000 mg/L As and arsenite percentages of up to 90% at Eh-values close to 0 mV [23], [31]. Often As-enriched waters also contain elevated concentrations of U, often linked to high radon gas concentrations [28], [29], [31].

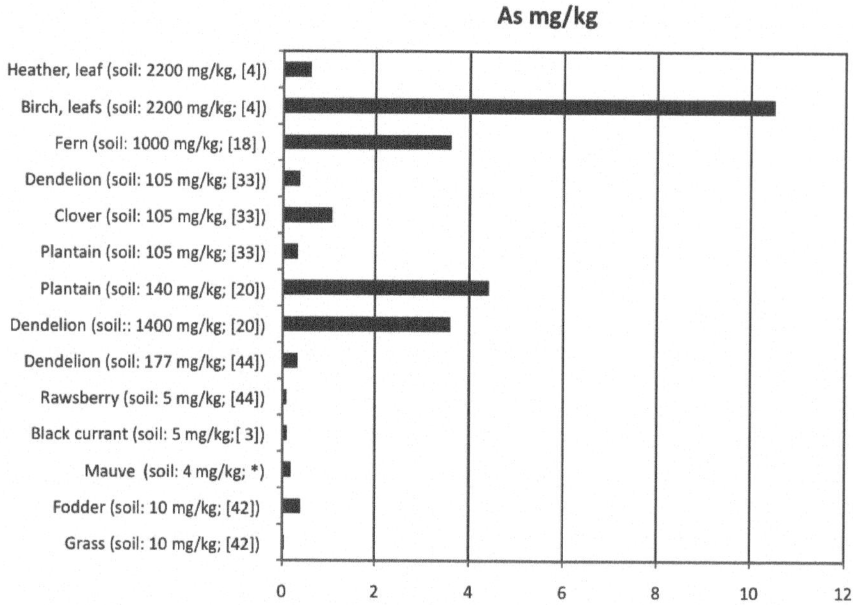

Figure. 2. Typical arsenic contents in plants from Switzerland (mg/kg dry plant). *: Pfeifer, unpublished

7. Plants

Typical As-concentrations of fodder plants grown on non-contaminated normal Swiss soils vary between 0.2 and 0.4 mg/kg As [42]. Studies on plants grown on As-enriched soils vary between 0.2 and 4.5 mg/kg (Figure. 2). Leaves of birch trees grown on As-rich mine waste can contain up to 11 mg/kg As [4], [20]. Natural ferns (Dryopteris filix-mas) grown on alpine soils of the Pétoudes area in the Trient valley/Wallis show leaf-concentrations between 0.3 and 3.5 mg/kg dry weight [18].

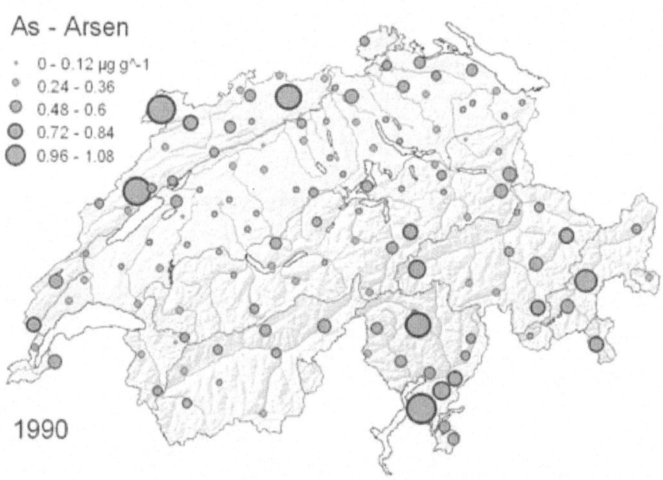

Figure. 3. As-concentrations in mosses in Switzerland in 1990, reflecting the contamination of the air (reproduced from [46]). Maps of 1995 and 2005 show in general lower values, but the overall picture remains the same.

8. Wetlands

Several authors found clear indication that organic matter accumulates As [23], [25], [45]. Sphagnum mosses of wetlands seem particularly efficient. The presence of As-free authigenic pyrite in the peatland of Gola di Lago (max. 350 mg/kg As) in Southern Switzerland [45] indicates that there is probably direct complexing between solid organic matter and arsenic. In wetlands of St. Moritz area in the Engadine valley of SE-Switzerland, As-contents between 110 [43] and 1350 mg/kg [31] have been measured.

9. Air

No direct measurements of arsenic-rich dust particles are known from Switzerland, however moss analyses for As repeated every 5 years between 1990 and 2005 [46], show a clear correlation between areas with known naturally elevated As-contents in rocks and soils and elevated moss analyses (Figure. 3), suggesting that a large part of the As in the air is natural.

10. Health risks

Only one study on the possible correlation between As-concentrations in drinking water is available for all of Switzerland [14]. It came to the conclusion that only small differences in tumor incidence between population exposed to higher and lower As-concentrations could be observed, but which were with one exception, statistically not significant. In terms of really observed health problems possibly related to As, only two women, having been exposed to spring waters with about 150 mg/l As for some time in the Engadine Valley, showed skin symptoms (small bald areas on their heads), (T.Peters, Univ. Bern, pers. comm. 2001).

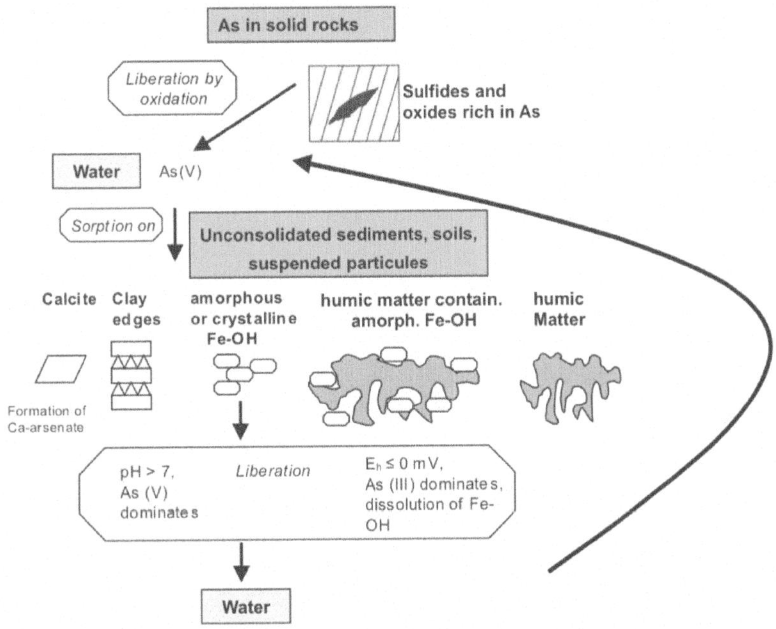

Figure. 4. Typical release and fixation mechanisms of As from primary to secondary solid phases in the presence of water (including solid organic matter, modified from [17])

11. Conclusions

Twenty years of research on As in the Swiss environment show that a largely natural contamination exists in several parts of the country (Jura, Alps), which is comparable to that of other European countries [47] : local occurrences of As-rich minerals in carbonate and silicate rocks are at the origin of As-anomalies in spring waters, soils, plants and wetlands. The surface area of these anomalies varies from a few hectares to 20 km2. The complete survey of public drinking water supplies finished in 2002 revealed that around 20'000 people were exposed to As-concentrations above 10 mg/l. In the meanwhile, in most of the touched communities, other springs providing drinking water with As < 2 mg/l are in use. However, only few privately used springs in the alpine areas have been monitored.

In terms of processes that are behind the observed As-distribution, all the known release and fixation mechanism in the presence of water from primary to secondary solid phases at different pH and Eh conditions have been identified (Figure. 4). In several cases, solid organic matter seem to have played an important role in fixing and releasing As. However the exact mechanism is still little known.

Acknowledgments

This research has been supported by the Swiss National Science Foundation (project no. 20-61860.00). We thank J.-C. Lavanchy (Univ. of Lausanne) for having provided numerous rock and soil analyses during the last 20 years and P.Y.Favarger, J. Poté (Univ. of Geneva), W.Halter, A.Ulianov (Univ. Lausanne), J.P.Dubois, J.-D.Teuscher (EPF-Lausanne) and V. Matera (Univ. Neuchâtel) for their help analyzing As in water and plants. The important contribution of more than 15 master students and 4 PhD candidates and postdocs involved in this study is acknowledged by citing the publications for which they have been co-authors or their thesis in the references. We also thank M.Berg, E.Hoehn, S.Hug, A.Johnson and J.Zobrist (EAWAG-ETH-Zürich), H.Surbeck (Cordast, FR), R.Hänny (Univ. Basel) and V. Lenoble (Univ. Toulon) for many stimulating discussions.

References

[1] Bovey, R. La défense des plantes cultivées. Traité pratique de phytopathologie et zoologie agrcicole. 6e edit. Payot, Lausanne, 1972.

[2] Gini, G., L'estrazione mineraria nel Malcantone et l'impatto delle sostanze nocive sul suolo. Internal report, Sezione Agricoltura, Cantone Ticino, Bellinzona,1992.

[3] Bondietti , G., Gex, P., Gini, G., Hansen, J., Hunziker, J. & Pfeifer, H.-R., Heavy metal contamination around the As-Pb-Zn-Au-mine at Astano (Malcantone, Ticino, Switzerland). Eclogae geol. Helv. 87, 487- 490, 1994.

[4] Rey, D., Arsenic dans les sols et eaux d' Astano (TI). Diplôma thesis, Sciences de la Terre, Univ. of Lausanne, 1996.

[5] Pfeifer, H.-R., Hansen, J., Hunziker, J., Rey, D., Schafer, M. & Serneels,V., Arsenic in Swiss soils and waters and their relation to rock composition and mining activity, in: Prost, R., ed., Contaminated soils: 3rd Internat. Conf. Biogeochemistry of Trace Elements, Paris, May 15-19, D:/data/communic/050.PDF, Colloque 85, INRA ed., Paris, 1996.

[6] Jäggli, M., Arsenico nel acqua di alcuni communi del Malcantone, 34- 36, Rapporto d'esercizio 1996. Laboratorio cantonale, Cantone di Ticino, Dip. delle opere sociale, 1996.

[7] Chatterjee, A., Das, D., Mandal, B.K., Roy Chowdhury, T., Samanta, G. and Chakraborti, D., Arsenic in groundwater in six districts of West Bengal, India, The biggest arsenic calamity in the world. Part-1. Arsenic species in drinking water and urine of the affected people. Analyst, 1995, 120, 643-650

[8] Das, D., Chatterjee, A., Mandal, B.K., Samanta, G., Chanda, B. and Chakraborti, D., Arsenic in ground water in six districts of West Bengal, India, The biggest arsenic calamity in the world. Part-2. Arsenic concentration in drinking water, hair, nail, urine, skin-scale and liver tissue (biopsy) of the affected people. Analyst, 1995,120, 917-924.

[9] Pfeifer H.-R. & Zobrist J., Arsenic in drinking water- also a problem in Switzerland? EAWAG News 2002, 53, pp. 15-17.

[10] Fournier, C., L'arsenic en Valais : Association entre anomalies et formation géologique, et étude de la contamination par le gisement de La Rasse. Diplôma thesis, Sciences Naturelles de l'Environnement, Univ. of Lausanne and Univ. of Geneva, 2004.

[11] Thétaz, C., Détermination d'arsenic dans l'eau potable. Laboratoire Cantonal, Valais, Rapport annuel. Département des Transports, de l'Equipement et de l'Environnement, 1999.

[12] Haldimann M.,Brüschwiler, B., Schlatter, J. & Dudler V., L'arsenic: une substance naturelle toxique,présente dans l'eau. Rapport annuel 2004, Securité alimentaire. Office fédérale de la santé publique, Bern, 53-54.

[13] Haldimann M., Pfammatter E., Venetz P. Studer P., Dudler V., Occurrence of arsenic in drinking water of the canton Valais. Mitt. Lebensm. Hygiene. 2005, 96, 89-105. Ed. Swiss Federal Office of Public Health.

[14] Brüschweiler, B. J., Schlatter, J.R., De Weck, D., Favre, F. & Luthi, J.-C., Occurrence of arsenic in drinking water of the canton of Valais. Part II: Epidemiological comparison between arsenic concentrations and cancer incidence rates. Mitt. Lebensm Hygiene, 2005, 96, 106-117, 2005. Ed. Swiss Federal Office of Public Health.

[15] Greppin, R., Géologie régionale et contamination environmental par arsenic de l'indice d'arsenopyrite de La Payanne, Bruson, Val de Bagnes, VS. Diploma thesis, Sciences de la Terre, Univ. of Lausanne, 1997.

[16] Henry G., Etude des dispersions de métaux traces autour des minéralisations du Col des Mines. Diploma thesis, Institut de Minéralogie et de Géologie, Université de Lausanne, 128p, 2003.

[17] Pfeifer, H.-R., Häussermann, A., Lavanchy, J.C. & Halter, W., Distribution and behavior of arsenic in soils and waters in the vicinity of the gold-arsenic mine of Salanfe, Western Switzerland. J. geochem. explor. 2007, 93, 121-134.

[18] Blanc, N., Contamination naturelle de l'environnement par l'arsenic en milieu alpin. Le cas des Petoudes, Val de Trient (VS). Master thesis, Fac. Geosci. Env., University of Lausanne, 2010.

[19] Hofmann B.A., Knill, M.D., Geochemistry and genesis of the Lengenbach Pb-Zn-As-Tl-Ba-mineralisation, Binn Valley, Switzerland. Mineralum Deposita, 1996, 31,319-339, 1996.

[20] Pfeifer, H.-R., Derron, M.-H., Rey, D., Schlegel, C., Dalla Piazza, R., Dubois, J.D. & Mandia, Y., Natural trace element input to the soil-water-plant system, examples of background and contaminated situations in Switzerland, Eastern France and Northern Italy. p. 33- 86, in: Markert, B. & Friese, K., eds., Trace elements- Their distribution and effects in the environment- Elsevier, Amsterdam, 2000.

[21] Beatrizotti, G., Berthoud,, J., De Rossa, M., Gueye-Girardet, A., Jäggli, M., Lavanchy, J.-C., Pfeifer, H.-R., Reymond, D., Schmit, V. & Temgoua, E., Contaminazione naturale da arsenico di acque superficiali et sottterraneee in Ticino (Svizzera meridionale). Geol. Insubr. 2002, 7, 1-16.

[22] Pfeifer H.-R, Beatrizotti G, Berthoud J, De Rossa M, Girardet A, Jäggli M, Lavanchy J.-C, Reymond D, Righetti G, Schlegel C, Schmit V and Temgoua E., Natural arsenic contamination of surface and ground waters in Southern Switzerland. Bulletin appl. Geol. 2002, 7, 83-105.

[23] Pfeifer , H.-R., Gueye-Girardet, A., Reymond, D., Schlegel. C., Temgoua, E., Hesterberg. D., Chou, J., Dispersion of natural arsenic in the Malcantone watershed, Southern Switzerland: Field evidence for repeated sorption-desorption and oxidation-reduction processes. Geoderma, 2004, 122, 205-234.

[24] Gonzalez A., Z. I., Krachler, M., Cheburkin, A. K. & Shotyk, W., Natural enrichment of arsenic in a minerotrophic peatland (Gola di Lago, Canton Ticino, Switzerland), and implications for the treatment of contaminated waters. In: Natural arsenic in groundwater; Bundschu, J., Bhattacharya, P., Chandrashekharam, D., Eds.; Taylor and Francis Group plc: The Netherlands, vol. 27, 205-210, 2005.

[25] Gonzales, Z., Krachler, M., Cheburkin, A. & Shotyk, W., Spatial Distribution of Natural Enrichments of Arsenic, Selenium, and Uranium in a Minerotrophic Peatland, Gola di Lago, Canton Ticino, Switzerland. Environ.Sci.Technol. 2006, 40, 6568-6674.

[26] Högl, O., Die Mineral-und Heilquellen der Schweiz. Edit. Haupt, Bern, 302p, 1980.

[27] Rafflenbeul J., Zur Hydrogeologie der arsenhaltigen Quellenwässer im Bereich St. Moritz, Diploma thesis, Geol. Institute, Swiss Federal Institute of Technology (ETH) Zürich, 2002.

[28] Deflorin O. Natürliche Radionuklide in Grundwässern des Kantons Graubünden.PhD thesis, Centre d'Hydrogéologie de Neuchâtel, Univ. Neuchâtel, 189p, 2004.

[29] Voirol, J.-M., Étude sur l'origine et la relation de l'arsenic avec les contaminants radiogéniques naturels dans les eaux du Val Poschiavo Les Grisons, GR. Master thesis, Fac. Geosci. Env. , University of Lausanne, 2006.

[30] Peters, T., Geologischer Atlas der Schweiz 1.25'000, Blatt 1257 St. Moritz. Erläuterungen. Bundesamt für Wasser und Geologie. Bern, 96p, 2005.

[31] Chiandussi, L., L'arsenic et ses interactions avec les communautés microbiennes dans une zone marécageuse de la Haute Engadine, Grisons. Master thesis, Fac. Geosci. Env., University of Lausanne, 2010.

[32] Donzel, P.-Y. , Arsenic dans les roches et sols du Haut-Jura suisse : distribution générale sur la chaîne et étude détaillées dans la région du Weissenstein (SO). Diploma thesis, Sciences de la Terre, Univ. de Lausanne, 80p., 2001.

[33] Donzel, P.-Y., Dubois, J.P., Lavanchy, J-C., Pfeifer, H.R. & Adatte, T., Arsenic dans l'environnement du Haut-Jura Suisse. Bull.soc.vaud.sci.nat. (in print), 2011.

[34] Degen, C., Etude de l'impact des lombriciens sur la dynamique de l'arsenic (disponibilité, transfert) dans les sols naturellement enrichis. Master thesis, Univ. Neuchâtel, 2005.

[35] Grisel, N., Etude du transfert de l'arsenic présent dans les sols naturellement enrichis vers différentes plantes. Master Univ. Neuchâtel, 2005.

[36] Bayon R.C., Matera V., Kohler-Milleret R., Degen C. and Gobat J.-M., The effects of earthworm activities (Aporrectodea giardi) on geogenic arsenic and soil nutrients dynamics in the Jura Mountains (Switzerland). Pédobilogia 2011 (in print).

[37] Hamel, M., Minéralogie et sédimentologie du sidérolithique et de son substratum Mésozoïque à Vicques-Courcelon (bassin de Delémont, Jura). Diploma thesis, Lausanne, University Lausanne, 1998.

[38] Hofmann, B., Erzmineralien in paläozoischen, mesozoischen und tertiären Sedimenten der Nordschweiz und Südwestdeutschlands, Schweiz. Min. und Petr. Mitt. 1989, 69, 345-357.

[39] Hofmann, B., A regional anomaly of arsenic and cesium in northern Switzerland and SW-Germany. Abstract. EUG, Strasbourg, April 1995.

[40] Grocolas, J., Investigation historique et modélisation SIG de l'ancienne décharge de Genève, 101p. Diploma thesis, Sciences de l'Environnement, Univ. of Lausanne and Univ. of Geneva, 2003.

[41] Looser, M., Méthode de détection et de caractérisation de pollutions du sous-sol par les sites contaminés à l'aide des traces inorganiques.PhD thesis, Swiss Federal Institute of Technology Lausanne (EPFL), 1996.

[42] Stünzi, H., Arsen im Rauhfutter. Kolloquium Analyt. Atomspektroscopie Leipzig, 1993, 457-462.

[43] Knecht K., Keller, T., Desaules, A., Arsen in Böden der Schweiz. Umweltschutz und Landwirtschaft, Bern, 37p, 1999.

[44] Girardet, A., Contamination en arsenic des sols de la région Sessa-Astano, Malcantone, Ti. Diploma thesis, Sciences naturelles de l'Environnement, Univ. of Lausanne and Univ. of Geneva, 101p., 2001.

[45] Rothwell, J.J., Taylor, K.G., Ander, E.L., Evans, M.G., Daniels, S.M., & Allott, T.E.H., Arsenic retention and release in ombrotrophic peatlands. Sc. Tot. Env. 2009, 407, 1405-1417.

[46] Thöni, L., Matthaei, D., Seitler, E. & Bergamini, A., : Deposition von Luftschadstoffen in der Schweiz, Moosanalysen 1990-2005. Bundesamt für Umwelt BAFU Bern, 2008.

[47] Bhattacharya P., Mukherjee A. B., Bundschuh J., Zevenhoven R. & Loeppert R. H., Arsenic in Soil and Groundwater Environment, Biological interactions, health effects and remediation, Trace metals and other contaminants in the environment, vol. 9 (Series Editor Nriagu J. O.), Elsevier, Amsterdam, 2007.

Heavy metal pollution of surface water sources of Konya Basin

M.E. Aydin, S. Ozcan, and S. Ucar
Department of Environmental Engineering, Selcuk University, Konya, Turkey
Corresponding E-mail: meaydin@selcuk.edu.tr

Abstract
Surface water continuously exposed to numerous environmental pollutants among which the most potentially hazardous are toxic chlorinated compounds, heavy metals, residual chemicals and radioactive compounds. Heavy metals can enter waters through natural and anthropogenic sources. Most heavy metal contaminants originate from different natural sources such as magmatic, sedimentary, and metamorphic rocks. The origin of heavy metals in surface and groundwater are also from anthropogenic sources due to human activities such as industrial production and agriculture. Many of heavy metals have been detected in different environmental compartments. Konya (in Turkey) watershed is a closed basin and has 4.52 billion m3 water capacity. Surface water sources are being polluted by anthropogenic sources such as domestic, agricultural and industrial activities. The determination of the water quality of surface water sources in Konya closed basin is very important. Because Konya closed basin is the biggest closed basin in Turkey and larger part of Turkey is in semi-arid climate area. In this work 32 monitoring stations were selected for investigation of heavy metal pollution within the closed basin. Water samples collected from these monitoring stations were analysed for arsenic (As), cadmium (Cd), lead (Pb), copper (Cu), chromium (Cr), cobalt (Co), nickel (Ni), zinc (Zn), iron (Fe), manganese (Mn), aluminum (Al), beryllium (Be), selenium (Se) using Inductively Coupled Plasma - Mass Spectrometry (ICP-MS). The results obtained were compared with drinking and irrigation waters guidance values given by the Turkish Regulations, the European Community Council Directive 98/83/EC, US Environmental Protection Agency and World Health Organization.

1. Introduction

Anthropogenic influences as well as natural processes degrade surface and groundwater, and impair their use for drinking, industrial, agricultural, recreation or other purposes. Metals enter the aquatic environment from a variety of sources. Although most metals are naturally occurring through the biogeochemical cycle, they may also be added to environment through anthropogenic sources, including industrial and domestic effluents, urban storm, water runoff, landfill leachate, atmospheric sources, coal-burning power plants, non-ferrous metal smelteries, iron and steel plants and dumping of sewage sludge [1,2].

Excess of some essential metals can damage human health, and nonessential metals can be toxic at even very low concentrations [3]. Health effects reported have included neurological, bone and cardiovascular diseases, renal dysfunction, and various cancers. The interest on the effects on humans and other animals of heavy metals has increased in recent years. The definition of the maximum admissible concentration (MAC) values for certain elements in spring, drinking, thermal and surface waters has been and still is the subject of chemical and biological research in several countries [4,5]. Humans can be exposed to high metal levels from the ingestion of contaminated drinking water, vegetables, fruits, fish and soil [6]. The permissible levels of toxic metals in drinking water and surface water sources used for irrigation recommended by various regulatory authorities, for example, Turkish Regulations (TR), the European Council Directive 98/83/EC (EU), US Environmental Protection Agency (EPA) and World Health Organization (WHO) [7-12].

Rocks and soils are the principal natural sources of heavy metals in the environment. Water chemistry of surface waters such as streams, rivers, springs, ponds, and lakes, is greatly influenced by the kind of soil and rock the water flows on or flows through. Heavy metals are also released into the environment by many human activities. They are also used in a large variety of industrial products, which in the long term have to be deposited as waste. Heavy metal release into the environment occurs at the beginning of the production chain, whenever ores are mined, during the use of products containing them, and also at the end of the production chain. The natural sources are dominated by parent rocks and metallic minerals, while the main anthropogenic sources are agricultural activities, where fertilizers, animal manures, and pesticides containing heavy metals are widely used, metallurgical activities, which include mining, smelting, metal finishing, and others, energy production and transportation, microelectronic products, and finally waste disposal. Heavy metals can be released into the environment in gaseous, particulate, aqueous, or solid form and emanate from both diffuse or point sources. Heavy metals are mainly introduced into groundwater by agricultural and industrial activities, landfilling, mining, and transportation. The ever growing world population requires intensive land use for the production of food, which includes repeated and heavy input of fertilizers, pesticides, and soil amendments. Phosphatic fertilizers contain various amounts of Zn, Cd, and other heavy metals depending from which parent rock the fertilizer has been produced. Those made from sedimentary rocks tend to have high levels of Cd, while those made from magmatic rocks have only small Cd concentrations. The differences in heavy metal content are caused by impurities coprecipitated with the phosphates. Therefore, Cd input into agricultural soils varies considerably according to the Cd concentration of the fertilizer used. Pesticides are used for insect and disease control for high-production in agriculture and can be applied as seed treatment, by spraying, dusting, or by soil application. Although metal-based pesticide are no longer in use, their former applications lead to increased accumulation of heavy metals, especially of Hg from methyl mercurials, of As, and of Pb from lead arsenate into soils and groundwater. Land application of waste water is widely used in industrialized countries for the last 50 to 100 years. Several reviews of hazards from heavy metal concentration in waste water have been conducted and phytotoxic symptoms when using waste water containing Cd, Zn, Cu, Ni, Pb, and especially B have been observed [13,14].

The discharge of effluents and associated toxic compounds into aquatic systems represents an ongoing environmental problem due to their possible impact on communities in the receiving aquatic water and a potential effect on human health. Further these materials enter the surface water and subsurface aquifers resulting in pollution of irrigation and drinking water. Urbanization increases in population density and the intensification of agricultural activities in certain area is among the main causes of water pollution. The main heavy metals of concern in sewage sludge are Cd, Zn, Cu, Pb, Se, Mo, Hg, Cr, As, and Ni. The concentration of heavy metals in animal wastes depends on a variety of factors such as class of animal, age of the animals, type of ration, housing type, and waste management practice. Heavy metals such as Cu, Co, and Zn originate from rations and dietary supplements fed to the animals. Although animal wastes are usually rather low in heavy metal content, input of excess N and salts as well as nutrient imbalance in plants poses a problem [13,14].

The most important industrial activities, by which heavy metals are introduced into the environment, are mining, coal combustion, effluent streams, and waste disposal. Most metals occurring in ore deposits have only low concentration. During the extraction process, large amounts of waste rock are produced, The waste rock is usually disposed of in mine tailings or rock spoils. In the case of pyrite, this mineral will weather in the tailing due to oxidizing environmental conditions and thus create acid mine drainage. The acid conditions also mobilize heavy metals form the waste rock. This mobilization can cause fatal environmental and health problems through respiration, drinking and cooking contaminated water, and eating food grown on soils influenced by irrigation. Numerous examples are known especially for the heavy metals As, Cd, Cu, Hg, and Pb. The combustion of fossil fuel contributes heavily to the release of heavy metals in the environment, especially into the atmosphere. Notable heavy metals in coal residues are As, Cd, Mo, Se, and Zn, especially compared to their mobilization due to natural weathering. Solid wastes are produced worldwide in millions of tons annually. The most important sources of heavy metals stem from wastes from industrial activities, especially energy generation, from mining, agricultural activities (animal manure), and domestic waste (e.g. batteries, tires, appliances, junked automobiles). These wastes are often disposed of without proper treatment at waste disposal sites, which do not meet the requirements necessary for a secure deposition [13,14].

Freshwater sources are unevenly distributed throughout the world, with much of the water located far from human populations. Dramatic increase in population resulted in an enormous consumption of the worlds water reserves. Today, 450 million people in 29 countries suffer from water shortages and water-related concerns are the most acute in arid or semi-arid areas. Water amount per person is declining since 1927 considering the population growth in Turkey. Moreover this water potential is not distributed evenly. Shortage of freshwater throughout the world can be attributed to human abuse which most commonly is in the form of pollution. Increasing water pollution causes not only the deterioration of water quality but also threatens human health and the balance of aquatic ecosystems, economic development and social prosperity [15,16]. Heavy metals have been found in potentially harmful concentrations in numerous drinking water systems naturally or due to human activities, such as agricultural practices, transport, industrial activities and waste disposal [17-19].

In recent years, protecting the quality of water resources has become important for everyone. To control pollution and protect water quality in exchange for the water quality must be determined. Konya closed basin exists at the central Anatolia Region and covers a region of 44841 km2 area corresponding to the 7% area of Turkey and has 4.52 billion m3 water capacity. In this work 32 sampling points were selected for determining heavy metal pollution of surface water sources throughout the closed basin. Water samples were collected and analysed for arsenic (As), cadmium (Cd), lead (Pb), copper (Cu), chromium (Cr), cobalt (Co), nickel (Ni), zinc (Zn), iron (Fe), manganese (Mn), aluminum (Al), beryllium (Be), selenium (Se). The results obtained were compared with the Turkish Regulations, the European Community Council Directive 98/83/EC, US Environmental Protection Agency and World Health Organization guidance values [7-12].

2. Materials and Methods

2.1 Study sites

Konya is a city with a population of a million, annual mean temperature is 11.5 oC and average precipitation is about 325 mm. City has semi-arid climate, limited water sources, this is especially problem for wide agricultural land and irrigation water demand. Water shortage is faced because increasing water demand due to the increasing population. Uncontrolled drilled bore-holes and water abstraction intensified water shortages. In the basin, three million people live, 45% in rural areas and 55% in urban areas. The basin is surrounded with the city centers of Konya, Aksaray, Karaman and Niğde cities. The basin is flat plain and altitute changing from 900 m to 1050 m. Konya closed basin is shown in Figure 1.

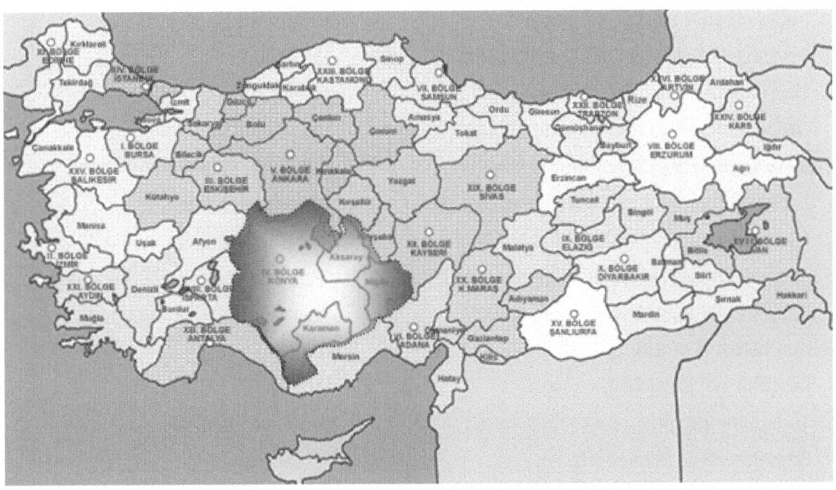

Figure 1. Konya closed basin

Konya closed basin is the country's largest closed basin. There are many lakes, and wetland areas such as Samsam, Kozanlı, Kulu, Beyşehir, Suğla, Bolluk, Tersakan and Tuz lakes, Hotamış, Eşmekaya and Ereğli wetlands. The water sources of Konya Closed Basin is only rainfall. Intensive agricultural activities carried out in the basin and water resources limited in the region. Beyşehir lake is the largest water source in the basin and there is no outlet to the sea. After completing the circulation from underground and from surface in the basin the water reaches Salt Lake. Due to its large grassy steppes, biological diversity and wetlands, it is one of the 200 most important ecological regions in the world [20]. The determination of the quality of existing water resources in Konya Closed Basin and identifying pollutant sources is of great importance. Konya closed basin, especially in recent years, under of pressure and negative effects. Lack of rainfall and water sources, climate change and drought, development of industry, and untreated domestic and industrial wastewater discharges, non efficient water consumption for agricultural purposes, drainage water from agriculture, drop in groundwater level, solid waste disposal problem are main sources of the effects. Important part (about 80%) of the basin water consumed for agricultural irrigation.

Improper irrigation techniques and agricultural practices negatively affect the water potential of the basin.

In this work 32 monitoring stations were selected for investigation of heavy metal pollution within the closed basin. Samples were taken during rain and dry periods. Names, resource type and purpose of the water sampling points are given in Table 1. Sampling points are contained in various sources such as stream, lake and spring water, dam outlet, irrigation and drainage channel. These waters are used for different purposes such as drinking water and general for example irrigation water or quality control. Sampling points on surface water sources of Konya closed basin are shown in Figure 2.

Table 1. Names, resource type and purpose of the water sampling points

No	Names	Resource type	Purpose
1	Karaman İbrala deresi	Stream	Drinking water sources
2	Kırkgözler kaynağı Ihlara	Spring	
3	Başarakavak çıkışı	Stream	
4	Tepeköy çıkışı-Meram çayı	Stream	
5	Mamasın barajı (Aksaray)	Dam outlet	
6	Bağbaşı barajı derivasyon tüneli girişi	Stream	
7	Bozkır barajı Gördürüp köprüsü	Stream	
8	Afşar Ilıcapınar deresi Sazak köprüsü	Stream	
9	Altınapa barajı	Dam	
10	Çavuşcu gölü çıkışı	Irrigation channel	Irrigation water sources
11	Peçeneközü deresi Şereflikoçhisar	Irrigation channel	
12	Beyşehir göl girişi soğuksu yeşildağ köprü	Stream	
13	Ekecik deresi ulukışla (Aksaray)	Stream	
14	Orhaniye köprüsü (Ilgın) Çavuşcu gölü çıkışı	Irrigation channel	
15	BSA kanalı İncesu Seydişehir giriş	Irrigation channel	
16	Zaferiye köprüsü (Ilgın şeker fabrikası çıkışı)	Irrigation channel	
17	BSA kanalı suğla çıkışı Seydişehir	Irrigation channel	
18	Aksaray T1 tahliye kanalı fidanlık yöresi	Drainage channel	
19	Beyşehir göl girişi Üstünler köprüsü	Stream	
20	Apa barajı (Çumra)	Dam	
21	Beyşehir göl girişi Çeltik kanalı	Stream	
22	1 nolu pompa girişi Apa tahliye kanalı	Drainage channel	
23	Beyşehir gölü tarihi köprü	Stream	
24	Beyşehir göl girişi Ilısu	Stream	
25	Beyşehir göl girişi Sarısu Eylikler	Stream	
26	Ereğli Akgöl girişi	Drainage channel	
27	İvriz barajı Ereğli	Dam outlet	
28	T1 T2 karışım öncesi	Drainage channel	
29	Gölyam Cihanbeyli	Drainage channel	
30	Niğde Akkaya baraj gölü	Lake	
31	Niğde çayı Niğde öncesi	Stream	
32	Niğde çayı Niğde sonrası	Stream	

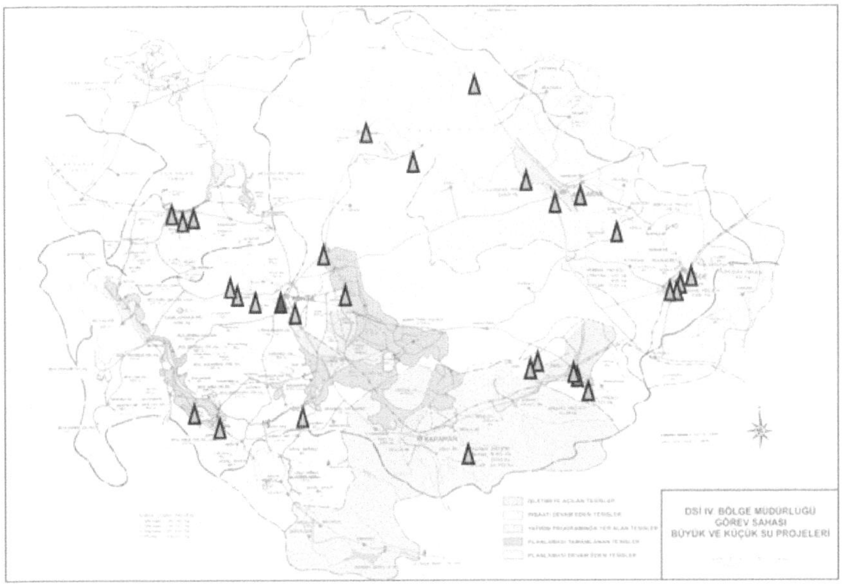

Figure 2. Water sampling points in Konya closed basin

2.2 Analytical methods

All samples were collected free of air bubbles in glass containers and they were stored in the dark at 4 oC. Water samples collected from these monitoring stations were analysed for arsenic (As), cadmium (Cd), lead (Pb), copper (Cu), chromium (Cr), cobalt (Co), nickel (Ni), zinc (Zn), iron (Fe), manganese (Mn), aluminum (Al), beryllium (Be), selenium (Se). Metal concentrations in water samples were measured by using Inductively Coupled Plasma - Mass Spectrometry (ICP-MS, Perkin Elmer).

3. Results and Discussion

LOD and R2 values of heavy metal compounds can be seen in Table 2. Heavy metal concentration levels of water samples taken from the surface water sources in Konya closed basin are given in Table 3. The results obtained for sources used drinking water purpose were compared with the Turkish Regulations, the European Community Council Directive 98/83/EC, US Environmental Protection Agency and World Health Organization guidance values. The results obtained for sources used irrigation water purpose were compared with the Turkish Regulations. The results were also evaluated according to the values of inland water resources quality (Table 4) and surface quality classes of the basin are determined.

Table 2. LOD and R2 values of heavy metal compounds

	As	Cd	Pb	Cu	Cr	Co	Ni	Zn	Fe	Mn	Al	Be	Se
LOD (μg/L)	0.0246	0.0071	0.0286	0.0076	0.1880	0.0047	0.0093	0.0178	0.2210	0.0257	0.8290	0.0149	0.0431
R^2	0.9990	0.9994	0.9999	0.9999	0.9983	0.9995	0.9999	0.9993	0.9998	0.9995	0.9990	0.9999	0.9999

Table 3. Heavy metal concentration levels of water samples (µg/L)

Sample No	As	Cd	Pb	Cu	Cr	Co	Ni	Zn	Fe	Mn	Al	Be	Se
1	1.02-1.83	<dl-0.17	<dl	0.70-0.81	3.16-15.54	0.12-0.47	2.30-2.47	<dl-3.66	165.5-266.1	0.23-0.83	<dl-1.59	<dl-0.16	1.60-2.11
2	6.09-17.54	0.01-0.09	<dl	1.84-5.63	5.49-13.37	0.05-0.16	0.74-1.56	2.25-9.00	98.6-150.4	0.04-1.04	<dl-3.44	<dl-0.01	1.00-1.64
3	4.72-8.37	<dl-0.009	<dl	0.89-2.19	2.36-19.73	0.19-0.20	1.86-3.42	<dl	193.3-298.8	0.24-1.78	<dl-4.50	<dl	0.78-1.34
4	2.25-13.53	<dl-0.01	<dl	2.72-4.63	5.34-26.12	0.54-1.43	6.38-15.46	<dl-29.48	589.0-1273.3	8.42-39.22	<dl-13.12	<dl	0.06-1.45
5	91.11	0.02	<dl	3.62	33.76	0.23	2.30	<dl	165.41	0.65	<dl	<dl	1.68
6	3.10-3.68	<dl-0.009	<dl	0.77-1.10	3.14-23.30	0.13-0.19	1.47-2.10	<dl	190.6-221.8	0.20-1.04	<dl-7.32	<dl	0.72-1.41
7	0.95-2.60	<dl	<dl	0.74-0.80	2.66-16.57	0.05-0.12	1.10-1.25	<dl	141.2-157.8	0.10-1.18	<dl-7.08	<dl	0.46-1.00
8	0.69-1.32	<dl	<dl	0.48-0.59	1.96-14.83	0.04-0.11	0.90-1.24	<dl	122.1-141.8	0.05-0.40	<dl-4.45	<dl	0.25-1.21
9	4.91-14.78	0.01-0.02	<dl	3.76-27.6	2.80-29.01	0.16-0.28	2.60-5.75	<dl-65.8	204.0-422.6	0.29-5.00	<dl-22.59	<dl	0.68-1.79
10	12.94	0.02	<dl	2.91	30.11	0.24	4.39	<dl	203.5	0.82	<dl	<dl	1.25
11	32.37	0.03	<dl	4.51	43.55	0.39	7.04	<dl	209.7	1.31	<dl	<dl	7.58
12	1.05-1.51	<dl-0.02	<dl	0.85-5.13	3.42-23.55	0.12-0.15	2.61-6.68	<dl	108.7-247.6	1.79-1.95	<dl-8.44	<dl	0.44-0.95
13	68.75	0.08	<dl	6.88	66.76	0.35	4.58	<dl	161.2	1.03	<dl	<dl	44.74
14	90.13	0.04	<dl	5.50	65.58	0.70	0.20	<dl	557.9	2.49	<dl	<dl	4.79
15	3.36-7.60	0.02-0.03	<dl	1.98-22.51	4.13-28.14	0.12-0.24	2.60-3.92	<dl-18.54	205.1-220.7	0.43-1.35	<dl-13.60	<dl	0.69-1.73
16	26.13	0.01	<dl	0.55	42.25	1.74	9.60	<dl	492.8	46.42	<dl	<dl	2.50
17	11.45	0.04	<dl	2.28	25.80	0.28	3.92	<dl	241.5	0.24	<dl	<dl	0.59
18	5.09	0.18	<dl	5.18	75.56	0.27	8.19	<dl	302.4	3.08	<dl	<dl	<dl
19	0.54-1.98	<dl	<dl	0.61-2.20	3.90-23.80	0.15-0.24	2.40-4.50	<dl	222.5-252.8	1.88-3.58	0.90-2.39	<dl	0.40-1.23
20	6.40	0.02	<dl	2.26	15.63	0.29	2.54	<dl	159.3	1.48	<dl	<dl	0.60
21	2.06-4.01	<dl-0.009	<dl	2.28-6.74	3.34-36.02	0.26-0.30	4.52-5.88	<dl	234.3-268.5	1.54-6.38	<dl-5.49	<dl	0.74-0.84
22	17.62	0.15	<dl	12.14	25.18	1.32	24.43	<dl	387.0	7.34	<dl	<dl	0.30
23	5.05-5.18	0.03-0.04	<dl	1.66-5.28	3.96-26.44	0.12-0.31	3.58-6.60	<dl-153.0	184.5-631.6	0.28-6.66	<dl-23.80	<dl	0.60-1.44
24	1.01	<dl	<dl	0.72	9.12	0.08	1.20	<dl	141.2	0.56	<dl	<dl	0.26
25	2.98-11.48	<dl-0.01	<dl	0.72-1.56	1.30-22.70	0.12-0.28	2.36-4.68	<dl	244.4-350.4	1.58-1.78	<dl-1.95	<dl	0.83-1.10
26	6.73	<dl	<dl	1.30	2.30	0.43	6.65	<dl	567.6	16.42	1.58	<dl	3.01
27	0.66	<dl	<dl	4.70	1.58	0.05	1.34	<dl	125.6	0.74	14.51	<dl	1.18
28	7.60	0.01	<dl	1.57	1.05	0.28	5.51	20.58	537.8	0.60	<dl	<dl	4.58
29	17.90	0.12	<dl	7.64	5.46	0.46	8.22	36.86	657.5	1.24	10.34	<dl	12.26
30	8.48	<dl	<dl	0.83	5.66	0.80	9.56	62.58	804.2	184.6	8.89	<dl	2.82
31	59.05	0.009	<dl	0.59	2.48	0.82	4.70	64.84	1044.4	179.8	3.98	<dl	2.26
32	6.07	<dl	<dl	0.14	3.57	0.45	8.11	<dl	422.8	135.5	4.48	<dl	1.50

According to inland water resources water quality criteria given in Table 4, surface water of Konya closed basin is in 1st class high quality water in terms of Cd, Pb, Co, Ni, Zn, Al, Be and Se metals. As, Cu, Cr, Fe, Mn are exceeded limit values given for 1st class (high quality water) or 2nd class (less contaminated water). Cu values surface water samples taken from sampling point 9 and 15 exceeded 20 μg/L limit values given for high quality water. Cr values were determined at sampling point 4, 5, 6, 9-19, 21-23 and 25 exceeded 1st class less contaminated water standard values (20 μg/L) given by Turkish Regulation. Fe was determined in surface water samples and as it can be seen in Table 2 Fe values were determined at sampling point 4, 9, 14, 16, 23, 25, 26, 28-32 exceeded 1st class high quality water standard values (300 μg/L).

Table 4. Quality criteria and water quality classes for inland water resources

Parameter (μg/L)	Water quality class			
	1st class (high quality water)	2nd class (less contaminated water)	3rd class (dirty water)	4th class (very dirty water)
As	20	50	100	> 100
Cd	3	5	10	> 10
Pb	10	20	50	> 50
Cu	20	50	200	> 200
Cr	20	50	200	> 200
Co	10	20	200	> 200
Ni	20	50	200	> 200
Zn	200	500	2000	> 2000
Fe	300	1000	5000	> 5000
Mn	100	500	3000	> 3000
Al	300	300	1000	> 1000
Se	10	10	20	> 20
Ba	1000	2000	2000	> 2000

Cd, Pb, Co, Cr, Ni, Mn, Zn, Al, Be and Se in water samples used for drinking water do not exceed the limit values by Turkish Regulation, EU, US EPA and WHO while As and Fe exceeded the limit values in some sampling point and as it can be seen in Table 3 some of them exceeded limit values given by regulations. As values exceeded the 10 μg/L level in water samples taken four sampling points while Fe values exceeded the 200 μg/L level in water samples taken five sampling points. The value of MAC of Arsenic in Turkey was reduced from 50 μg/L to 10 μg/L in 2005. The EPA, the National Research Council (NRC) and several research groups stated that chronic effects on humans may be caused by prolonged consumption of water with a concentration of As as low as 5 μg/L (EPA) or even 3 μg/L (NRC). The lowest level of As of water samples of Konya closed basin some drinking water sources was observed with 0.69 μg/L level. As concentration determined in some sampling points exceed 3 or 5 μg/L levels. Fe was also determined in all sampling point and their concentrations were changed between 98.6 μg/L and 1273.3 μg/L levels.

As is widely distributed in the environment and is known to be highly toxic to humans. Both natural and anthropogenic activities result in the significant input of As to the environment. Natural processes like erosion and weathering of crustal rocks lead to the breakdown and translocation of arsenic from the primary sulfide minerals, and the background concentrations of arsenic in soils are strongly related to the nature of parent rocks. An extensive range of anthropogenic sources may enhance concentration of As in the environment. Some of these activities include industrial processes that contribute to both atmospheric and terrestrial depositions, such as mining and metallurgy, wood preservation, urban and industrial wastes, and applications of sewage sludge and fertilizer. Among the two modes of As input, the environment is mostly threatened by anthropogenic activities. The fate of As accumulated in the surface environment depends essentially on its retention and mobility in the host medium, soil and groundwater, and is most vulnerable for biota [21,22]. Arsenic may cause skin lesions, gangrene in leg, lung, bladder, liver and renal cancer, laryngitis, tracheae bronchitis, rhinitis, phrryngitis, shortness of breath and nasal congestions. US EPA and several researchers stated that chronic effects on humans may be caused by prolonged consumption of water with a concentration of Arsenic [23-27]. Copper is widely used for wire production and in the electrical industry. Its main alloys are brass (with zinc) and bronze (with tin). Other applications are kitchenware, water delivery systems, fertilizers, bactericides and fungicides, feed additives and growth promoters, and as an agent for disease control in livestock and poultry production. The main sources of copper are copper fertilizers, which are widely used in agriculture. If the manure is applied to soils, this may lead to potential accumulation and toxic effects, e.g. to sheep. Cu is also emitted by metallurgical processing for Cu, iron, and steel production, and coal combustion. Heavy metal contamination caused by industrial emissions is well documented. The deposition rate of heavy metals from smelters is a function of distance. Copper is one of the seven well known micro nutrients (Zn, Cu, Mn, Fe, B, Mo, and Cl), which are essential for plant nutrition, although it is only needed in small amounts of 5 to 20 mg/l. Concentrations of <4 mg/L are considered deficient, and concentrations >20 mg/l are considered toxic [13,14].

Some heavy metals accumulate in environment and are toxic to plants and animals. Their presence may limit the suitability of the surface water for irrigation or other uses. Aluminum can cause nonproductivity in acid soils but soils at 5.5 to 8.0 will precipitate the ion and eliminate toxicity. Toxicity to plants varies widely for Beryllium, ranging from 5 mg/L for kale to 0.5 mg/L for bush beans. Cadmium is toxic to beans, beets and turnips at concentrations as low as 0.1 mg/L in irrigation water. Cobalt is toxic to tomato plants at 0.1 mg/L in irrigation water. Fluoride is inactivated by neutral and alkaline soils. Copper is toxic to a number of plants at 0.1 to 1.0 mg/L in irrigation water. Iron is not toxic to plants in aerated soils, but can contribute to soil acidification and loss of essential phosphorus and molybdenum. Lead can inhibit plant cell growth at very high concentrations. Lithium is tolerated by most crops at up 5 mg/L mobile in soil and toxic to citrus at low doses recommended limit is 0.075 mg/L. Manganese is toxic to a number of crops at a few mg/L in acid soils. Nickel is toxic to a number of plants at 0.5 to 1.0 mg/L and reduced toxicity at neutral or alkaline pH. Selenium is toxic to plants at low concentrations and to livestock if forage is grown in soils with low levels of added selenium. Zinc is toxic to many plants at widely varying concentrations, reduced toxicity at increased at pH (6 or above) and in fine-textured or organic soils [28,29]. Maximum allowable heavy metals and toxic elements for irrigation waters are presented in Table 5. These all toxic elements are not exceeding limit values for continuous irrigation for any type of soil and for clayey soils for irrigation less than 24 years (pH between 6.0-8.5).

Table 5. Maximum allowable heavy metals and toxic elements for irrigation waters

Elements	Maximum allowable concentrations	
	Limit values for continuous irrigation for any type of soil, µg/L	Limit values for clayey soils for irrigation less than 24 years (pH between 6.0-8.5), µg/L
As	100	2000
Be	100	500
Cd	10	50
Cr	100	1000
Co	50	5000
Cu	200	5000
Fe	5000	20000
Pb	5000	10000
Mn	200	10000
Ni	200	2000
Se	20	20
Zn	2000	10000

4. Conclusions

Konya closed basin is in a semi arid region and has very wide farm land. High temperature, low precipitation due to global climate change increased water demand in the region. Konya surface water falls in 1th or 2nd class inland water according to water pollution control regulation. Cd, Pb, Cu, Cr, Ni, Mn, Al and Se metals measured in this work in surface water samples used for drinking water were complying the limit values by Turkish Regulation, US EPA and WHO except for As and Fe. Possible origin of these metals in surface water samples in the basin may be residential, industrial and agricultural activities. Maximum allowable heavy metals and toxic elements limits for irrigation waters are given by water pollution control regulation. Konya closed basin surface water heavy metal and toxic element contents are below the maximum levels for continuous irrigation for any type of soil and for less than 24 years irrigation for any type of land. Surface water sources used for drinking water supply should strictly be controlled and for heavy metal contaminants a detailed monitoring program should be established initially and then the optimum sampling program should be determined considering individual pollutant sources and local circumstances.

Acknowledgments

This investigation was carried out by support of Selcuk University Scientific Research Projects (BAP) Foundation (Project number: 10201046).The authors thank General Directorate of State Hydraulic Works (DSI) for helping the water sampling.

References

[1] Özmen, H., Külahçı, F., Çukurovalı, A., Doğru, M., Chemosphere, 2004, 55, 401.
[2] Zarazua, G., Ávila-Pérez, P., Tejeda, S., Barcelo-Quintal, I., Martínez, T., Spectrochimica Acta Part B, 2006, 61, 1180.
[3] Rajaratnam, G., Winder, C., An, M., Environmental Research, 2002, 89, 165.
[4] Calderon, R.L., Food and Chemical Toxicology, 2000, 38, 13.
[5] Tamasi, G., Cini, R., Science of the Total Environment, 2000, 327, 41.
[6] Miller, J.R., Hudson-Edwards, K.A., Lechler, P.J., Preston, D., Macklin, M.G., Science of the Total Environment, 2004, 320, 189.
[7] Ministry of Environment and Forestry, Turkish Official Gazette, Regulations of waters for human consumption, Date: 17.02.2005, Number: 25730, 2005.
[8] Ministry of Environment and Forestry, Turkish Official Gazette, Water pollution control regulation, Date: 31.12.2004, Number: 25687.
[9] Ministry of Environment and Forestry, Turkish Official Gazette, Water pollution control regulation technical procedure, Date: 7.01. 1991, Number: 20748.
[10] US EPA, 2006 Edition of the Drinking Water Standards and Health Advisories, EPA 822-R-06-013, Washington, DC, USA, 2006.
[11] Council Directive of 15 July 1980 Relating to the Quality of Water Intended for Human Consumption, 80/778/EEC.
[12] World Health Organization, Guidelines for drinking-water quality [electronic resource]: incorporating first addendum. Vol. 1, Recommendations. - 3rd ed, 2006.
[13] Sarkar, B., Heavy Metals in the Environment, The Hospital for Sick Children and University of Toronto, Ontorio, Canada, Marcel Dekker, Inc., New York, 2002.
[14] Bradl, H.B., Heavy Metals in the Environment, University of Applied Sciences Trier Neubrucke, Germany, Elsevier Academic Pres, 2005.
[15] Buschmann, J., Berg, M., Stengel, C., Winkel, L., Sampson, M.L., Tranh, P.T.H., Viet, P.H., Environmental International, 2008, 34, 756.
[16] Krishna, A.K., Satyanarayanan, M., Govil, P.K., Journal of Hazardous Materials, 2009, 167, 366.
[17] Oehmen, A., Viegas, R., Velizarov, S., Reis, M.A.M., Crespo, J.G., Desalination, 2006, 199, 405.
[18] Abollina, O., Aceto, M., Malandrino, M., Sarzinini, C., Mentasti, E., Water Research, 2003, 37, 1619.
[19] Lin, S.H., Juang, R.S., Journal of Hazardous Materials, 2002, 315.
[20] WWF-Turkey, Project Final Report of Turkey's last, World Wildlife Fund, 2010.
[21] Leung, C.M., Jiao, J.J., Water Research, 2006, 40, 753.
[22] Sorme, L., Lagerkvist, R., The Science of the Total Environment, 2002, 298, 131.
[23] Virkutyte, J., Sillanpaa, M., Environmental International, 2006, 32, 80.
[24] World Health Organization, Environmental Health Criteria, Arsenic, Geneva, 1981.
[25] Anawar, H.M., Akai, J., Mostofa, K.M.G., Safiullah, S., Tareq, S.M., Environment International, 2002, 27, 597.
[26] Wyatt, C.J., Fimbres, C., Romo, L., Mendez, R.O., Grijalva, M., Environmental Research, 1998, 76, 114.
[27] Guo, Q., Journal of Hazardous Materials, 1997, 56, 181.
[28] US EPA, Manual Guidelines for Water Reuse", EPA/625/R-92/004, 1992.
[29] US EPA, Municipal Wastewater Reuse, Selected Readings on Water Reuse, EPA430/09-91-022, 1991.

Geochemical evidences in the release processes of Arsenic into the groundwater in a part of Brahmaputra Floodplains

C. Mahanta[1] , P. Bhattacharya[2] , Bibhas Nath[3] and L. Sailo[1]

[1] Department of Civil Engineering, Indian Institute of Technology Guwahati 781039, India
[2] KTH-International Groundwater Arsenic Research Group, Department of Land and Water Resources Engineering, Royal Institute of Technology (KTH), SE-100 44 Stockholm, Sweden
[3]School of Geosciences, University of Sydney, Sydney NSW 2006, Australia
Corresponding author E-mail: mahantaiit@gmail.com

Abstract
To understand the sources and mobilization processes responsible for arsenic enrichment in groundwater in the Brahmaputra Basin where higher arsenic concentration have been reported, the geochemical features of the aquifer sediments were studied. Six boreholes were drilled near the tubewells (1 and 2) where aqueous arsenic concentration varies between 250 - 350 µg/l. The soil sediment was collected at 3 m (10 ft) interval and it was drilled to the depth of 45 m (150 ft) which is the common depth of the tubewell installed in the study area. The bulk chemical studies on the sediments show that the pH of soils varies from 4.2 to 5.2 with a mean value of 4.75. The groundwater composition in the study area is of Na-HCO3-. The major anions HCO3- is likely from the decomposition of organic matter and originates from weathering of silicate and calcite minerals by atmospheric or respired CO2. Selective sequential extraction (SSE) method proposed by Wenzel et al., (2001) for extraction of arsenic from soil was used. Results of sequential extraction experiment show that solid-phase arsenic is present predominantly in the reducible fraction (Ext_5 and Ext_6), and residual fraction (Ext_7) contributes to highest fraction in many soil sediment. The major processes of arsenic mobilization probably linked to desorption of As from Fe oxides/oxyhydroxides and the reductive dissolution of Fe rich phases in the aquifers sediments under reducing and alkaline conditions.

1. Introduction

Naturally occurring arsenic contamination in groundwater has become a major environmental globally, affecting large human population in Bangladesh, India, Nepal, Vietnam and Taiwan [1]. The Arsenic (As) contamination in Asia, derived probably from the Himalayan region [2] has exposed tens of millions of individuals to drinking water with hazardous levels of this metalloid. The existence of arsenic contamination [3] in the upper Brahmaputra plain in Assam has already been reported. Concentration is often higher than the drinking water guideline values of World Health Organization (WHO) and the Bureau of Indian Standards (BIS). In a recent study, concentrations beyond 50 ppb have been confirmed in 72 blocks out of 192 blocks in 22 districts of Assam [3]. Very few studies on As has been done and reported in Assam plains of the Brahmaputra Basin, and no in depth research work has yet been carried out for deducing sources and mechanisms of As release despite strong indication of its wider spread [3].

The processes controlling the release of As to the groundwater have been studied intensively but still they remain a subject of dispute. The reductive dissolution of Fe oxides, which are common in sedimentary environments, is widely accepted as a key process for the release of As into the groundwater [4]. The reduction of Fe oxides alone cannot explain the large range of groundwater As concentrations encountered in similarly reducing aquifers. Further, processes which could lead to higher As concentration in groundwater are sulfide oxidation which has been postulated as a source of As, especially in West Bengal, but this mechanism is now largely abandoned or precipitation and dissolution of secondary mineral phases (e.g. siderite, magnetite, amorphous phases incorporating As), competition with other dissolved anions such as PO43- or HCO3- [5]. Assam has a great deal of similarities to the Bangladesh plains located downstream of the Brahmaputra in terms of sedimentology. The problem of As enrichment is likely therefore to be of a similar magnitude in Assam, and possibly can be investigated within similar source and process mechanisms to begin with [3].

2. Materials and Methods

2.1 Sampling sites

The study area is located at Titabor, Jorhat district situated in the eastern part of Assam state of India between Latitude 26°27.3′ North and 26°30.8′ N, Longitude 94°6.3′ E and 94°9.8′ E as show in Figure 1. Based on the results of groundwater and soil sample analysis, it was decided to drill a fresh wells in this high arsenic contaminated location i.e. Titabor. Six boreholes were drilled during latter part of January 2010 with the help of local drillers provided by the PHED Assam.

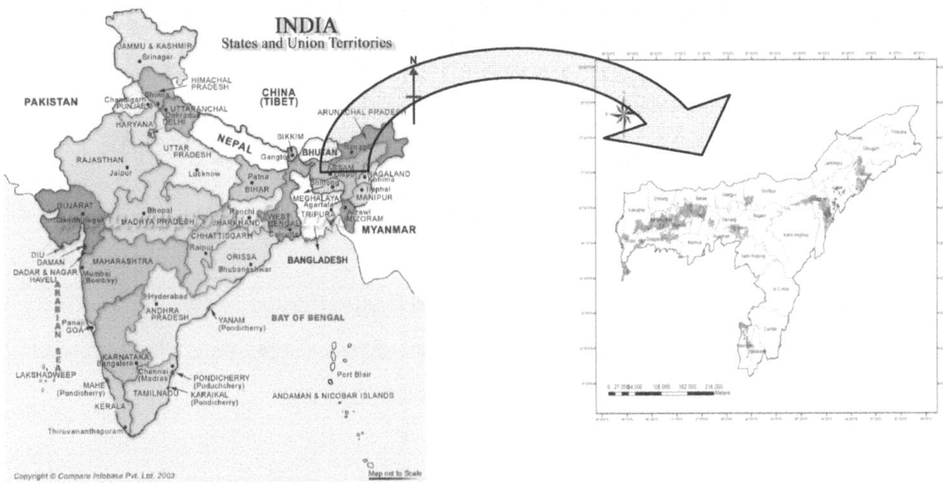

Figure 1. The location of the study area, Titabor, Assam

2.2 Selective sequential extraction methods

A selective sequential extraction (SSE) method was used to assess trace metals of differential liability with reagents of increasing dissolution strength. Since As is found mainly in anionic form the Selective Sequential Extraction (SSE) procedures were adopted to follow those that have been used for the studies of phosphorous retention. A sequential extraction procedure proposed by [6] in combination with extraction method for carbonates by [7] was used for the sediment extraction. This methods involves seven steps i.e. from 1_Ext to 7_Ext each Ext steps targeting selectively specific phase. The brief steps along with target phase and extractant used are given in table 1.

Table1: Sequential extraction scheme for sediments [6] and [7]

Step	Target phase	Extractant	Extraction conditions	SSR[a]
Ext_1	Non-specifically bound *As*	0.05 M $(NH_4)_2SO_4$	4 hour shaking, 20°C	1:25
Ext_2	Specifically bound *As*	0.05 M $(NH_4)H_2PO_4$	16 hour shaking, 20°C	1:25
Ext_3	Carbonate bound *As*	1M *NaOAc+HOAc*, pH 5	6 hour shaking, 20°C	1:25
Ext_4	Resorbed *As*, release from carbonates	0.05 M $(NH_4)H_2PO_4$	4 hour shaking, 20°C	1:25
Ext_5	Amorphous hydrous oxide-bound *As*	NH_4-*Oxalate* buffer (0.2M), pH 3.25	4 hour shaking 20°C in the dark	1:25
Ext_6	Crystalline hydrous oxide-bound *As*	NH_4-*Oxalate* buffer (0.2M) + ascorbic acid (0.1M) pH 3.25	30 minutes in a water basin at 96±3°C	1:25
Ext_7	*As* in sulfide minerals	16N HNO_3	Microwave digestion	1:50

3. Results and Discussion

Selective sequential extractions were carried out to fractionate the target elements in the solid materials to assess their potential effects. The sequential extraction scheme proposed by [6] was used and the results are shown in Figure 2. The results show that step 1 (Ext_1) using $(NH_4)_2SO_4$ which target non-specifically bound As is very low (less than 5%) for arsenic and iron. The step 2 (Ext_2) with $(NH_4)H_2PO_4$ targeting specifically bound As is significant with average of 10 % of sum of total. The extraction steps 3 and 4 (Ext_3 and Ext_4) targets arsenic bound to carbonate phases is very low (< 1%), this shows that carbonate dissolution does not contribute to As leaching. The step 5 and 6 (Ext_5 and Ext_6) extract As from Fe-amorphous hydrous oxide and crystalline Fe-oxide. From the Figure 2 (c and d) the extraction of Fe shows that majority of Fe (iron) is in the form of sulfide and silicate minerals extracted from Ext_7. Whereas the amorphous Fe extracted from Ext_5 is lesser than Ext_4 (crystalline Fe-oxide) in most of the sediment samples. Some caution is warranted in the interpretation of extracted Fe in terms of amorphous and crystalline Fe-oxides as even small amount of Fe^{2+}in combination with oxalate will catalyze the dissolution of more crystalline Fe-oxides like goethite and hematite.

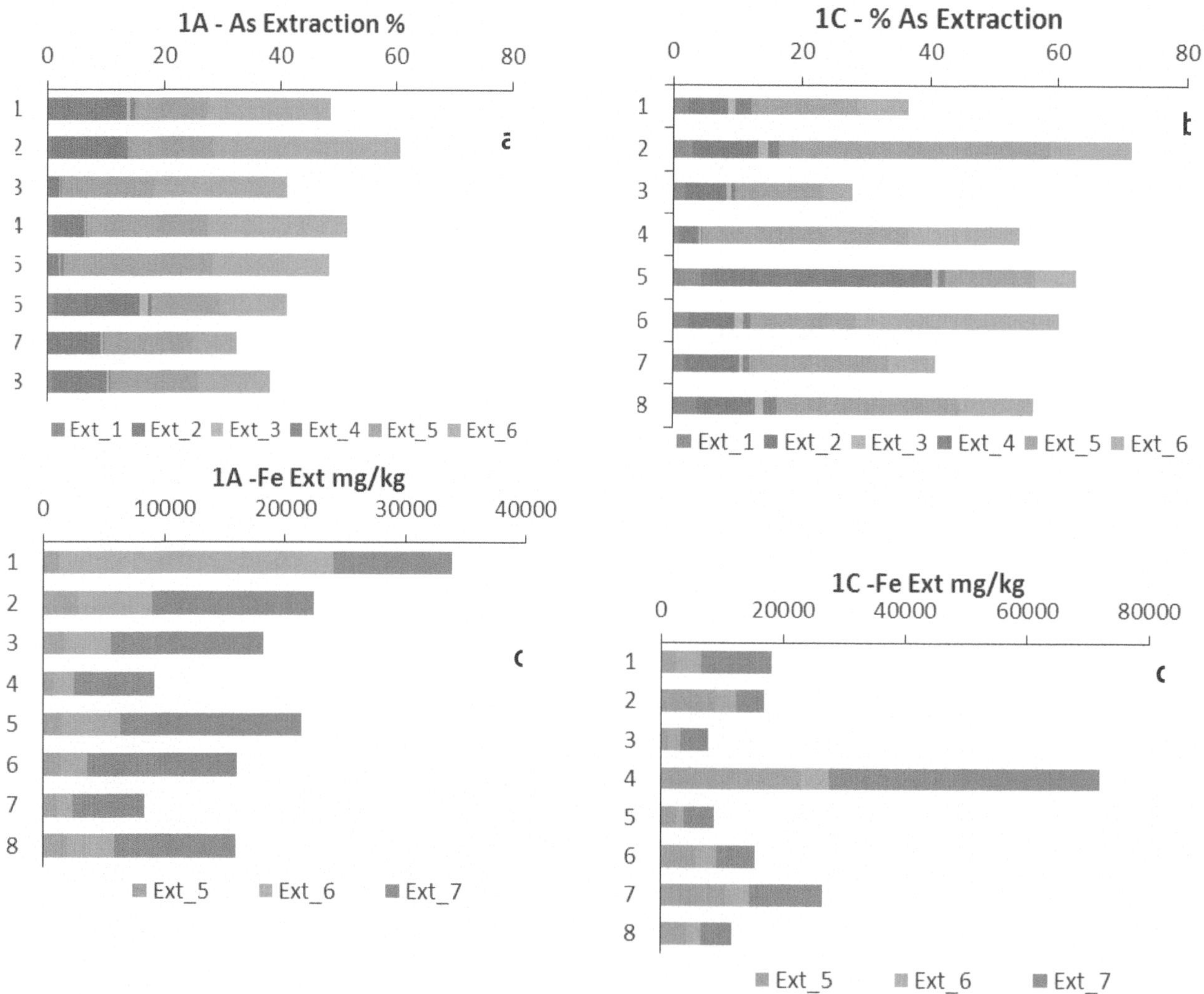

Figure 2.(a) and (b) As extraction; Figure 2 (c) and (d) Fe leaching from the sediment of 1A and 1C.

From the sequential leaching the sorbed As i.e.(Ext_1 + Ext_2) constitute only a small fraction of extracted As. Ext_2 contributes to about 10 % of As extracted this suggests that competitive anion exchange with PO_4^{3-} because of smaller charge and higher charge density [6] is responsible for higher As in few tube wells. The Ext_5 and Ext_6 contributes highest fraction of arsenic which targets the Fe-oxides. An additional As is extracted in Ext_7 from pyrite and sulfide. The As fractions in individual soil depends strongly on the extractant and extraction procedure employed [6]. Small modification of the sequential extraction procedure could yield significantly variable results among laboratories. But the trends in the characterization of distribution into the main soil fractions should agree regardless of the extraction procedure [6]. The same sequential extraction procedure was done in Dan Phuong, Vietnam and reported total As between 7 - 20 μg/g and total Fe around 20 mg/g [8] and in Bangladesh, using extraction procedure proposed by Koen et al., (2001) reported total As concentration generally to be less than 3 μg/g and total Fe about 20 mg/g [10]. In our study area total arsenic concentration was between 6 - 33 μg/g with an average of 14 μg/g and total Fe about 25 mg/g. The amorphous Fe extracted contributes to about 10 % and the crystalline Fe contributes to 25 % of the total Fe extracted in Ext_7. A study in Bangladesh reported to extract 10% of total Fe associated with Fe-oxide and 30% extraction associated with crystalline Fe-oxide [10]. In Vietnam more than 50% of total iron was associated with Fe-oxides [8].

4. Conclusions

The hydrogeochemical characteristics and arsenic contamination of groundwater are evaluated in Tokobari (Titabor), Jorhat district. The groundwater composition is Na-HCO3- type water. Major cations Na+, Ca2+ and Mg2+ have likely resulted from the weathering of silicates, dissolution of carbonates and cation-exchange enhanced by respired CO2 from organic matter degradation, which is also responsible for HCO3- forming a major anion.

Sediment extraction studies revealed that arsenic concentration of 6 to 33 μg/g with an average of 14 μg/g, is sufficiently high to mobilize As above acceptable drinking water quality standard. The sequential extraction results suggest that predominant part of arsenic is expected in sorbed or co-precipitated state in Fe-bearing mineral phases.

The main mechanism for As mobilization seems to be the reductive dissolution of Fe-oxyhydroxides in the present case. The reduction of Fe-oxyhydroxide is coupled to the degradation of organic matter in the sediments. Competitive anions exchange between As with PO43- is also a likely mechanism for the enrichment of As in some tubewells.

References

[1] Smedley, P.L., Kinniburgh, D.G., Applied Geochemistry, 2002, **17**:517-68.

[2] McArthur, J.M., Banarjee, D.M., Hudson-Edwards, K.A., Mishra, R., Purohit, R., Ravenscroft, P., Cronin A, Howarth RJ, Chatterjee A, Lowry D, Houghton, S. & Chadha, D.K., Applied Geochemistry,2004,**19**: 1255-1293.

[3] Mahanta, C., Pathak, N., Borah, P. Choudhury, R and Alam, W., Proceedings of World Environmental and water Resources Congress, 2009, **342**: 180.

[4] Nickson, R.T., McArthur, J.M., Ravenscroft, P., Burgess, W.G., and Ahmed, K.M., Applied Geochemistry, 2000, **15**: 403 - 413.

[5] Harvey, C.F., Swartz, C.H., Badruzzaman, A.B.M., Keon-Blute, N., Yu, W., Ashraf Ali, M., Jay, J., Beckie, R., Niedan, V., Brabander, D., Oates, P.M., Ashfaque, K.N., Islam, S., Hemond, H.F., Ahmed, M.F., Science, 2002, **298**: 1602-1606.

[6] Wenzel, W. W., Kirchbaumer, N., Prohaska, T., Stingeder, G., Lombic, E. and Adriano D. C., Analytical Chimica Acta, 2001, **436**: 309-323.

[7] Tessier, A., Campbell, P.G.C., and Bisson, M., Analytical Chemistry, 1979, 51: 844-851.

[8] Postma, D., Larsen, F., Minh Hue, N.T., Thanh Duc, M., Viet, P.H., Nhan, P.Q., Jessen, S., Geochimica et Cosmochimica Acta, 2007, **71**: 5054-5071.

[9] Keon, N. E., Swartz, C. H., Brabander, D. J., Harvey C., and Hemond, H. F., 2001, Environ. Sci. Technol. **35**: 2778-2784.

[10] Swartz C. H., Blute N. K., Badruzzman B., Ali A., Brabander D., Jay J., Besancon J., Islam S., Hemond H. F. and Harvey C., Geochimica et Cosmochimica Acta, 2004, **68**, 539-4557.

Sustainable Arsenic Mitigation (SASMIT): An approach for developing a color based tool for targeting arsenic-safe aquifers for drinking water supply

M. Hossain[1,2], P. Bhattacharya [2], K.M. Ahmed[3], M.A. Hasan [3], M. von Brömssen[4], M.M. Islam [1], G. Jacks[2], M.M. Rahman [3], M. Rahman[1], A. Sandhi[2] and S.M.A. Rashid[1]

[1] NGO Forum for Drinking Water Supply and Sanitation, Lalmatia, Dhaka, 1207, Bangladesh

[2] KTH-International Groundwater Arsenic Research Group, Department of Land and Water Resources Engineering, Royal Institute of Technology (KTH), Teknikringen 76, SE-10044 Stockholm, Sweden

[3] Department of Geology, University of Dhaka, Dhaka, 1000, Bangladesh

[4] Ramböll Sweden AB, Box 4205, SE-102 65 Stockholm,, Sweden

Abstract

Presence of high concentration of geogenic arsenic (As) in water and soil become a big health risk towards millions of people in various magnitudes through drinking water. To minimize arsenic interaction with human considered as a global challenge. The main objective of this research is to develop a simple, easy and cost-effective arsenic identification tool which would be easily acceptable by the inhabitants and local well drillers. The relationship of sediment color and corresponding As concentrations in water has already been demonstrated and is being further studied under SASMIT project. A total of 1920 sediment samples from 15 locations bored up to a depth of 250 m have been scientifically evaluated according to the color codes using Munsell Color Chart. A total of 60 varieties observed and simplified into four color groups viz. black, white, off-white and red. It is revealed that red and off-white sands can be targeted for As-safe water. White sands can also be safe but uncertainty is high and black sediments produce water with highest As concentration, although Mn content in waters sampled from white and black sediments is relatively low. Further refinement is going on for improving the tool for targeting aquifers which can be safe for both arsenic and manganese.

1. Introduction

In last millennium, presence of high level of arsenic in sediment and aquifer was recognized one of the biggest environmental disasters in South Asia and new arsenic contaminated areas are growing exponentially *[1]* all over the world. According to the survey performed by British Geological Society (BGS),the arsenic concentration in groundwater in Bengal delta found above than global arsenic standard (10µg L^{-1}) in last *[2]* millennium. Geogenic arsenic (As) exposed to millions of people in Bangladesh through drinking water, collected from the groundwater sources or tubewells. A wide range of diseases caused in human body by drinking this carcinogenic metalloid associated water *[3,4]* and mitigation of this problem is a major challenge for ensuring safe drinking water for the population. The development of a suitable, simple to understand and cost-effective arsenic identification tool for local communities was a big challenge in ongoing arsenic research fields.

Under the Sustainable arsenic mitigation (SASMIT) project, the integrated approach taken will ensure installation of the project wells in those locations that are sustainable in the context of water quantity and quality; and where the demand for safe water is high. a permanent, color comparator chart *[5]* could be considered for using field arsenic test kit. In order to provide safe aquifers for local inhabitants, development of an identifying tool based on color due to easy to understandable by the rural people. Previous studies showed, a sediment color based identification tool *[6]* could be utilized for recognizing safe water collection point in Matlab, Bangladesh. The basic outcome of this color based tool to educate the regional drillers and resident people how to identify a safe aquifer using the locally affordable technology. Successful completion of Matlab area, this strategy initiated by SASMIT would be replicable in other As affected areas within and beyond the country where the geological conditions are similar.

Therefore, the main aim of this research was to develop a color based tool for identification of safe aquifers from various sediments on the basis of their color. The local well drillers were the main target groups those can utilize this tool. It will play a significant role in As mitigation approach, as tubewells are considered the major source of drinking water in the rural areas of Bengal delta.

2. Materials and Methods

For comprehensive hydrogeological investigations through determining the geological characteristics of the sediments and monitoring of hydraulic head and water quality, 15 piezometer nests have been installed in the study area. Most of these nests are constructed of 5 piezometers - 4 in shallower reach (within 100 m) and the deepest one is at a depth of around 250 m.

At each of these 15 locations (Figure. 1), samples were collected for each of 5 feet and visual inspection was made for construction of borelogs. Wash boring sediment samples collected in the field were described first in terms of their grain size and color. 1920 sediment samples collected from all the 15 sites up to a depth of 250 meter were then standardized using Munsell colour chart and categorized into four groups namely, red, off-white, white and black.

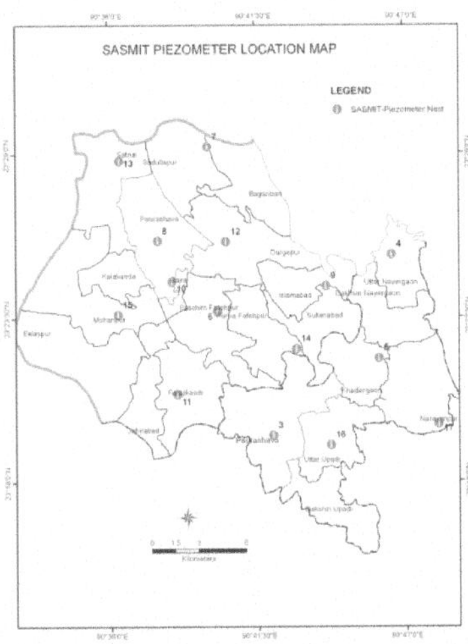

Figure. 1: Location of test borings and piezometer nests

Groundwater analyses in study area, include analysis of on-site parameters (pH, Eh, Temperature and EC), major ions (cations and anions), trace elements in aquifer and speciation of As (data not shown),. A total number of 492 groundwater samples from 197 stations covering the period from 2004 to Premonsoon 2010 have been taken into account for this assesment.

Finally, harmonization was done involving classification of all sediment and groundwater samples according to four-colour concept developed *[6]* by KTH-GARG, Dept. of LWR, KTH, Sweden.

3. Results

Based on visual examination, all the 1920 samples were simplified into the following groups. Different categories and their proportion were given in the (Table 1). Due to abundance, medium sand was found in the most cases and was followed by fine sand and silty sand.

In case of color based observation, all 1920 samples were categorized and simplified into the following five colors; grey, light grey, olive, yellow and brown (Table 2). Among the color components, grey was the top most one. Other frequently visible colors observed were brown and yellow.

In order to amplify the relationship between arsenic content in groundwater and the well depth, data of the 492 groundwater samples harmonization was performed (Figure. 2). From the depth vs As concentration curve showed, highest arsenic concentration was found in the sediments which was black and the lowest in red color.

Table 1: Grain size variety for tool development

<table>
<tr><th colspan="3">Primary observation (description of individual sample)</th><th colspan="3">Generalized observation (simplified)</th></tr>
<tr><td>Category name (primary observation)</td><td>No. of samples found in each category</td><td>% (by number)</td><td>Name to be used</td><td>No. of samples found in each category</td><td>% (by number)</td></tr>
<tr><td>Clay</td><td>6</td><td>0.31</td><td>CLAY</td><td>6</td><td>0.31</td></tr>
<tr><td>Silty clay</td><td>36</td><td>1.88</td><td rowspan="4">Silty CLAY</td><td rowspan="4">53</td><td rowspan="4">2.76</td></tr>
<tr><td>Sandy clay</td><td>1</td><td>0.05</td></tr>
<tr><td>Clayey Silt</td><td>15</td><td>0.78</td></tr>
<tr><td>Silt</td><td>1</td><td>0.05</td></tr>
<tr><td>Clayey Very Fine Sand</td><td>24</td><td>1.25</td><td rowspan="9">Silty SAND</td><td rowspan="9">174</td><td rowspan="9">9.06</td></tr>
<tr><td>Very Fine Sand with clay</td><td>5</td><td>0.26</td></tr>
<tr><td>Very FINE SAND with Silt</td><td>29</td><td>1.51</td></tr>
<tr><td>Clayey Fine Sand</td><td>44</td><td>2.29</td></tr>
<tr><td>Silty Fine Sand</td><td>3</td><td>0.16</td></tr>
<tr><td>Fine sand with Silt</td><td>2</td><td>0.10</td></tr>
<tr><td>Silty Fine to Medium Sand</td><td>4</td><td>0.21</td></tr>
<tr><td>Clayey Sand</td><td>18</td><td>0.94</td></tr>
<tr><td>Silty Sand</td><td>45</td><td>2.34</td></tr>
<tr><td>Very Fine Sand</td><td>29</td><td>1.51</td><td rowspan="6">Fine SAND</td><td rowspan="6">692</td><td rowspan="6">36.40</td></tr>
<tr><td>Fine Sand</td><td>404</td><td>21.04</td></tr>
<tr><td>Fine sand with rare clay nodules</td><td>1</td><td>0.05</td></tr>
<tr><td>Fine to medium Sand with clay</td><td>1</td><td>0.05</td></tr>
<tr><td>Fine to medium Sand</td><td>227</td><td>11.82</td></tr>
<tr><td>Fine Sand to Medium Sand</td><td>30</td><td>1.56</td></tr>
<tr><td>Medium Sand with clay</td><td>1</td><td>0.05</td><td rowspan="9">Medium SAND</td><td rowspan="9">977</td><td rowspan="9">50.89</td></tr>
<tr><td>Medium Sand (with clay coating)</td><td>1</td><td>0.05</td></tr>
<tr><td>Medium sand with Silt</td><td>1</td><td>0.05</td></tr>
<tr><td>Medium to Fine Sand</td><td>8</td><td>0.42</td></tr>
<tr><td>Medium Sand to Fine Sand</td><td>17</td><td>0.89</td></tr>
<tr><td>Medium Sand</td><td>883</td><td>45.99</td></tr>
<tr><td>Medium Sand (with organic matter)</td><td>1</td><td>0.05</td></tr>
<tr><td>Medium Sand To Coarse Sand</td><td>8</td><td>0.42</td></tr>
<tr><td>Medium to Coarse Sand</td><td>57</td><td>2.97</td></tr>
<tr><td>Coarse Sand to Medium Sand</td><td>8</td><td>0.42</td><td rowspan="2">Coarse SAND</td><td rowspan="2">18</td><td rowspan="2">0.94</td></tr>
<tr><td>Coarse Sand</td><td>10</td><td>0.52</td></tr>
<tr><td></td><td>1920</td><td>100.00</td><td></td><td>1920</td><td>100.00</td></tr>
</table>

Table 2: Color variety for tool development

Primary observation (description of individual sample)			**Generalized observation (simplified)**		
Color category (primary observation	No. of samples found in each category	% (by number)	Simplified Color	No. of samples found in each category	% (by number)
Dark gray	225	11.72	**Dark GREY**	**225**	**11.72**
Gray	670	34.90	**GREY**	**878**	**45.73**
Greenish gray	94	4.90			
Olive grey	87	4.53			
Brownish gray	27	1.41			
Whitish gray	195	10.16	**Light GREY**	**646**	**33.64**
Light gray	111	5.78			
Greyish white	58	3.02			
Light olive gray	70	3.65			
Light brownish gray	12	0.63			
Brownish white	8	0.42			
Pale yellow	109	5.68			
Very dark greyish brown	2	0.10			
Dark grayish brown	9	0.47			
Grayish brown	39	2.03			
Light olive brown	7	0.36			
Brown	26	1.35			
Pale olive	28	1.46	**OLIVE**	**44**	**2.29**
Olive	16	0.83			
Yellow	35	1.82	**YELLOW**	**57**	**2.97**
Olive yellow	11	0.57			
Brownish yellow	11	0.57			
Very pale brown	1	0.05	**BROWN**	**70**	**3.65**
Pale brown	6	0.31			
Light brown	39	2.03			
Dark brown	1	0.05			
Light yellowish brown	14	0.73			
Yellowish brown	6	0.31			
Reddish brown	3	0.16			
Total	**1920**	**100.00**		**1920**	**100.00**

4. Conclusions

To ensure safe drinking water for people was the primary objective of this research. The data showed that developing a color based tool may be possible to deploy for searching a safe aquifer.It has been observed that red and off-white sediments, which were relatively more oxidized are characteristically low in As. On the other hand, low arsenic content could not ensure safe drinking water, as high Mn concentration revealed in some aquifers. Therefore, presence of both As and Mn in aquifers could not be counted as safe water source. Considering the toxicity of As and thereby the relative importance compared to Mn, red and off-white sands should get priority for installing water wells if there is no other alternative sources are available. This research needs continuous refinement and would be continued until find the ultimate solution for providing safe water to the people. As providing safe water is already selected as one of the millennium development goals (MDGs) for developing countries.

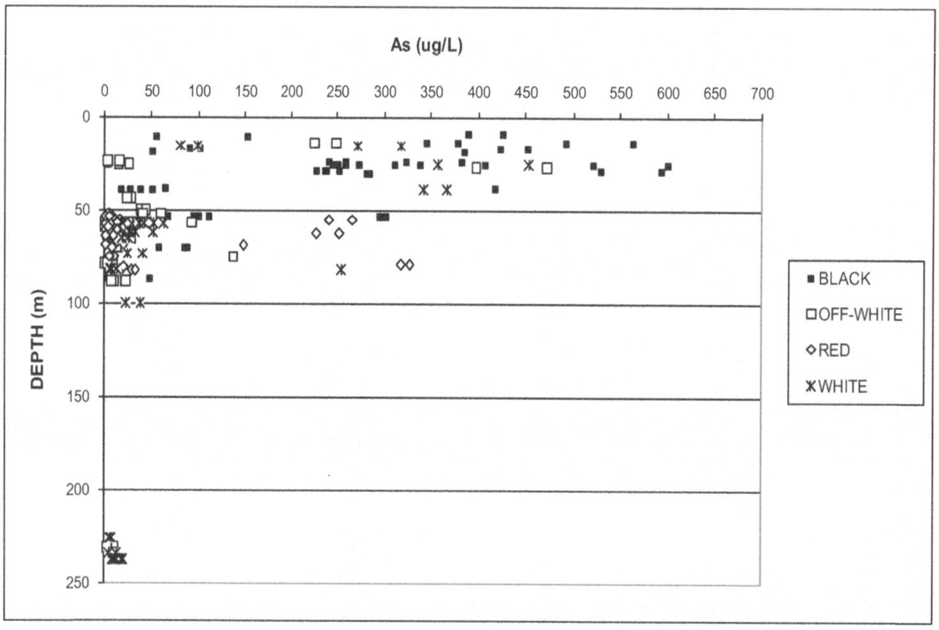

Figure-2: As concentration vs. sediment color

The median values for As in black observed was 241 µg/L and for off-white, red and white were 11 µg/L, <5.6 µg/L, and 32 µg/L respectively. Manganese concentration in off-white and red sediments is relatively higher *[7]* than in black and white sediments.

5. Acknowledgments

This action research cum implementation project is being carried out with the support of Swedish International Development Agency (SIDA) in the form of grant.

6. References

[1] Bundschuh, J., Litter, M.,Bhattacharya, P. (2010) Targeting arsenic-safe aquifers for drinking water supplies. Environ. Geochem. & Health 32: 307-315.

[2] BGS and DPHE (British Geological Survey and Department of Public Health Engineering). Arsenic contamination of groundwater in Bangladesh. Technical report WC/00/19. UK: Keyworth; 2001.

[3] Bhattacharya, P., Welch, A. H., Stollenwerk K. G., McLaughlin M.J., Bundschuh, Panaullah, G. (2007) Arsenic in the environment: biology and chemistry. Science of the total environment. 379. 109-120.

[4] Argos, M., Kalra, T., Rathouz,P.J., Chen, Y., Pierce, B., Parvez, F., Islam,T. Ahmed, A., Rakibuz-Zaman, M., Hasan, R., Sarwar, G., Slavkovich, V., van Geen, A., Graziano, J., Ahsan, H. (2010) Arsenic exposure from drinking water, and all-cause and chronic-disease mortalities in Bangladesh (HEALS): a prospective cohort study. The Lancet, DOI:10.1016/S0140-6736(10)60481-3

[5] Deshpande, L.S. and Pande, S. P. (2005) Development of arsenic testing field kit -a tool for rapid on-site screening of arsenic contaminated water sources Environmental Monitoring and Assessment. 101: 93-101.

[6] von Brömssen, M., Jakariya, Md., Bhattacharya, P., Ahmed, K. M., Hasan, M.A., Sracek, O., Jonsson, L., Lundell, L., Jacks G. (2007) Targeting low-arsenic aquifers in groundwater of Matlab Upazila, Southeastern Bangladesh. Science of the Total Environment 379: 121-132.

[7] Mozumder, M.R.H., Bhattacharya, P.,Ahmed, K.M., von Brömssen, M.,Hasan, M.A., Islam, M. M., Hossain, M., Jacks, G. (2011) Identifying low-As aquifers in terms of redox-facies delineation by correlating sediment color with groundwater chemistry in Matlab, Upazila, Southeastern Bangladesh. 11th ICOBTE conference proceedings. Florence, Italy.

Section 7

Bottled water

The Elemental Composition and Taste of Bottled Water

Helle Marcussen, Hans Chr. B. Hansen and Peter E. Holm
Department of Basic Sciences and Environment, Strategic Water Initiative ViVa, Faculty of Life Sciences, University of Copenhagen, DK-1871 Frederiksberg C, Denmark.
Corresponding author e-mail: hma@life.ku.dk

Abstract
The worldwide consumption of bottled water is increasing. Bottled water is normally divided into different categories of waters depending on its origin; and the regulation with respect to bottling, microbiological and chemical quality differs between the different categories. Taste is often mentioned as the main reason for choosing bottled water over tap water but there is little scientific evidence supporting that bottled water has superior taste over tap water. This paper reviews the linkages between taste and the elemental composition of bottled waters. The elemental content of bottled water is relatively well investigated at least for the major elements and ions. Through a literature review of 40 studies we collect information about the elemental content of one to 67 parameters for a total of 1935 bottled waters. We conclude that the content of different elements varies up to 6 orders of magnitude for bottled water and that the relation between the taste and the chemical content in general is poorly understood.

1. Introduction

The worldwide consumption of bottled water is increasing. Bottled water is up to 10.000 more expensive than tap water and the understanding of why bottled water has become such a commercial success is limited. Different authors speculate that the increase in bottled water consumption is due to lack of trust in tap water, beliefs that bottled water has superior health promoting properties, that bottled water is marketed as trendy and a signal of youth and has better taste qualities compared to tap water.

Bottled water is normally divided into different categories of waters depending on its origin; and the regulation with respect to bottling, microbiological and chemical quality differs between the different categories. In the EU the three following categories are used: 1) natural mineral water, 2) spring water, and 3) other types of bottled drinking water e.g. bottled tap water. The latter category is regulated under the same directive as tap water namely the EC Drinking Water Directive (98/83/EEC). It must therefore conform to all its standards comprising 48 quality parameters including limit values for anions and cations, conductivity, taste, color, organoleptic and microbial quality. Natural mineral water is regulated under the EC Mineral Water Directive (80/777/EEC). This directive was adopted to allow for taste differences and the higher mineral content seen in some mineral waters which is not acceptable under the Drinking Water Directive. Bottled water drinkers often cite taste as the main reason for choosing bottled water over tap water but there is little scientific evidence supporting that bottled water has superior taste over tap water. Bottled water may in many cases be tap water filled on bottles possible after treatment. This paper reviews the elemental compositions of bottled waters with focus on possible links between composition and taste.

2. Review of the chemical composition of bottled water

The chemical content of bottled water is relatively well investigated at least for the major elements and ions. Through a literature review of 40 studies we collected information about the chemical content of one to 67 major parameters (data not shown) for a total of 1935 bottled waters. A preliminary compilation of selected minor elements is shown in Table 1.

The difference between the minimum and maximum concentrations of minor and major elements and ions varies with 2 to 6 orders of magnitude. The chemical content of bottled water depends on its classification. Contamination during the bottling process or from the bottles may also have influence on the chemical content. Much less is known about the organic content of bottled water compared to the inorganic content and such information is normally not given on bottle labels.

Organic compounds in bottled water may originate from the natural compounds or anthropogenic contaminants in the aquifer, contamination during bottling or migration from bottles during storage.

Table 1. Concentration ranges (µg L^{-1}) of selected minor elements in bottled water. The number of studies which included the parameter and the total number of bottled water brands analyzed is shown.

Element	Range (µg L^{-1})	Studies	Brands
Ag	0.0004-92	9	615
Al	<0.06-2500	15	1203
As	0.006-49	12	899
Cd	0.0006-24	21	1215
Co	0.0009-11	13	708
Cr	0.001-102	19	996
Cu	<0.01-761	18	1005
Fe	0.07-56000	15	743
Mn	<0.005-1150	16	1016
Mo	0.003-179	11	701
Ni	<0.01-420	14	792
P	0.21-718	4	213
Pb	<0.001-74	17	1095
Sb	0.001-5	11	880
U	0.0002-92	8	306
V	0.0006-93	13	736
Zn	<0.001-980	19	992

3. Taste of drinking water

3.1 Inorganic parameters

Few studies have investigated the taste of drinking water in relation to the content of inorganic parameters and the present knowledge of water taste caused by inorganic components is predominantly based on studies of tastes and taste thresholds for single salts dissolved in distilled water. Such studies have been carried out for the major cations Ca and Mg, the halides Li, Na, K, Ru and Cs, and trace metals such as Fe, Zn and Cu. Through these studies it has become widely accepted that cations primarily are responsible for the taste properties of salts but anions may modify the strength of the taste. There is a lack of studies examining the taste in more complex solutions as spring, mineral or tap water where the taste of salt mixtures not necessarily will be the sum of the tastes of the single salts.

Koseki et al. [1] found that the taste quality of a bottled mineral water decreased if more calcium was added to the water. At lower calcium concentrations (10-20 mg l-1) the water was perceived as sweet and tasted significantly better compared to a calcium concentration of 100 mg l-1 which resulted in a bitter taste.

3.2 Iron

Trace elements may have a strong effect on the taste of water when it is present above the taste detection threshold. Iron has been the focus in many sensory studies because Fe has high taste intensity and foods are Fe fortified due to the high occurrence of Fe deficiency especially in developing countries. The tastes of different ferrous salts have been found to be predominantly metallic, bitter and astringent and to a lesser degree sour, sweet and salty. Studies of ferrous salts have shown that taste qualities of metals are complex and consist of several gustatory qualities, retronasal smell and likely also weak tactile sensations (mouthfell). The reported mean taste detection threshold for iron ranges from 0.4 to 49 mg l-1 for different ferrous salts. This large range in determined taste detection thresholds may be due to different sensory or calculation methods, the application of different iron salts such a sulphate and chloride. However, the taste detection threshold may vary several orders of magnitude between people with different taste sensitivity and it is therefore also likely that the variation in taste detection threshold is a result of the different panel members taste sensitivity.

3.3 Copper and Zinc

The taste of copper (Cu) has been described as both bitter, astringent, sour, salty or metallic. Cuppett et al. [2] found that persons who are sensitive to the Cu taste can taste it at concentrations down to 0.4 mg l-1. Zacarías et al. [3] determined the taste detection threshold for Cu defined as the lowest concentrations which could be tasted by 50% of a taste panel. The taste threshold ranges between 2.4 and 2.6 mg l-1 for sulphate and chloride salts in distilled and tap water. For a mineral water the threshold was 3.5 - 3.8 mg l-1 for Cu sulphate and Cu chloride, respectively. Cuppett et al. [2] found much lower threshold taste levels for Cu in distilled and mineral water which were in the range of 0.4-0.7 mg l-1. The study demonstrated that free and complexed Cu ions could be readily tasted whereas particulate copper only added slightly to the copper taste. This may be the reason to the higher threshold values observed by others since they typically tested concentrations above the solubility limit.

Zinc salts in concentrations of 327-3270 mg l-1 have a bitter, sour, salty, tingling/stinging and astringent taste with astringent being the most dominant taste [4]. Further, Zn seemed to inhibit the sweet taste of other substances. Cohen et al. [5] found Zn taste detection thresholds in distilled water between 17.6 and 24.8 mg l-1. The threshold was determined as the lowest concentration which could be detected by 50% of a sensory panel. The detection threshold increased in the following order: sulphate < nitrate < chloride.

A few older studies have focused on the taste of drinking water as a result of the total inorganic content. These studies applied water from the consumers tab, water works or produce water with different mineral content from distilled water and a range of salts. Through these studies the relation between the taste quality and content of major inorganic components Ca2+, Mg2+, Na+, HCO3-, Cl-, NO3-, SO42- and total dissolved solids (TDS) was investigated. This research has indicated that the taste of water gradually increase with increasing TDS when the concentration is below 50 mg L-1. At higher concentrations the taste will decreases with increasing TDS and that water with a TDS ≤ 450 mg L-1 results in good quality water [6].

4. Conclusions

From review of the literature it can tentatively be concluded that the chemical content of different elements varies up to 6 orders of magnitude for bottled water. Trace elements like Fe, Cu and Zn may have a strong effect on the taste of water when it is present above the taste detection threshold. However, the taste detection threshold is unlikely to be exceeded in most bottled drinking waters. The taste of a cation can be modified both by the cation concentration and by the presence of other cations or anions, e.g. through complexation reactions. There is a lack of studies examining the taste in more complex solutions as spring, mineral or tap water where the taste of salt mixtures not necessary will be the sum of the tastes of the single salts. We conclude that the relation between the taste and the elemental content in general is poorly understood.

References

[1] Koseki, M., Makagawa, A., Tanaka, Y., Noguchi, H., Omochi, T. Journal of Food Science, 2003, 68, 354-358.

[2] Cuppett, J.D., Duncan, S.E., Dietrich, A.M. Chemical Senses, 2006, 31, 689-697.

[3] Zacarías, I., Yáñez, C.G., Araya, M., Oraka, C., Olivares, M., Uauy, R., Chemical Senses 2001, 26, 85-89.

[4] Keast, R.S.J. Journal of Food Science, 2003, 68, 1871-1877.

[5] Cohen, J.M., Kamphake, L.J., Harris, E.K., Woodward, R.L. American Water works Association Journal, 1960, 52, 660-670.

[6] Bruvold, W.H., Daniels, J.I., American Water Works Association Journal, 1990, 82, 59-65.

Elucidating the Parameters Involved with Antimony and Phthalates Co-leaching in Bottled Water

Syam S. Andra and **Konstantinos C. Makris***

Water and Health Laboratory, Cyprus International Institute for Environmental and Public Health in association with the Harvard School of Public Health, Cyprus University of Technology, Cyprus

Corresponding author e-mail: konstantinos.makris@cut.ac.cy

Plastic bottles of varying volumes (0.5, 0.75 and 1.5L) are used worldwide to package drinking-water. Most of the small-sized (by volume) plastic bottles are typically made out of polyethylene terephthalate (PET), which contains an appreciable amount of antimony (Sb) used as a catalyst in the polymerization process. Antimony has been charged with diaphoretic, cardiac and respiratory adverse health effects to humans, and phthalates are associated with endocrine disrupting effects, but no toxicological data exist for mixtures of Sb and phthalates in drinking-water. Adverse health effects associated with the co-presence of arsenic and antimony in water have been documented, but there is no human exposure study, or epidemiologic dataset utilizing data on the simultaneous presence of Sb and phthalates in the finished water of PET bottles. Our preliminary data collected in Cyprus suggest that the average daily dose of water may significantly increase if further use of bottled water for preparation of hot or cold coffees, tea and juices is taken into account. Recent studies report that heating and microwaving may increase Sb concentrations in bottled water to values > maximum contaminant level in the EU (5 ppb). However, little, if any, information is available on the co-leaching of Sb and phthalates into the finished water of PET bottles. The objectives of this study were to: i) assess the effect of leached phthalates on the possible concomitant release of Sb and ii) investigate the environmental factors, i.e., storage time, temperature and UV radiation that maximize co-leaching of Sb and phthalates in the finished water. Twenty different samples of PET bottled water in the Boston, USA and Limassol, Cyprus will be used in our study. Antimony will be measured in water samples using an inductively coupled plasma mass spectrometer equipped with a dynamic reaction cell technology, while phthalates analyses (Diethyl phthalate, Dibutyl phthalate, Di(2-ethylhexyl) phthalate, Dimethyl phthalate, and benzyl butyl phthalate) will be conducted with solid phase microextraction coupled to gas chromatography mass spectrometry. Results will illustrate the main and interaction effects of environmental factors, such as, storage time, temperature and UV radiation on the co-leaching of Sb and phthalates into finished water. Expected results will elucidate the environmental conditions needed to minimize human exposure to both Sb and phthalates via the oral ingestion of bottled water.

Element Composition of Mineral Waters and Different Beverages

Bengt Nihlgård[1] and Ingegerd Rosborg[2]

[1]Section of Plant Ecology and Systematics, Lund University, Sweden
[2]Royal Institute of Technology, KTH, Stockholm, Sweden

Corresponding author e-mail: bengt.nihlgard@me.com

The mineral composition of 21 different beverages (four wines, two beers, three mineral waters, six different soft drinks, two apple juice and cider, and four vichy and soda waters), all being sold in South Sweden or Poland, with contributions also from Denmark, France and Germany, were analysed concerning 34 elements. Both red and white wines showed very high levels of many nutrient elements, especially K (600-1100 mg/L), Mg (40-60 mg/L), P (150-180 mg/L), Fe (1-5 mg/L), Mn (0,7-2,4 mg/L), and also of many micro elements, e.g. As (9-40 µg/L), Co (4.8-6.5 µg/L), Cr (6-33 µg/L), Ni (50-300 µg/L) and V (7-110 µg/L). The nutritional values of the microelements may be questioned at least at the highest concentrations. They also showed high levels of some toxic elements like Be (0,0-2,4 µg/L), Cd (0,2-0,4 µg/L), Hg (0,0-0,4 mg/L) and Pb (7-32 µg/L). Also the analysed beers, apple cider and apple juice showed similar high values of many of these mineral nutrients and toxic microelements, except for Cd, Hg and Pb. The mineral waters from Poland showed very high contents of Ca (170-370 mg/L) and Mg (70-250 mg/L), and unusually high Li-values (420-750 µg/L). Most soft drinks contained on the other hand low values of almost all elements, including a so called "Sport drink" and a "Soda" water, that both, however, had very high levels of Na (490-640 mg/L) from added sodiumbicarbonate. High levels of both Na and K appeared in "vichy" waters. A conclusion is that especially wines and beers contain lots of minerals, much higher than recommended for ordinary drinking water. So does also juices and natural ciders, while exceptions are the soft drinks, often containing minerals below recommended nutritional levels if Na and K is not added. Wines may show intolerable levels of toxic elements.

Mineral balance in bottled waters

Ingegerd Rosborg[1], Prosun Bhattacharya[1] and Jimmie Parkes[2]

[1]Royal Institute of Technology, KTH, Stockholm, Sweden.
[2]Inter-Euro Technology Ltd., Carlow, Ireland.

Corresponding author e-mail: rosborg@spray.se

The mineral content of bottled water varies greatly throughout Europe, even within each country, depending on the brand and the source. The pH levels vary from around 4 in highly carbonated waters to above 8 in naturally alkaline waters. It is of utmost importance to consider the amounts of minerals present and their ratios and also the additions to some bottled waters because more and more people are now drinking bottled water when travelling and at home too, in preference to their normal water supply. In many cases tap water is healthier than bottled water, since well and municipal water supplies are controlled to a large extent by EU and other regulations. Bottled waters seem to have fewer controls. On the other hand, some bottled waters are very rich in minerals, while if these waters were municipal or private well waters, they would have been treated with softening filter or other treatments that reduce or eliminate the water hardness.

33 different brands of bottled waters from Sweden were analyzed on about 60 elements and ions. In addition, Information was collected from labels on a number of brands from most European countries.

In both cases, we considered the following factors,

The origins of bottled waters - bottled waters are collected from many different types of regions, e.g. from limestone regions or from barren districts.

Amounts of minerals present in different bottled waters - There was a large variation with regards concentrations of Ca, Mg, Na, K and Cl. Two of the Swedish waters had additions of Na_2CO_3, K_2CO_3 and NaCl, which lead to increased concentrations of Na, K and Cl, as well as decreased ratios of Ca/Na and larger ratios of Na/K.

Levels of pH and how this can differ depending on addition of CO_2 or natural water. Depending on the type of container that the water is stored in; glass, plastic or cans different substances may be dissolved from the material, e.g. aluminum cans are less suited for storage of carbonated waters, as the lowered pH-values may dissolve Al.

Benefits or dangers to health of levels of various minerals in bottled water. - e.g. the waters added with Na_2CO_3, K_2CO_3 and NaCl may contribute significantly to the daily intake of Na and/or Cl in high water consumers. 2 litres of bottled water per day may be harmful for persons suffering from e.g. high blood pressure, but good for persons who need salt. The ratios of Ca/Na and Mg/Na which vary a lot in some brands of bottled waters are of great importance especially for patients with heart diseases. HCO_3 from drinking water may raise the pH of the gastrointestinal tract. The variation of SO_4 levels were 3-320 mg/L in Croatian bottled waters. SO_4 is good against constipation, while too high levels can cause diarrhea._

Comparisons with WHO, EU and National guidelines

Some unusual results were found also in our study, e.g. levels of radioactivity found in some Hungarian and Polish bottled water; Li concentration in some Italian bottled water; importance of Reverse Osmosis treatment in bottled waters, as levels of Ca, Mg and HCO_3 in Portuguese bottled waters were extremely low and a bottled water from Great Britain, poor in minerals, had 280 mg/L of NO_3, which seems odd, since the water was regarded as one of "The best of British". The WHO guideline value is 50 mg/L. K was also high in this bottle brand at 77 mg/L.

Author Index

Arslan, G. : *Department of Chemistry, Selcuk University, Konya 42075, Turkey*
Achene , L. : *Section of Inland waters, Department of Environment and primary prevention, Istituto Superiore di Sanità, Rome, Italy*
Ahmed, K. M. : *Department of Geology, University of Dhaka, Dhaka, 1000, Bangladesh*
Akin, I. : *Department of Chemistry, Selcuk University, Konya 42075, Turkey*
Andra, S. S. : *Water and Health Laboratory, Cyprus International Institute for Environmental and Public Health in association with the Harvard School of Public Health, Cyprus University of Technology, Cyprus.*
Karam, P.A. : *New York City Department of Health and Mental Hygiene, Bureau of Environmental Emergency Preparedness and Response, New York, USA*
Annadotter, H. : *Regito AB, SE-28 022, Vittsjö, Sweden*
Åström, M. : *Geochemistry Research Group, Linnaeus University, SE-391 82 Kalmar, Sweden*
Beduk, F. : *Selcuk University, Environmental Engineering Department, Konya-Turkey*
Benodiel, M.J. : *Empresa Portuguesa das Águas Livres, S.A., Rua do Alviela, 12, 1170-012, Lisboa, Portugal*
Berger, T.: *Geochemistry Research Group, Linnaeus University, SE-391 82 Kalmar, Sweden*
Bergqvist, C. : *Department of Botany, Stockholm university, 106 91 Stockholm, Sweden*
Bhattacharya, P. : *KTH-International Groundwater Arsenic Research Group, Department of Land and Water Resources Engineering, Royal Institute of Technology (KTH), SE-100 44 Stockholm, Sweden*
Błaszczuk, A. : *Częstochowa University of Technology, Faculty of Environmental Protection and Engineering, Brzeznicka 60a, 42-200 Częstochowa, Poland*
Boxwall, J. : *Pennine Water Group, University of Sheffield, UK*
Bratkowski, J. : *National Institute of Public Health - National Institute of Hygiene, Department of Environmental Hygiene, 24 Chocimska Str., 00-791 Warsaw, Poland*
Breach, B. : *Water Quality and Environmental Consultancy, UK*
Brenner, A. : *Dept. of Environmental Engineering, Ben-Gurion University, Beer-Sheva 84105, Israel*
Buchheit, D.: *AIT Austrian Institute of Technology GmbH, Seibersdorf, Austria*
Cengeloglu, Y.: *Department of Chemistry, Selcuk University, Konya 42075, Turkey*
Colavita, P. E. : *School of Chemistry, Trinity College Dublin, College Green, Dublin 2, Ireland*
Collivignarelli, C. : *Department of Civil Engineering, Architecture, Land and Environment, University of Brescia -Via Branze 43, 25123 Brescia, Italy*
Cosma, C. : *National Research and Development Institute for Industrial Ecology - INCD ECOIND, 90-92 Panduri Avenue, 050663 Bucharest-- 5, Romania*
Criscuoli, A. : *Institute of Membrane Technology (ITM-CNR), 87030 Rende (CS), Italy*
Croll, B.T. : *Consultant UK*
Cruceru, L. : *National Research and Development Institute for Industrial Ecology - INCD ECOIND, 90-92 Panduri Avenue, 050663 Bucharest 5, Romania*
Cullen, R. : *School of Chemistry, Trinity College Dublin, College Green, Dublin 2, Ireland*
de Jongh, C. : *KWR Watercycle Research Institute, Nieuwegein, the Netherlands*
Dean, T. : *School of Health Sciences and Social Work, University of Portsmouth, Portsmouth, UK*
Deowan, S. A. : *Karlsruhe University of Applied Sciences, Moltkestr.30, 76133 Karlsruhe, Germany*
Dimitric, M. : *Institute of Public Health of Zrenjanin, Zrenjanin, Serbia*
Dinu, C. : *Instumental Analysis Laboratory, National Research and Development Institute for Industrial Ecology - INCD ECOIND, Romania*
Doniecki,T. : *Częstochowa University of Technology, Faculty of Environmental Protection and Engineering, Brzeznicka 60a, 42-200 Częstochowa, Poland*
Dragana, J. : *Institute for public health, Belgrade, Serbia*
Drake, H. : *Geochemistry Research Group, Linnaeus University, SE-391 82 Kalmar, Sweden*
Duffy, P. : *School of Chemistry, Trinity College Dublin, College Green, Dublin 2, Ireland*
Edwards, M. : *Virginia Tech, Blacksburg VA, 24061, USA*
Eike, B. : *STIFTELSEN SINTEF, Department of Water and Environment, Strindveien 4, 7034 Trondheim, Norway*
Emin Aydin, M. : *Selcuk University, Environmental Engineering Department, Konya-Turkey*

Ersoz, M. : *Department of Chemistry, Selcuk University, Konya 42075, Turkey*
Figoli, A. : *Institute of Membrane Technology (ITM-CNR), 87030 Rende (CS), Italy*
Filipov, R. : *Institute of Public Health of Zrenjanin, Zrenjanin, Serbia*
Forssblad, J. : *Regito AB, SE-28 022, Vittsjö, Sweden*
Fröjdö, S. : *Dept. of Geology and Mineralogy, Åbo Akademi, FI-20500 Åbo, Finland*
Frost, A. : *UK Water Treatment Association, Loughborough, UK*
Ganguli, B.: *Department of Statistics, University of Calcutta, 35 Bullygunge Circular Road, Kolkata - 700 019, West Bengal, India*
Garboś, S. : *National Institute of Public Heath - National Institute of Hygiene;Departament of Environmental Hygiene; 24 Chocimska Str., 00-791 Warsaw, Poland*
Gari, D.W. : *National Institutes of Public Health, Department of Environmental Health, CZ-10042 Prague, Czech*
Gheorghe, D. : *AQUATIM Company, Street Gheorghe Lazăr no 11A, code 300081, Timisoara, Timis County, Romania*
Gialdini, F. : *Department of Civil Engineering, Architecture, Land and Environment, University of Brescia -Via Branze 43, 25123 Brescia, Italy*
Giri, A. : *Molecular and Human Genetics Division, Indian Institute of Chemical; Biology, 4, Raja S.C. Mullick Road, Kolkata-700 032, West Bengal, India*
Gordana, P. : *Department of Clinical Nephrology and HD Unit, Zemun Clinical Hospital, Belgrade, Serbia*
Górski, J. : *Department of Hydrogeology and Water Protection; Adam Mickiewicz University; 16 Maków Polnych Str., 61-606 Poznań, Poland*
Greger, M. : *Department of Botany, Stockholm university, 106 91 Stockholm, Sweden*
Gyliene, O. : *Institute of Chemistry of the Center for Physical Sciences and Technology, Lithuania*
Sach, T.H. : *School of Pharmacy, University of East Anglia, Norwich, UK*
Pallett, I.H. : *British Water, 1 Queen Anne's Gate, London SW19BT, UK*
Hansen, H.C.B. : *Strategic Water Initiative ViVa, Department of Basic Sciences and Environment, Faculty of Life Sciences, University of Copenhagen, Denmark*
Hasan, M. A. : *Department of Geology, University of Dhaka, Dhaka, 1000, Bangladesh*
Hayes, C. : *Swensa University, UK*
Hem, L. J. : *STIFTELSEN SINTEF, Department of Water and Environment, Strindveien 4, 7034 Trondheim, Norway*
Hoinkis, J. : *Karlsruhe University of Applied Sciences, Moltkestr.30, 76133 Karlsruhe, Germany*
Holm, P. E. : *Strategic Water Initiative ViVa, Department of Basic Sciences and Environment, Faculty of Life Sciences, University of Copenhagen, Denmark*
Hossain, M. : *NGO Forum for Drinking Water Supply and Sanitation, Lalmatia, Dhaka, 1207, Bangladesh*
Hug, S. : *EAWAG, Zurich, Switzerland*
Islam, M.M. : *NGO Forum for Drinking Water Supply and Sanitation, Lalmatia, Dhaka, 1207, Bangladesh*
Jacks, G. : *Dept Land & Water Resources Eng., KTH, SE 100 44 Stockholm, Sweden*
Jakóbczyk, S. : *Department of Hydrogeology and Engineering Geology, University of Silesia, Sosnowiec, Poland*
Jelea, M. : *North University of Baia Mare, Faculty of Science, 76 Victoriei Street, 430072.Baia Mare, ROMANIA*
Jeligova, H. : *National Institute of Public Health, Department of Environmental Health, CZ-10042 Prague, Czech Republic*
Jez-Walkowiak, J. : *Poznan University of Technology,Institute of Environmental Engineering, Piotrowo 3a, 60-950,Poznan,POLAND*
Johnsson, S. : *Sydvatten AB, Skeppsgatan 19, 211 19 MALMÖ, Sweden*
Jovanovic, D. : *Institute of Public Health of Serbia "Dr Milan Jovanovic Batut", Belgrade, Serbia*
Kantorová, J. : *Institute of Public Health, CZ-70200 Ostrava, Czech Republic Republic*
Khattak, S. : *School of Earth, Atmospheric and Environmental Sciences, University of Manchester, M13 9PL, UK; National Centre For Excellence in Geology, University of Peshawar, Peshawar, 25120, NWFP, PAKISTAN*
Kmiecik, E : *AGH University of Science and Technology, Faculty of Geology, Geophysics and Environmental Protection, Krakow, Poland*
Knezevic, T. : *Institute of Public Health of Serbia "Dr Milan Jovanovic Batut", Belgrade, Serbia*
Koller, K. : *Centre of Evidence Based Dermatology, University of Nottingham, Nottingham, UK*

Kowalczyk, A. : *Department of Hydrogeology and Engineering Geology, University of Silesia, Sosnowiec, Poland*

Kozisek, F. : *National Institute of Public Health, Department of Environmental Health, CZ-10042 Prague, Czech Republic; Department of General Hygiene, Third Faculty of Medicine, Charles University in Prague*

Landi, D. : *AQUATIM Company, Street Gheorghe Lazăr no 11A, code 300081, Timisoara, Timis County, Romania*

Lazarova, Z. : *AIT Austrian Institute of Technology GmbH, Seibersdorf, Austria*

Leventon, J. : Central European University, Budapest, Hungary; Currently at the Technical University of Crete, Chania, Greece

Ljiljana, B. : *Biochemical Laboratory, Clinical Hospital, Belgrade, Serbia*

Lucaciu, I. : *National Research and Development Institute for Industrial Ecology - INCD ECOIND, 90-92 Panduri Avenue, 050663 Bucharest-- 5, Romania,*

Lucentini, L. : *Section of Inland waters, Department of Environment and primary prevention, Istituto Superiore di Sanità, Rome, Italy*

Jelea, S.M. : *North University of Baia Mare, Faculty of Science, 76 Victoriei Street, 430072.Baia Mare, ROMANIA*

Mahanta, C. : *Department of Civil Engineering, Indian Institute of Technology Guwahati 781039, India*

Mandic-Miladinovic, M. : *Public Health Institute, Belgrade*

Makris, K.C. : *Water and Health Laboratory, Cyprus International Institute for Environmental and Public Health in association with the Harvard School of Public Health, Cyprus University of Technology, Cyprus*

Marcussen, H. : *Strategic Water Initiative ViVa, Department of Basic Sciences and Environment, Faculty of Life Sciences, University of Copenhagen, Denmark*

Melini, V. : *Section of Inland waters, Department of Environment and primary prevention, Istituto Superiore di Sanità, Rome, Italy*

Milce, C. : *Institute of blood transfusion, Belgrade, Serbia*

Mirana, A. : *Empresa Portuguesa das Águas Livres, S.A., Rua do Alviela, 12, 1170-012, Lisboa, Portugal*

Mondal , D. : *School of Earth, Atmospheric and Environmental Sciences, University of Manchester, M13 9PL, UK; Molecular and Human Genetics Division, Indian Institute of Chemical; Biology, 4, Raja S.C. Mullick Road, Kolkata-700 032, West Bengal, India; London School of Hygiene & Tropical Medicine, Keppel Street, London WC1E 7HT, UK*

Mons, M. : *Prorail, Utrecht, the Netherlands*

Murphy , D. : *School of Chemistry, Trinity College Dublin, College Green, Dublin 2, Ireland*

Nemcova, V. : *Institute of Public Health, CZ-70200 Ostrava, Czech Republic*

Nicolau, M. : *National Research and Development Institute for Industrial Ecology - INCD ECOIND, 90-92 Panduri Avenue, 050663 Bucharest 5, Romania,*

Nihlgård, B. : *Section of Plant Ecology and Systematics, Lund University, Sweden*

Nordstrand, D. : Department of Botany, Stockholm university, 106 91 Stockholm, Sweden

Okoniewska, E. : *Częstochowa University of Technology, Faculty of Environmental Protection and Engineering, Brzeznicka 60a, 42-200 Częstochowa, Poland*

Ormachea, M. : *Department of Land and Water Resources Engineering, KTH, SE-100 44 Stockholm, Sweden*

Ottaviani, M. : *Section of Inland waters, Department of Environment and primary prevention, Istituto Superiore di Sanità, Rome, Italy*

Ozcan, S. : *Selcuk University, Environmental Engineering Department, Konya-Turkey*

Parkes, J. : *Inter-Euro Technology Ltd., Carlow, Ireland.*

Paunivic, K. : *Institute of Hygiene and Medical Ecology, School of Medicine, Belgrade, Serbia*

Peltola, P. : *Geochemistry Research Group, Linnaeus University, SE-391 82 Kalmar, Sweden*

Persson, K. : *Sydvatten AB, Skeppsgatan 19, 211 19 Malmö, Sweden*

Perunicic, G. : *Departments of Endocrinology, University Hospital Zemun, Belgrade, Serbia*

Petre, J. : *National Research and Development Institute for Industrial Ecology - INCD ECOIND, 90-92 Panduri Avenue, 050663 Bucharest-- 5, Romania,*

Pfeifer, H. : *IMG-Centre d'Analyse Minérale, Faculté de Géosciences et de l'Environnement, Université de Lausanne, CH-1015 Lausanne, Switzerland*

Phawadee, N. : *School of Earth, Atmospheric and Environmental Sciences, University of Manchester, M13 9PL, UK ;CEDS, National University of Laos, Dongdok Campus, Vientiane, Lao PDR*

Plavsic , S. : *Institute of Public Health of Serbia "Dr Milan Jovanovic Batut", Belgrade, Serbia*
Polya, D. : *School of Earth, Atmospheric and Environmental Sciences, University of Manchester, M13 9PL, UK*
Postawa, A. : *AGH University of Science and Technology, Faculty of Geology, Geophysics and Environmental Protection, Krakow, Poland*
Pott, B. : *Sydvatten AB, Skeppsgatan 19, 211 19 Malmö, Sweden*
Prandini, F. : *Department of Civil Engineering, Architecture, Land and Environment, University of Brescia -Via Branze 43, 25123 Brescia, Italy*
Pruss, A. : *Poznan University of Technology, Institute of Environmental Engineering, Piotrowo 3a, 60-950,Poznan,POLAND*
Radosavljevic, M. : *Institute of Public Health of Serbia "Dr Milan Jovanovic Batut", Belgrade, Serbia*
Rahman, M. : *NGO Forum for Drinking Water Supply and Sanitation, Lalmatia, Dhaka, 1207, Bangladesh*
Rahman, M. M. : *Department of Geology, University of Dhaka, Dhaka, 1000, Bangladesh*
Ramos, O. : *Department of Land and Water Resources Engineering, KTH, SE-100 44 Stockholm, Sweden*
Rapp, T. : *Federal Environment Agency, Bad Elster, Germany*
Rashid, S. M. A. : *NGO Forum for Drinking Water Supply and Sanitation, Lalmatia, Dhaka, 1207, Bangladesh*
Rasic-Milutinovic, Z. : *Departments of Endocrinology, University Hospital Zemun, Belgrade, Serbia; University Hospital Zemun, Belgrade*
Ristanovic-Ponjavic, I. : *Public Health Institute, Belgrade*
Rosborg, I. : *Royal Institute of Technology, KTH, Stockholm, Sweden.*
Rubin, H. : *Department of Hydrogeology and Engineering Geology, University of Silesia, Sosnowiec, Poland*
Rubin, K. : *Department of Hydrogeology and Engineering Geology, University of Silesia, Sosnowiec, Poland*
Russell, L. : *REED International LTD, USA*
Thomas, K.S. : *Centre of Evidence Based Dermatology, University of Nottingham, Nottingham, UK*
Salio, L. : *Department of Civil Engineering, Indian Institute of Technology Guwahati 781039, India*
Sandhi, A. : *Department of Botany, Stockholm university, 106 91 Stockholm, Sweden*
Siepak, M. : *Department of Hydrogeology and Water Protection; Adam Mickiewicz University; 16 Maków Polnych Str., 61-606 Poznań, Poland*
Skotak, K. : *National Institute of Public Health - National Institute of Hygiene - 00-791 Warsaw, 24 Chocimska str., Poland*
Soldi, L. : *School of Chemistry, Trinity College Dublin, College Green, Dublin 2, Ireland*
Sorlini, S. : *Department of Civil Engineering, Architecture, Land and Environment, University of Brescia - Via Branze 43, 25123 Brescia, Italy*
Sovann, C. : *School of Earth, Atmospheric and Environmental Sciences, University of Manchester, M13 9PL, UK; Royal University of Phnom Penh, Russian Federation Boulevard, Toul Kork, Phnom Penh, Cambodia*
Sozanski, M.M. : *Poznan University of Technology,Institute of Environmental Engineering, Piotrowo 3a, 60-950,Poznan,Poland*
Staniloae, D. : *Instumental Analysis Laboratory, National Research and Development Institute for Industrial Ecology - INCD ECOIND, Romania*
Staniloae, D. : *National Research and Development Institute for Industrial Ecology - INCD ECOIND, 90-92 Panduri Avenue, 050663 Bucharest-5, Romania,*
Stefanescu, A.: *AQUATIM Company, Street Gheorghe Lazăr no 11A, code 300081, Timisoara, Timis County, Romania*
Svensson , M. : *Dept. of Soil and Water Environment, Ramböll Sweden, Box 4205, SE-102 65 Stockholm, Sweden*
Swiatczak, J. : *Expert, Poland*
Święcicka, D. : *National Institute of Public Heath - National Institute of Hygiene;Departament of Environmental Hygiene; 24 Chocimska Str., 00-791 Warsaw, Poland*
Tenne, A. : *Head of Desalination Division, Israel Water Authority, Tel-Aviv 61203, Israel*
Thorvalden, G. : *STIFTELSEN SINTEF, Department of Water and Environment, Strindveien 4, 7034 Trondheim, Norway*
Tor, A. : *Department of Environmental Engineering, Selcuk University, Konya 42075, Turkey*
Triantafyllidou, S.: *Virginia Tech, Blacksburg VA, 24061, USA*

Uçar, S.: *Selcuk University, Environmental Engineering Department, Konya,-Turkey*
Van Wezel, A. : *KWR Watercycle Research Institute, Nieuwegein, the Netherlands*
Vasile, G. : *National Research and Development Institute for Industrial Ecology ECOIND, Road Panduri no. 90-92, code 050663, Bucharest, Romania*
Veschetti, E.: *Section of Inland waters, Department of Environment and primary prevention, Istituto Superiore di Sanità, Rome, Italy*
von Brömssen, M. : *Ramböll Sweden AB, Box 4205, SE-102 65 Stockholm,, Sweden*
Vukcevic, S. : *Public Health Institute, Belgrade*
Wator, K: *AGH University of Science and Technology, Faculty of Geology, Geophysics and Environmental Protection, Krakow, Poland*
Williams, H.C. : *Centre of Evidence Based Dermatology, University of Nottingham, Nottingham, UK*
Yilmaz , A. : *Department of Chemistry, Selcuk University, Konya 42075, Turkey*
Zabochnicka-Świątek, M. : *Częstochowa University of Technology, Faculty of Environmental Protection and Engineering, Brzeznicka 60a, 42-200 Częstochowa, Poland*
Zorica, R. : *Department of Endocrinology, Zemun Clinical Hospital, Belgrade, Serbia*

Kristianstad, Sweden

Kristianstad is one of Scandinavian cities which founded during 16th century. The city is divided by a canal which flowing throughout the city. The large square of the city (*Swedish, Stora Torg*) is the icon for administrative part whereas Small Square (*Swedish, Lilla Torg*) plays a gathering point of craftsmen and trade.

On the countryside, a vast natural green landscape accompanied with a large number of manors and castles in the Kristianstad area. Of the oldest, Bishop Eskil's castle in Åhus, there are only ruins left while one of Scandinavia's most magnificent renaissance manors.

The Kristianstad plain around the lower reaches of the Helgeån river has historically been farming land with good pastures around the reed lined waters. The area is bounded in the north, west and south by forests and ridges.

Source: http://www.kristianstad.se/en/Tourism/History/Kristianstad/